T0207370

Lecture Notes in Computer Science 13298

More information about this series at https://link.springer.com/bookseries/558

Merav Parter (Ed.)

Structural Information and Communication Complexity

29th International Colloquium, SIROCCO 2022
Paderborn, Germany, June 27–29, 2022
Proceedings

Editor
Merav Parter
Weizmann Institute
Rehovot, Israel

ISSN 0302-9743 ISSN 1611-3349 (electronic)
Lecture Notes in Computer Science
ISBN 978-3-031-09992-2 ISBN 978-3-031-09993-9 (eBook)
https://doi.org/10.1007/978-3-031-09993-9

This Springer imprint is published by the registered company Springer Nature Switzerland AG
The registered company address is: Gewerbestrasse 11, 6330 Cham, Switzerland

Preface

This volume contains the papers presented at SIROCCO 2022: 29th International Colloquium on Structural Information and Communication Complexity held during June 27–29, 2022, in Paderborn, Germany.

SIROCCO is devoted to the study of the interplay between structural knowledge, communication, and computing in decentralized systems of multiple communicating entities. Special emphasis is given to innovative approaches leading to better understanding of the relationship between computing and communication.

SIROCCO 2022 received 28 submissions. Each submission was reviewed by at least three Program Committee (PC) members. The committee decided to accept 16 papers. The program also included four keynote talks and four invited talks. Of the regular papers, "Proof Labeling Schemes for Reachability-Related Problems in Directed Graphs" by Yoav Ben Shimon, Orr Fischer, and Rotem Oshman received the Best Paper Award and the Best Student Paper Award.

I would like to thank the authors who submitted their work to SIROCCO this year and the PC members and subreviewers for their valuable reviews and comments. I would also like to thank the keynote speakers, George Giakkoupis, Maurice Herlihy, and Gillat Kol, and the invited talk speakers, Michal Dory, Laurent Feuilloley, Sebastian Forster and Goran Zuzic, for their insightful talks. I am grateful to Christian Scheideler for his featured talk as the recipient of the 2022 SIROCCO Innovation in Distributed Computing Prize. The SIROCCO Steering Committee, chaired by Magnús Halldórsson, and the local arrangements chair, Christian Scheideler, provided help and guidance throughout the process. The EasyChair system was used to handle the submission of papers and to manage the review process.

I hope that you enjoy reading these papers and find inspiration to further your own research.

May 2022 Merav Parter

Organization

Program Committee

Dan Alistarh	IST Austria, Austria
Silvia Bonomi	Sapienza University of Rome, Italy
Sebastian Brandt	CISPA Helmholtz Center for Information Security, Germany
Trevor Brown	University of Waterloo, Canada
Gregory Chockler	University of Surrey, UK
Giuseppe Antonio Di Luna	Sapienza University of Rome, Italy
Michael Dinitz	Johns Hopkins University, USA
Shlomi Dolev	Ben-Gurion University of the Negev, Israel
Michal Dory	ETH Zurich, Switzerland
Thomas Erlebach	Durham University, UK
Sebastian Forster	University of Salzburg, Austria
Pierre Fraigniaud	Université Paris Cité and CNRS, France
Rati Gelashvili	University of Toronto, Canada
Olga Goussevskaia	Federal University of Minas Gerais, Brazil
Tomasz Jurdzinski	University of Wroclaw, Poland
Othon Michail	University of Liverpool, UK
Rotem Oshman	Tel Aviv University, Israel
Merav Parter (Chair)	Weizmann Institute of Science, Israel
Andrzej Pelc	Universite du Quebec en Outaouais, Canada
Sriram Pemmaraju	University of Iowa, USA
Seth Pettie	University of Michigan, USA
Giuseppe Prencipe	Università di Pisa, Italy
Ivan Rapaport	Universidad de Chile, Chile
Stefan Schmid	University of Vienna, Austria
Lewis Tseng	Boston College, USA
Jennifer Welch	Texas A&M University, USA

Additional Reviewers

Almethen, Abdullah
Castaneda, Armando
Connor, Matthew
Davies, Peter
Fischer, Orr
Gańczorz, Adam
Miyano, Eiji
Montealegre, Pedro
Pai, Shreyas
Skretas, George
Smith, Tyler
Vargas Godoy, Karla
Wang, Wanliang

Contents

Local Mending

Alkida Balliu[1], Juho Hirvonen[2], Darya Melnyk[2(✉)], Dennis Olivetti[1], Joel Rybicki[3], and Jukka Suomela[2]

[1] Gran Sasso Science Institute, L'Aquila, Italy
{alkida.balliu,dennis.olivetti}@gssi.it
[2] Aalto University, Espoo, Finland
{juho.hirvonen,darya.melnyk,jukka.suomela}@aalto.fi
[3] IST Austria, Klosterneuburg, Austria
joel.rybicki@ist.ac.at

Abstract. In this work we introduce the graph-theoretic notion of *mendability*: for each locally checkable graph problem we can define its *mending radius*, which captures the idea of how far one needs to modify a partial solution in order to "patch a hole." We explore how mendability is connected to the existence of efficient algorithms, especially in distributed, parallel, and fault-tolerant settings. It is easy to see that $O(1)$-mendable problems are also solvable in $O(\log^* n)$ rounds in the LOCAL model of distributed computing. One of the surprises is that in paths and cycles, a converse also holds in the following sense: if a problem Π can be solved in $O(\log^* n)$, there is always a restriction $\Pi' \subseteq \Pi$ that is still efficiently solvable but that is also $O(1)$-mendable. We also explore the structure of the landscape of mendability. For example, we show that in trees, the mending radius of any locally checkable problem is $O(1)$, $\Theta(\log n)$, or $\Theta(n)$, while in general graphs the structure is much more diverse.

Keywords: Mendability · Fault tolerance · LCL problems · Distributed algorithms · Parallel algorithms · Dynamic algorithms

1 Introduction

Naor and Stockmeyer [41] initiated the study of the following question: given a problem that is locally *checkable*, when is it also locally *solvable*? In this paper, we explore the complementary question: given a problem that is locally checkable, when is it also locally *mendable*?

Warm-Up: Greedily Completable Problems. There are many graph problems in which *partial solutions can be completed greedily, in an arbitrary order.* Several classic problems from the field of distributed computing fall into this category: the canonical example is vertex coloring with $\Delta + 1$ colors in a graph with maximum degree Δ. Any partial coloring can be always completed; the neighbors of a node can use at most Δ distinct colors, so a free color always exists for any uncolored node. This simple observation has far-reaching consequences:

M. Parter (Ed.): SIROCCO 2022, LNCS 13298, pp. 1–20, 2022.
https://doi.org/10.1007/978-3-031-09993-9_1

- Any such problem *can be solved efficiently* not only in the centralized sequential setting, but also in distributed and parallel settings. For example, for maximum degree $\Delta = O(1)$ any such problem can be solved in $O(\log^* n)$ communication rounds in the usual LOCAL model [40, 43] of distributed computing (we will explain this in detail in Sect. 5).
- Any such problem admits simple *fault-tolerant and dynamic algorithms*. One can, for example, simply clear the labels in the immediate neighborhood of any point of change and then greedily complete the solution.

Classic symmetry-breaking problems such as maximal matching and maximal independent set also fall in this class of problems. However, there are problems that admit efficient distributed solutions even though they are not greedily completable.

In this work, we introduce the notion of *local mendability* that captures a much broader family of problems, and that has the same attractive features as greedily completable problems: it implies efficient centralized, distributed, and parallel solutions, as well as fault-tolerant and dynamic algorithms. Local mendability can be seen as a formalization and generalization of the intuitive idea of greedily completable problems.

Informal Example: Mending Partial Colorings in Grids. Let G be a large two-dimensional grid graph; this is a graph with maximum degree 4. As discussed above, 5-colorings in such a graph can be found greedily; any partial solution can be completed. However, 4-coloring is much more challenging. Consider, for example, the partial 4-coloring in Fig. 1a: the unlabeled node in the middle does not have any free color left; this is not greedily completable. Also, the four neighbors of the node do not have any other choices.

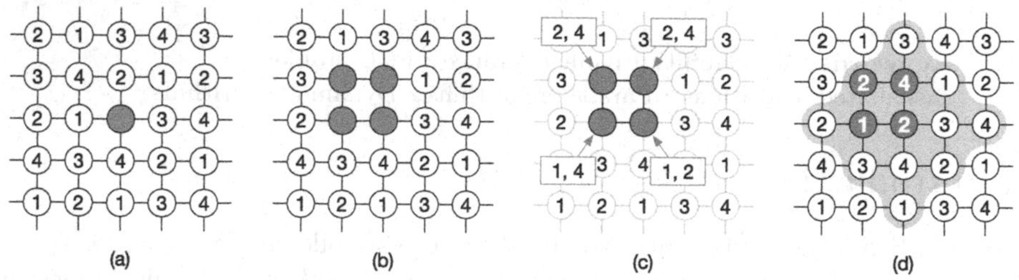

Fig. 1. Mending of 4-colorings in a two-dimensional grid: (a) A partial solution given as input, with one hole (blue). (b) The hole enlarged. (c) This leads to a list coloring instance with lists of size 2 in a 4-cycle. (d) As 4-cycles are 2-choosable, we can always complete the coloring in the hole. The orange shading indicates the radius-2 neighborhood of the hole, and by following this procedure, it is sufficient to modify the partial input in this region to patch the hole; therefore we say that 4-coloring in grids is 2-mendable. (Color figure online)

However, one can make the empty region a bit larger and create a 2×2 hole in the grid, as shown in Fig. 1b. This way, we will always create a partial coloring that is completable—this is a simple corollary of more general results by Chechik and Mukhtar [19]. To see this, notice that each node in the 2×2 part has got at least two possible choices left; hence we arrive at the task of list coloring of a 4-cycle with lists of size at least 2 (Fig. 1c). It is known that any even cycle is 2-choosable [25], i.e., list-colorable with lists of size 2. We can therefore find a feasible solution, e.g., the one shown in Fig. 1d.

Note that to complete the partial coloring in this setting, we had to change the label of a node at distance 2 from the hole, but we never needed to modify the given partial labeling any further than that. Therefore we say that **4-coloring in grids is 2-mendable**; informally, *we can mend any hole in the solution with a "patch" of radius at most* 2. We can contrast this with the 5-coloring problem, which is 0-mendable (i.e., greedily completable without touching any of the previously assigned labels), and the 3-coloring problem, which turns out to be not T-mendable for any constant T (i.e., there are partial 3-colorings that no constant-radius patch will mend).

Remark 1. The above example is informal. We will later formalize the notion of mendability so that it is well-defined for *any* locally checkable problem, not just for the particularly convenient problem of graph coloring. This definition will also change the precise mending radius of graph coloring by an additive $+1$. Both the above informal view and the formalism are suitable for understanding our results; we are, after all, primarily concerned about the asymptotics. We return to this aspect later in Sect. 4.

Consequences of Local Mendability. Problems that are locally mendable have several attractive properties. In particular, as the notion of mendability does not depend on any specific computational model, it naturally lends itself to algorithm design paradigms in centralized, parallel, and distributed systems. We now discuss a few motivating examples.

Consider any graph problem Π that is T-mendable for a constant T; we formally define this notion in Sect. 4, but for now, the informal idea that holes can be mended with radius-T patches will suffice. To simplify the discussion, we will work in bounded-degree graphs, i.e., $\Delta = O(1)$, and assume that we have a discrete graph problem in which nodes are labeled with labels from some finite set of labels; 4-coloring grids is a simple example of a problem that satisfies all these properties, with $T = 2$.

We can now make use of *mendability as an algorithm design paradigm.* First, we obtain a very simple centralized, sequential linear-time algorithm for solving Π: process the nodes in an arbitrary order and whenever we encounter a node that has not been labeled yet, mend the hole with a constant-size patch. Mendability guarantees that such a patch always exists; as there are only constantly many candidates, one can find a suitable patch e.g. by checking each candidate.

A key observation is that we can *mend two holes simultaneously in parallel* as long as the distance between them is at least $2T + 1$; this ensures that the patches do not overlap. This immediately suggests efficient parallel and distributed algorithms: find a set of holes that are sufficiently far from each other, and patch all of them simultaneously in parallel. Concretely, one can first find a distance-$(2T + 1)$ coloring of the graph, and then proceed by color classes, mending all holes in one color class simultaneously in parallel. For example, in the usual LOCAL model [40,43] of distributed computing, this leads to an $O(\log^* n)$-time algorithm for solving Π: first use e.g. Linial's [40] graph coloring algorithm to find the constant-distance coloring in $O(\log^* n)$ communication rounds, and then there are constantly many color classes, each of which can be processed in constant time (see Sect. 5).

Local mendability also directly leads to very efficient *dynamic graph algorithms* that can maintain a feasible solution in constant time per update: simply clear the solution around the point of change and patch the hole locally. This approach holds not only in the classic setting of centralized dynamic graph algorithms [22,30], but also in so-called distributed input-dynamic settings [26], where the communication topology remains static but the local inputs may dynamically change.

Furthermore, local mendability can be used to design algorithms that can *recover from failures*. It can, for example, be used as a tool for designing *self-stabilizing algorithms* [23,24]: in brief, nodes can repeatedly check if the solution is locally correct, switch to an empty label if there are local errors, and then patch any holes in a parallel manner as sketched above.

Case Study: Mending as an (automatic) Algorithm Design Tool. So far, we have seen that mendability can be used as a tool to design efficient algorithms; but is this a powerful tool that makes algorithm design *easier*? The problem of 4-coloring grids provides a rather convincing example. Directly designing an efficient 4-coloring algorithm e.g. for the LOCAL model is challenging; none of the prior algorithms [13,31,42] are easy to discover or explain, but the above algorithm that makes use of the concept of mending is short and simple.

In the full version of this work [9], we give another example: $\{1, 3, 4\}$-*orientation*. In this problem, the task is to orient the edges in a two-dimensional grid such that each (internal) node has indegree 1, 3, or 4. This problem served as an example of computational algorithm synthesis in prior work [13]. It is a nontrivial problem for algorithm designers. In fact, it is the most constrained orientation problem that is still $O(\log^* n)$-round solvable—for any $X \subsetneq \{1, 3, 4\}$ the analogous X-orientation problem is *not* solvable in $O(\log^* n)$ rounds [13].

The prior $O(\log^* n)$-round algorithm for $\{1, 3, 4\}$-orientation was discovered with computers [13]; this algorithm is in essence a large lookup table, and as such not particularly insightful from the perspective of human beings. It moreover made additional assumptions on the input (specifically, oriented grids).

With the help of mending we can design a much simpler algorithm for the same problem. We show that $\{1, 3, 4\}$-orientations are locally mendable, and we can patch them as follows:

- Let G be a two-dimensional grid, and let H be the subgraph induced by 3×3 nodes in G.

- Regardless of how the edges of G outside H are oriented, we can always orient the edges of H such that all nodes of H have indegree 1, 3, or 4.

Note that we can mend any partial orientation by simply patching the orientations in a 3×3 square around each hole. This way, we arrive at a very simple $O(\log^* n)$-round algorithm for the orientation problem. To show that this works, it is sufficient to show the *existence* of a good orientation of H. And since there are only 12 edges in H, it is quick to *find* a suitable orientation for any given scenario.

We give the proof of the mendability of the $\{1, 3, 4\}$-orientation problem in the full version of this work [9]. While we give a human-readable proof, we also point out that one can quickly verify that $\{1, 3, 4\}$-orientations are mendable using a simple computer program that enumerates all possible scenarios (there are only 1296 cases to check). This demonstrates the power of mending also as an *algorithm synthesis* tool: for some problems we can determine automatically with computers that they are locally mendable, and hence we immediately also arrive at simple and efficient algorithms for solving them.

2 Contributions and Key Ideas

Concept. Our first main contribution is the new notions of mendability and the mending radius. While ideas similar to mendability have been used in numerous papers before this work (see Sect. 3), we are not aware of prior work that has given a definition of mendability that is as broadly applicable as ours:

- Our definition is *purely graph-theoretic*, without any references to any model of computation.
- Our definition is applicable to the study of *any locally verifiable graph problem*; this includes all LCL problems (locally checkable labelings), as defined by Naor and Stockmeyer [41].

We emphasize that with our definition it makes sense to ask questions like, for example, what is the mendability of the minimal total dominating set problem [21] or the locally optimal cut problem [8]. Moreover, we aim at a simple definition that is as robust and universal as possible. As it does not refer to any model of computation, it is independent of the numerous modeling details that one commonly encounters in distributed computing (deterministic or randomized; private or public randomness; whether the nodes know the size of the network; if there are unique identifiers and what their range is; whether the message sizes are limited; synchronous or asynchronous; fault-free or fault-tolerant). The formal definition of mendability is presented in Sect. 4.

From Solvability to Mendability. As our second main contribution, we explore whether efficient mendability is *necessary* for efficient solvability. As we have already discussed in Sect. 1, it is well-known and easy to see that *local mendability implies efficient solvability*, not only in the centralized setting but also in parallel and distributed settings. One concrete example is the following result (here LCLs are a broad family of locally verifiable graph problems and LOCAL is the standard model of distributed computing; see Sect. 4.1 for the details): Let us consider an LCL problem Π. Then, if Π is $O(1)$-mendable, it is also solvable in $O(\log^* n)$ communication rounds in the LOCAL model. The proof of this result is presented in Sect. 5.

However, we are primarily interested in exploring the converse: given e.g. an LCL problem Π that is solvable in $O(\log^* n)$ rounds in the LOCAL model, what can we say about the mendability of Π? Is local mendability not only sufficient but also necessary for efficient solvability?

It turns out that the answer to this is not straightforward. We first consider the simplest possible case of *unlabeled paths and cycles*. This is a setting in which local solvability is very well understood [13,18,41]. We show the following connections between solvability and mendability:

1. There are LCL problems that are $O(\log^* n)$-round solvable but $\Omega(n)$-mendable in paths and cycles.
2. However, for every LCL problem Π that is $O(\log^* n)$-round solvable in paths and cycles, there exists a *restriction* $\Pi' \subseteq \Pi$ such that Π' is $O(\log^* n)$-round solvable and also $O(1)$-mendable. There is an efficient algorithm that constructs Π'.

The second point states that we can always turn any locally solvable LCL problem into a locally mendable one, without making it any harder to solve. In this sense, in paths and cycles, local mendability is *equivalent* to efficient solvability. These results are discussed in Sect. 6.

Let us next consider a more general case. The first natural step beyond paths and cycles is rooted trees; for problems similar to graph coloring (more precisely, for so-called edge-checkable problems) their local solvability in rooted trees is asymptotically equal to their local solvability in directed paths [18], and hence one might expect to also get a very similar picture for the relation between local mendability and local solvability. However, the situation is much more diverse:

1. First, the above idea of restrictions does not hold in rooted trees. We show that for 3-coloring there is no restriction that is simultaneously $O(\log^* n)$-round solvable and $O(1)$-mendable.
2. However, one can always augment a problem Π with auxiliary labels to construct a new problem Π^* that is $O(\log^* n)$-round solvable if Π is; furthermore, any solution of Π^* can be projected to a feasible solution of Π, and Π^* is $O(1)$-mendable. This works not only in rooted trees but also in general graphs.

Mendability on rooted trees is discussed in more detail in the full version of this work [9]. The key open question in this line of research is the following:

Open question 1. Can we develop *efficient, general-purpose techniques* that can turn any locally solvable problem into a concise and natural locally mendable problem?

Landscape of Mendability. Our third main contribution is that we initiate the study of the *landscape* of mendability, in the same spirit as what has been done recently for the local solvability of LCL problems [7,12,16,17]. We ask what are possible functions T such that there exists an LCL problem that is $T(n)$-mendable in graphs with n nodes. In this work, we show the following results:

1. In cycles and paths, there are only two possible classes: $O(1)$-mendable problems and $\Theta(n)$-mendable problems.
2. In trees, there are exactly three classes: $O(1)$-mendable, $\Theta(\log n)$-mendable, and $\Theta(n)$-mendable problems.
3. In general bounded-degree graphs, there are additional classes; we show that e.g. $\Theta(\sqrt{n})$-mendable problems exist.

These results are presented in Sect. 7. The key open question in this line of research is to complete the characterization of mendability in general graphs:

Open question 2. Are there LCL problems that are $\Theta(n^{\alpha})$-mendable for all rational numbers $0 < \alpha < 1$? For all algebraic numbers $0 < \alpha < 1$?

Open question 3. Are there LCL problems that are $T(n)$-mendable for some $T(n)$ that is between $\log^{\omega(1)} n$ and $n^{o(1)}$?

An Application. We highlight one example of nontrivial corollaries of our work: in trees, any $o(n)$-mendable problem can be solved in $O(\log n)$ rounds in the LOCAL model (see Corollary 6).

3 Related Work

The underlying idea of local mending is not new. Indeed, similar ideas have appeared in the literature over at least three decades, often using terms such as *fixing*, *correcting*, or, similar to our work, *mending*. However, none of the prior work that we are aware of captures the ideas of mendability and mending radius that we introduce and discuss in this work.

Despite some dissimilarities, our work is heavily inspired by the extensive prior work. The graph coloring algorithm by Chechik and Mukhtar [19] serves as a convincing demonstration of the power of the mendability as an algorithm design technique. Discussion on the maximal independent set problem in Kutten and Peleg [38] and König and Wattenhofer [32] highlights the challenges of choosing appropriate definitions of "partial solutions" for locally checkable problems.

Mending as an Ad-Hoc Tool. The idea of mendability has often been used as a tool for designing algorithms for various graph problems of interest. However, so far the idea has not been introduced as a general concept that one can define for any locally checkable problem, but rather as specific algorithmic tricks that can be used to solve the specific problems at hand.

For example, the key observation of Panconesi and Srinivasan [42] can be phrased such that Δ-coloring is $O(\log n)$-mendable. In their work, they show how this result leads to an efficient distributed algorithm for Δ-coloring. Aboulker et al. [1] and Chierichetti and Vattani [20] study list coloring from a similar perspective. Barenboim [10] uses the idea of local correction in the context of designing fast distributed algorithms for the $(\Delta + 1)$-coloring problem.

Harris et al. [29] show that an edge-orientation with maximum out-degree $a^*(1 + \varepsilon)$ can be solved using "local patching" or, in other words, mending; a^* here denotes the pseudo-arboricity of the considered graph. The authors first show that this problem is $O(\log n/\varepsilon)$-mendable. They then use network decomposition of the graph and the mending procedure to derive an $O(\log^3 n/\varepsilon)$ round algorithm for $a^*(1 + \varepsilon)$-out-degree orientation.

Chechik and Mukhtar [19] present the idea of "removable cycles", and show how it leads to an efficient algorithm for 4-coloring triangle-free planar graphs (and 6-coloring planar graphs). When one applies their idea to 2-dimensional grids, one arrives at the observation that 4-coloring in grids is 2-mendable, as we saw in the warm-up example of Sect. 1.

Chang et al. [15] study similar ideas from the complementary perspective: they show that edge colorings with few colors are *not* easily mendable, and hence mending cannot be used (at least not directly) as a technique for designing efficient algorithms for these problems.

Mending with Advice. The work by Kutten and Peleg [37,38] is very close in spirit to our work. They study the *fault-locality* of mending: the idea is that the cost of mending only depends on the number of failures, and is independent of the number of nodes. However, one key difference is that they assume that the solution of a graph problem is augmented with some auxiliary precomputed data structure that can be used to assist in the process of mending. They also show that this is highly beneficial: *any* problem can be mended with the help of auxiliary precomputed data structures. We note that the addition of auxiliary labels is similar in spirit to the brute-force construction discussed in the full version of this work [9] that turns any locally solvable problem into an equivalent locally mendable one.

Censor-Hillel et al. [14] also comes close to our work with their definition of locally-fixable labelings (LFLs). In essence, LFLs are specific LCLs that can be (by construction) mended fast. In our work we seek to understand which LCLs (other than these specific constructions) can be mended fast, and also look beyond the case of constant-radius mendability.

Mending with the help of advice is also closely connected to distributed verification with the help of *proofs*. The idea is that a problem is not locally verifiable as such, but it can be made locally verifiable if we augment the solution with a small number of proof bits; see e.g. Korman and Kutten [33,34], Korman et al. [35,36], and Göös and Suomela [28].

Mending Small Errors. Another key difference between our work and that of e.g. Kutten and Peleg [37,38] is that we want to be able to mend holes in *any* partial solution, including one in which there is a linear number of holes. In particular, mending as defined in this work is something one can use to efficiently compute a correct solution from scratch.

In the work by Kutten and Peleg [37,38], the cost of fixing can be very high if the number of faults is e.g. linear. The work by König and Wattenhofer [32] is similar in spirit: they assume that holes are small and well-separated in space or time.

Making Algorithms Fault-Tolerant. There is an extensive line of research of systematically turning existing efficient distributed algorithms into dynamic or fault-tolerant algorithms; examples include Afek and Dolev [2], Awerbuch et al. [4], Awerbuch and Sipser [5], Awerbuch and Varghese [6], Ghosh et al. [27], and Lenzen et al. [39]. This line of research is distinct from our work, where the starting point is a property of the problem (mendability); then both efficient distributed algorithms and fault-tolerant algorithms follow.

4 Defining Mendability

Starting Point: Edge-Checkable Problems. In Sect. 1, we chose vertex coloring as an introductory example. Vertex coloring is an *edge-checkable* problem, i.e., the problem can be defined in terms of a relation that specifies what label combinations at the endpoints of each edge are allowed. For such a problem the notion of *partial solutions* is easy to define.

More precisely, let $G = (V, E)$ be a graph, and let Π be the problem of finding a k-vertex coloring in G; a feasible solution is a labeling $\lambda\colon V \to \{1, 2, \ldots, k\}$ such that for each edge $\{u, v\} \in E$, we have $\lambda(u) \neq \lambda(v)$. We can then easily *relax* Π to the problem Π^* of finding a *partial* k-vertex coloring as follows: a feasible solution is a function $\lambda\colon V \to \{\bot, 1, 2, \ldots, k\}$ such that for each edge $\{u, v\} \in E$, we have $\lambda(u) = \bot$ or $\lambda(v) = \bot$ or $\lambda(u) \neq \lambda(v)$. Here we used $\bot$ to denote a missing label. The same idea could be generalized to *any* problem Π that we can define using some edge constraint C:

- Π: function $\lambda\colon V \to \Gamma$ such that each edge $\{u, v\} \in E$ satisfies $\big(\lambda(u), \lambda(v)\big) \in C$.
- Π^*: function $\lambda\colon V \to \{\bot\} \cup \Gamma$ such that each edge $\{u, v\} \in E$ with $\bot \notin \{\lambda(u), \lambda(v)\}$ satisfies $\big(\lambda(u), \lambda(v)\big) \in C$.

Note that our original definition Π was edge-checkable, and we arrived at a definition Π^* of partial solutions for Π such that Π^* is still edge-checkable. This is important—the problem definition itself remained as local as the original problem.

However, most of the problems that we encounter e.g. in the theory of distributed computing are *not* edge-checkable; we will next discuss how to deal with any locally verifiable problem. While most of our results deal with LCL problems, which are a specific family of locally verifiable problems (see Sect. 4.1), we believe mendability will find applications also outside this family, and therefore we define mendability in full generality for any locally verifiable problem.

Locally Verifiable Problems. In general, a *locally verifiable problem* Π is defined in terms of a set of input labels Σ, a set of output labels Γ, and a *local verifier* ψ. In problem Π we are given an input graph $G = (V, E, \sigma)$ with a vertex set V, an edge set E, and some input labeling $\sigma\colon V \to \Sigma$, and the task is to find an output labeling or *solution* $\lambda\colon V \to \Gamma$ that makes the verifier ψ *happy* at each node $v \in V$. In general a verifier ψ is a function that maps a (G, λ, v) to "happy" or "unhappy", and we say that ψ *accepts* (G, λ) if $\psi(G, \lambda, v) = \textit{happy}$ for all $v \in V$.

Finally, verifier ψ is a *local* verifier with verification radius $r \in \mathbb{N}$, if $\psi(G, \lambda, v)$ only depends on the input and output within radius-v neighborhood of v. That is, $\psi(G, \lambda, v) = \psi(G', \lambda', v')$ if the radius-r neighborhood of v in G (together with the input and output labels) is isomorphic to the radius-r neighborhood of v' in G'. Note that we can also generalize the definitions in a straightforward manner from node labelings to edge labelings. The verification radius of a locally verifiable problem Π is the smallest $r \in \mathbb{N}$ such that Π admits a verifier with verification radius r.

Now edge-checkable problems are clearly locally verifiable problems, with verification radius $r = 1$. But there are also numerous problems that are locally verifiable yet not edge-checkable; examples include maximal independent sets (note that independence is edge-checkable while maximality is not) and more generally ruling sets, minimal dominating sets, weak coloring (at least one neighbor has to have a different label), distance-k coloring (nodes of the same color must have distance more than k), and many other constraint satisfaction problems.

Partial Solutions of Locally Verifiable Problems. To capture the mendability of a locally verifiable problem, we first need to have an appropriate notion of partial solutions. Ideally, we would like to be able to handle *any* locally verifiable problem Π and define a relaxation Π^* of Π with all the desirable properties as what we had in the graph coloring example:

(P1) Problem Π^* captures the intuitive idea of partial solutions for Π, and it serves the purpose of forming the foundation for the notion of mendability and mending radius of Π.
(P2) An empty solution (all nodes labeled with $\perp$) is a feasible solution for Π^*.
(P3) Problem Π^* is a relaxation of Π: any feasible solution of Π is a feasible solution for Π^*.
(P4) A feasible solution for Π^* without any empty labels is also a feasible solution for Π.
(P5) The definition of Π^* is exactly as local as the definition of Π: if Π is defined in terms of labelings in the radius-r neighborhoods, so is Π^*.

It turns out that there is a definition with all of these properties, and it is surprisingly simple to state. Let Π be a locally verifiable problem with the set Γ of output labels and local verifier ψ with verification radius r. By definition, $\psi(G, \lambda, v)$ only depends on the radius-r neighborhood of v.

We can define a new verifier ψ^* that extends the domain of ψ in a natural manner so that $\psi^*(G, \lambda, v)$ is well-defined also for partial labelings $\lambda\colon V \to \Gamma^*$, where $\Gamma^* = \{\bot\} \cup \Gamma$, as follows:

> If there is a node u within distance r from v with $\lambda(u) = \bot$, let $\psi^*(G, \lambda, v) =$ *happy*. Otherwise let $\lambda'\colon V \to \Gamma$ be any function that agrees with λ in the radius-r neighborhood of v, and let $\psi^*(G, \lambda, v) = \psi(G, \lambda', v)$.

Note that such a λ' always exists, and ψ^* is independent of the choice of λ', so ψ^* is well-defined. Furthermore, if $\lambda\colon V \to \Gamma$ is a complete labeling, then ψ^* and ψ agree everywhere, and an empty labeling makes ψ^* happy everywhere. Finally, $\psi^*(G, \lambda, v)$ only depends on the radius-r neighborhood of v. If we now define problem Π^* using the local verifier ψ^*, it clearly satisfies properties (P2)–(P5). Let us now see how to use it to formalize the notion of mendability, and this way also establish (P1).

Mendability of Local Verifiers. We first define mendability with respect to a specific verifier ψ; as before, ψ^* is the relaxation that also accepts partial labelings. Our definition is minimalistic: we only require that we can make *one unit of progress* by turning any given empty label into a non-empty label—this will be both sufficient and convenient. The key definition is this:

> Let $\lambda\colon V \to \Gamma^*$ be a partial labeling of G such that ψ^* accepts λ. We say that $\mu\colon V \to \Gamma^*$ is a *t-mend of λ at node v* if:
>
> 1. ψ^* accepts μ,
> 2. $\mu(v) \neq \bot$,
> 3. $\mu(u) = \bot$ implies $\lambda(u) = \bot$,
> 4. $\mu(u) \neq \lambda(u)$ implies that u is within distance t of v.

That is, in μ we have applied a radius-t patch around node v. If there are some other empties around v, they can be left empty (note that this will never make mending harder, as empty nodes only help to make ψ^* happy, and this will not make mending substantially easier, either, as we will need to be able to eventually also patch all the other holes).

Fix a graph family $\mathcal{G}$ of interest. We define the mending radius of ψ for the entire family:

Definition 1. *Let $T\colon \mathbb{N} \to \mathbb{N}$ be a function. We say that a verifier ψ is T-mendable if for all graphs $G \in \mathcal{G}$ and all partial labelings λ accepted by ψ^*, there exists a $T(|V|)$-mend of λ at v for any $v \in V$.*

In general, the mending radius may depend on the number of nodes in the graph; it is meaningful to say that, for example, ψ is $\Theta(\log n)$-mendable, i.e., the mending radius of ψ is $\Theta(\log n)$. We use $n = |V|$ throughout this work to denote the number of nodes in the input graph.

Mendability of Locally Verifiable Problems. So far we have defined the mendability of a particular verifier ψ. In general, the same graph problem Π may be defined in terms of many equivalent verifiers ψ (for example, vertex coloring can be locally verified with a local verifier ψ that checks the coloring in the radius-7 neighborhoods, even if this is not necessary). Indeed, there are problems for which it is not easy to define a "canonical" verifier, and it is not necessarily the case that the smallest possible verification radius coincides with the smallest possible mending radius. Hence we generalize the idea of mendability from local verifiers to locally verifiable problems in a straightforward manner:

Definition 2. *A problem Π is T-*mendable *if for some $r \in \mathbb{N}$ there exists a T-mendable radius-r verifier for the problem Π. The* mending radius *of a problem Π is T if Π is T-mendable but not T'-mendable for any $T' \neq T$ such that $T'(n) \leq T(n)$.*

Now we have formally defined the mendability of any locally verifiable problem, in a way that makes it applicable to any locally checkable problem. The definition is a natural generalization of the informal idea of mendability for vertex coloring discussed in the introduction. We refer to the full version of this work [9] for some further discussion and examples.

4.1 Additional Definitions

LCL Problems. We have defined mendability for any locally verifiable problems, but a particularly important special case of locally verifiable problems is the family of *locally checkable problems* (LCL problems), as defined by Naor and Stockmeyer [41]. We say that a locally verifiable problem Π on a graph family $\mathcal{G}$ is an LCL problem if we have that

1. the set Σ of input labels and the set Γ of output labels are both finite,
2. $\mathcal{G}$ is a family of bounded-degree graphs, i.e., there is some constant Δ such that for any node v in any graph $G \in \mathcal{G}$ the degree of v is at most Δ.

Note that an LCL problem always has a *finite description*: we can simply list all possible non-isomorphic labeled radius-r neighborhoods and classify them based on whether they make the local verifier ψ happy.

LOCAL Model. Mendability is independent of any model of computing. However, the key applications for the concept are in the context of distributed graph algorithms. For concreteness, we use the LOCAL model [40,43] of distributed computing throughout this work. In this model, the distributed system is represented as a graph $G = (V, E)$, where each node $v \in V$ denotes a processor and every edge $\{u, v\} \in E$ corresponds to a direct communication link between nodes u and v. At the start of the computation, every node receives some local input.

The computation proceeds synchronously in discrete communication rounds and each round consists of three steps: (1) all nodes send messages to their neighbors, (2) all nodes receive messages from their neighbors, and (3) all nodes update their local state. In the last round, all nodes declare their local output. An algorithm has running time T if all nodes can declare their local output after T communication rounds. The bandwidth of the communication links is unbounded, i.e., in each round nodes can exchange messages of any size. We say that a problem is *T-solvable* if it can be solved in the LOCAL model in $T(n)$ communication rounds.

5 From Local Mendability to Local Solvability

In this section, we show that, in some cases, a bound on the mending radius implies an upper bound on the time complexity of a problem in the LOCAL model. Hence, the concept of mendability can be helpful in the process of designing algorithms in the distributed setting.

We start by proving a generic result, that relates mendability with network decomposition; we will make use of the following auxiliary lemma.

Lemma 1. *Let Π be an LCL problem with mending radius $f(n)$ and verification radius $r = O(1)$. Then we can create a mending procedure that only depends on the nodes at distance $f(n) + r$ from the node u that needs to be mended, and it does not even need to know n.*

Proof. We first show that it is sufficient to inspect the $(f(n) + r)$-radius neighborhood of u. We start from u and we inspect its $(f(n)+r)$-radius neighborhood, where $r = O(1)$ is the verification radius of Π. Since we know that there exists a $f(n)$-mend at u, and since the output of a node v may only affect the correctness of the outputs of the nodes at distance at most r from v, then it is possible to find a correct mend by brute force. We now remove the dependency on n as follows. We start by gathering the neighborhood of u at increasing distances and at each step we check if there is a feasible mend by brute force. This procedure must stop after at most $f(n)+r$ steps, since we know that such a solution exists.

We now show that *network decompositions* can be used to relate the mending radius of a problem with its distributed time complexity. A *(c, d)-network decomposition* is a partition of the nodes into c color classes such that within each color class, each connected component has diameter at most d [3]. Also, recall that G^i, the i-th power of $G = (V, E)$, is the graph (V, E') satisfying that $\{u, v\} \in E'$ if and only if u and v are at distance at most i in G.

Theorem 1. *Let Π be an LCL problem with mending radius k and verification radius r. Then Π can be solved in $O(cd(k+r))$ rounds in the LOCAL model if we are given a (c,d)-network decomposition of G^{2k+r}.*

Proof. We prove the claim by providing an algorithm for solving Π. We start by temporarily assigning $\bot$ to all nodes. Then, we process the nodes in c phases. In phase i, we mend all nodes that are in components of color i, denoted C_i, in parallel as follows. By Lemma 1, for each node v, we do not need to see the whole graph to find a valid k-mend at v, but only nodes that, in G, are at distance at most $k+r$ from v. This implies that we can find a valid mend for all nodes of each component by gathering the whole component and the nodes at a distance of at most $k+r$ from it. This mend only needs to modify the solution of nodes inside the component and nodes at distance at most k from it. Since we are given a network decomposition of G^{2k+r}, in G, nodes of different components are at distance strictly larger than $2k+r$ from each other. This implies that the mend applied on some component C_1 does not modify the temporary solution of nodes at distance of at most $k+r$ from some other component $C_2 \neq C_1$.

Hence, we obtain the same result that we would have obtained by mending each component of color i sequentially. Since we process all color classes and perform a valid mending, at the end, no node is labeled $\bot$, and hence the temporary labeling is a valid solution for Π. Each connected component has diameter at most $O(d(k+r))$ in G, so each phase requires $O(d(k+r))$ rounds. The total running time is $O(cd(k+r))$.

As a corollary of this theorem, we show that in order to prove an $O(\log^* n)$ upper bound on the time complexity of a problem, it is enough to prove that a solution can be mended by modifying the labels within a constant distance.

Corollary 1. *Let Π be an LCL problem with constant mending radius. Then Π can be solved in $O(\log^* n)$ rounds in the LOCAL model.*

Proof. We prove the claim by providing an algorithm running in $O(\log^* n)$. Let $k = O(1)$ be the mending radius of Π and r be its verification radius.

We start by computing a distance-$(2k+r)$ coloring using a palette of $c = \Delta^{2k+r} + 1 = O(1)$ colors. This can be done in $O(\log^* n)$ rounds by applying a c-coloring algorithm to the $(2k+r)$-th power of G; there is a wide variety of fast $(\Delta+1)$-coloring algorithms that we can use here, see, e.g., Barenboim and Elkin's book [11] on distributed graph coloring for more information.

Note that such a coloring is a $(c,1)$ network decomposition of G^{2k+r}, and we can hence apply Theorem 1 to solve Π in constant time.

The bound $O(\log^* n)$ in the above corollary is the best possible: for example, vertex coloring with $\Delta+1$ colors is an LCL problem with a constant mending radius, and solving this problem requires $\Omega(\log^* n)$ rounds in the LOCAL model [40]. Indeed, there is a broad class of LCL problems with complexity exactly $\Theta(\log^* n)$ rounds, and next we start to study their connection with mendability.

6 From Local Solvability to Local Mendability

In Sect. 5, we saw that local mendability implies local solvability. In this section, we consider the converse: does local solvability imply local mendability? First, we show that mending can be much harder than solving by considering an edge-checkable problem on undirected paths.

Theorem 2. *There are LCL problems that are $\Theta(\log^* n)$-round solvable and $\Theta(n)$-mendable.*

Proof. Consider the following LCL problem Π on undirected paths: nodes can either 2-color the path using labels $\{A, B\}$, or 3-color the path using labels $\{1, 2, 3\}$, but they cannot mix the labels in the solution.

Solving this problem requires $\Theta(\log^* n)$ time, as it is necessary and sufficient to produce a 3-coloring [40]. We now prove that this LCL is $\Theta(n)$-mendable. Consider a path $P = (p_0, p_1, \ldots)$ of length $n = 2k + 1$ for an even k, and the following partial solution on this path:

$$\underbrace{A, B, A, B, \ldots, A, B}_{k}, \bot, \underbrace{A, B, A, B, \ldots, A, B}_{k}.$$

Note that there are two regions that are labeled with a valid 2-coloring, these regions are separated by a node labeled $\bot$, and the two 2-colorings are not compatible, meaning that it is not possible to assign a label to p_k such that the LCL constraints are satisfied on both its incident edges.

We argue that mending this solution requires linear distance. Observe first that mending this solution using labels $\{1, 2, 3\}$ would require us to undo the labelings of all nodes. This is because the LCL constraints require that no node of the graph should be labeled $\{A, B\}$ in order to use labels $\{1, 2, 3\}$. Since there are nodes labeled $\{A, B\}$ at distance $k = \Theta(n)$ from p_k, changing the labels of these nodes would require linear time. The remaining option is to mend the solution by only using labels $\{A, B\}$. In this case, at least half of the nodes of the path need to be relabeled in order to produce a valid 2-coloring and satisfy the constraints.

6.1 Cycles

In Theorem 2, we showed that local solvability does not imply local mendability. The main idea of the counterexample was to use two different sets of labels $\{A, B\}$ and $\{1, 2, 3\}$ that cannot be both part of the same solution. In order to make this problem efficiently mendable, we could remove A and B from the set of possible labels and only allow the labels $\{1, 2, 3\}$. The restricted problem would still be $O(\log^* n)$-solvable and, in fact, the new problem would be $O(1)$-mendable.

We will now generalize this idea and show how to find a locally mendable restriction for any LCL problem in the case of unlabeled paths and cycles. To find such a restriction, we will make use of so-called *diagram representations* of the LCL problems [13, 18].

Let us first consider the simplest case: directed cycles with no inputs. In this case we can represent any LCL problem Π as a directed graph D, where each node represents a feasible neighborhood (a radius-r neighborhood that makes the verifier happy) and a directed edge (x, y) represents that the neighborhoods x and y are compatible (in a directed cycle, it is possible to have an edge (u, v) such that the radius-r neighborhood of u is x and the radius-r neighborhood of v is y) [13]. Here D is the *diagram* of Π.

Example 1. Consider first the problem of 3-coloring directed cycles. The feasible neighborhoods are $121, 123, 131, 132, 212, \ldots, 323$; for example, 123 represents a node with color 2 whose predecessor has color 1 and successor has color 3. Compatible neighborhoods include e.g. $(123, 231)$ and $(123, 232)$. The diagram has 12 nodes and 24 edges.

Next consider the problem of 2-coloring directed cycles. The only feasible neighborhoods are 121 and 212, and the only compatible neighborhoods are $(121, 212)$ and $(212, 121)$. The diagram has 2 nodes and 2 edges.

A key property of diagram representations is that directed walks in the diagram capture feasible solutions: If we consider a feasible solution of Π in a directed cycle, and walk along the cycle in the positive direction, the sequence of radius-r neighborhoods that we encounter corresponds to a walk in D. Conversely, each walk in D represents one possible way to label a fragment of a cycle.

Another key idea that we need is the notion of a *flexible node* [13]: we say that node x in D is flexible if there is a natural number K such that there exists a self-returning walk $x \rightsquigarrow x$ of any length $k \geq K$. Here the K is the *flexibility parameter* of x.

Example 2. In the diagram of 3-coloring, all states are flexible, while in the diagram of 2-coloring, none of the states are flexible (there is no self-returning walk of any odd length).

It turns out that Π is $O(\log^* n)$-solvable if and only if there is a flexible state in the diagram [13]. Now we are ready to prove the following result:

Theorem 3. *Suppose Π is an LCL problem on directed cycles with no input. If Π is $O(\log^* n)$-solvable, we can define a new LCL problem Π' with the same round complexity, such that a solution for Π' is also a solution for Π, and Π' is $O(1)$-mendable.*

Proof. Let D be the diagram of Π. Since Π is $O(\log^* n)$-solvable, D contains at least one flexible state f. Let D' be the strongly connected component of D that contains f. Let Π' be the LCL induced by D'.

Note that all self-returning walks from f to f in D are also contained in D'; hence f is flexible also in D', and Π' can be solved in $O(\log^* n)$ rounds. We still need to show that Π' is $O(1)$-mendable.

Let K be the flexibility parameter of f in D', let L be the number of nodes in D', and define $R = K + 2L$. Consider any two nodes x and y of D'. We claim that there always exists a walk of length exactly R from x to y in D'. To see this, recall that D' is strongly connected, and therefore we can find a path $x \rightsquigarrow f$ of some length ℓ_1, and a path $f \rightsquigarrow y$ of some length ℓ_2; these paths cannot have repeated nodes, so we have $\ell_1 < L$ and $\ell_2 < L$. Node f is flexible, so we can find a walk $f \rightsquigarrow f$ with exactly $R - \ell_1 - \ell_2 > K$ edges. Put together, we have a walk $x \rightsquigarrow f \rightsquigarrow f \rightsquigarrow y$ of length R.

Now we show how to mend the solution at some node u that is labeled with $\bot$. First consider the case that there are nodes s and t such that s is before u along the cycle, t is after u along the cycle, both s and t are labeled with $\bot$, and the distance from s to t is at most $2R + 2r$. Then we can relabel the fragment between s and t arbitrarily (by following some walk in D').

Otherwise we can find nodes s and t such that s is before u along the cycle, t is after u along the cycle, the distance between s and t is exactly R, and radius-r neighborhoods of s and t do not contain any empty labels. Then we can find a walk W of length R from x to y in diagram D', and re-label the path between s and t according to W.

In each case we have a T-mend at u with $T \leq 2R + 2r = O(1)$.

We can prove similar results also for directed paths, undirected cycles, and undirected paths; we refer to the full version of this work [9] for the details, as well as for discussion on the case of rooted trees.

7 Landscape of Mendability

In this section, we analyze the structure of the landscape of mendability in three settings: (1) cycles and paths, (2) trees, and (3) general bounded-degree graphs. We state the main results here; we refer to the full version of this work [9] for the proofs.

We start with a general result that we can use to prove a gap in all of these three settings.

Theorem 4. *Let $\mathcal{G}_{G_\Delta}$, $\mathcal{G}_{T_\Delta}$, and $\mathcal{G}_C$, be, respectively, the family general graphs of maximum degree Δ, the family of trees of maximum degree Δ, and cycles (in this case $\Delta = 2$). Let $d(n) = n$ if $\Delta \leq 2$, and $d(n) = \log n$ otherwise. Let $\mathcal{G}$ be one of the above families. There is no LCL problem Π defined on $\mathcal{G}$ that has mending radius between $\omega(1)$ and $o(d(n))$.*

Now the case of cycles and paths is fully understood. Theorem 4 implies the following corollaries:

Corollary 2. *There is no LCL problem with mending radius between $\omega(1)$ and $o(n)$ on cycles.*

Corollary 3. *There is no LCL problem with mending radius between $\omega(1)$ and $o(n)$ on paths.*

That is, there are only two possible classes: $O(1)$-mendable problems and $\Theta(n)$-mendable problems.

In the case of trees, Theorem 4 implies a gap between $\omega(1)$ and $o(\log n)$:

Corollary 4. *There is no LCL problem with mending radius between $\omega(1)$ and $o(\log n)$ on trees.*

One cannot make the gap wider: there are problems that are $\Theta(\log n)$-mendable. We can however prove another gap above $\Theta(\log n)$:

Theorem 5. *There is no LCL problem with mending radius between $\omega(\log n)$ and $o(n)$ on trees.*

The mendability landscape on general graphs looks different from the one on cycles and trees. Theorem 4 again implies a gap between $\omega(1)$ and $o(\log n)$:

Corollary 5. *There are no LCL problems with mending radius between $\omega(1)$ and $o(\log n)$ on general graphs.*

However, there is no gap between $\omega(\log n)$ and $o(n)$:

Theorem 6. *There exist $\Theta(\sqrt{n})$-mendable LCL problems on general bounded-degree graphs.*

Finally, by putting together Theorems 1 and 5, we can derive the following result:

Corollary 6. *Let Π be an LCL problem defined on trees with $o(n)$ mending radius. Then Π can be solved in $O(\log n)$ rounds in the LOCAL model.*

Acknowledgements. This project has received funding from the European Union's Horizon 2020 research and innovation programme under the Marie Skłodowska-Curie grant agreement No 840605. This work was supported in part by the Academy of Finland, Grants 314888 and 333837. The authors would also like to thank David Harris, Neven Villani, and the anonymous reviewers for their very helpful comments and feedback on previous versions of this work.

References

1. Aboulker, P., Bonamy, M., Bousquet, N., Esperet, L.: Distributed coloring in sparse graphs with fewer colors. Electr. J. Comb. **26**(4) (2019). https://doi.org/10.37236/8395
2. Afek, Y., Dolev, S.: Local stabilizer. In: PODC (1997). https://doi.org/10.1145/259380.259505
3. Awerbuch, B., Berger, B., Cowen, L., Peleg, D.: Fast distributed network decompositions and covers. J. Parallel Distrib. Comput. **39**(2), 105–114 (1996). https://doi.org/10.1006/jpdc.1996.0159
4. Awerbuch, B., Patt-Shamir, B., Varghese, G.: Self-stabilization by local checking and correction. In: FOCS (1991). https://doi.org/10.1109/SFCS.1991.185378
5. Awerbuch, B., Sipser, M.: Dynamic networks are as fast as static networks. In: FOCS (1988). https://doi.org/10.1109/SFCS.1988.21938

6. Awerbuch, B., Varghese, G.: Distributed program checking: a paradigm for building self-stabilizing distributed protocols. In: FOCS (1991). https://doi.org/10.1109/SFCS.1991.185377
7. Balliu, A., Hirvonen, J., Korhonen, J.H., Lempiäinen, T., Olivetti, D., Suomela, J.: New classes of distributed time complexity. In: STOC (2018). https://doi.org/10.1145/3188745.3188860
8. Balliu, A., Hirvonen, J., Lenzen, C., Olivetti, D., Suomela, J.: Locality of not-so-weak coloring. In: Censor-Hillel, K., Flammini, M. (eds.) SIROCCO 2019. LNCS, vol. 11639, pp. 37–51. Springer, Cham (2019). https://doi.org/10.1007/978-3-030-24922-9_3
9. Balliu, A., Hirvonen, J., Melnyk, D., Olivetti, D., Rybicki, J., Suomela, J.: Local mending (2022). https://arxiv.org/abs/2102.08703
10. Barenboim, L.: Deterministic $(\Delta + 1)$-coloring in sublinear (in Δ) time in static, dynamic, and faulty networks. J. ACM **63**(5) (2016). https://doi.org/10.1145/2979675
11. Barenboim, L., Elkin, M.: Distributed graph coloring: fundamentals and recent developments. Morgan Claypool (2013). https://doi.org/10.2200/S00520ED1V01Y201307DCT011
12. Brandt, S., et al.: A lower bound for the distributed Lovász local lemma. In: STOC (2016). https://doi.org/10.1145/2897518.2897570
13. Brandt, S., et al.: LCL problems on grids. In: PODC (2017). https://doi.org/10.1145/3087801.3087833
14. Censor-Hillel, K., Dafni, N., Kolobov, V.I., Paz, A., Schwartzman, G.: Fast deterministic algorithms for highly-dynamic networks. In: OPODIS (2021). https://doi.org/10.4230/LIPIcs.OPODIS.2020.28
15. Chang, Y.J., He, Q., Li, W., Pettie, S., Uitto, J.: The complexity of distributed edge coloring with small palettes. In: SODA (2018). https://doi.org/10.1137/1.9781611975031.168
16. Chang, Y.J., Kopelowitz, T., Pettie, S.: An exponential separation between randomized and deterministic complexity in the LOCAL model. In: FOCS (2016). https://doi.org/10.1109/FOCS.2016.72
17. Chang, Y.J., Pettie, S.: A time hierarchy theorem for the LOCAL model. SIAM J. Comput. **48**(1), 33–69 (2019). https://doi.org/10.1137/17M1157957
18. Chang, Y.-J., Studený, J., Suomela, J.: Distributed graph problems through an automata-theoretic lens. In: Jurdziński, T., Schmid, S. (eds.) SIROCCO 2021. LNCS, vol. 12810, pp. 31–49. Springer, Cham (2021). https://doi.org/10.1007/978-3-030-79527-6_3
19. Chechik, S., Mukhtar, D.: Optimal distributed coloring algorithms for planar graphs in the LOCAL model. In: SODA (2019). https://doi.org/10.1137/1.9781611975482.49
20. Chierichetti, F., Vattani, A.: The local nature of list colorings for graphs of high girth. SIAM J. Comput. **39**(6), 2232–2250 (2010). https://doi.org/10.1137/080732109
21. Cockayne, E.J., Dawes, R.M., Hedetniemi, S.T.: Total domination in graphs. Networks **10**(3), 211–219 (1980). https://doi.org/10.1002/net.3230100304
22. Demetrescu, C., Eppstein, D., Galil, Z., Italiano, G.F.: Dynamic graph algorithms. In: Algorithms and Theory of Computation Handbook: General Concepts and Techniques, chap. 9 (2010)
23. Dijkstra, E.W.: Self-stabilizing systems in spite of distributed control. Commun. ACM **17**(11), 643–644 (1974). https://doi.org/10.1145/361179.361202

24. Dolev, S.: Self-Stabilization. MIT Press, Cambridge (2000)
25. Erdős, P., Rubin, A.L., Taylor, H.: Choosability in graphs. In: Proceedings West Coast Conference on Combinatorics, Graph Theory and Computing (1980)
26. Foerster, K.T., Korhonen, J.H., Paz, A., Rybicki, J., Schmid, S.: Input-dynamic distributed algorithms for communication networks. In: SIGMETRICS (2021). https://doi.org/10.1145/3410220.3453923
27. Ghosh, S., Gupta, A., Herman, T., Pemmaraju, S.V.: Fault-containing self-stabilizing distributed protocols. Distrib. Comput. **20**(1), 53–73 (2007). https://doi.org/10.1007/s00446-007-0032-2
28. Göös, M., Suomela, J.: Locally checkable proofs. In: PODC (2011). https://doi.org/10.1145/1993806.1993829
29. Harris, D.G., Su, H.H., Vu, H.T.: On the locality of Nash-Williams forest decomposition and star-forest decomposition. In: PODC (2021). https://doi.org/10.1145/3465084.3467908
30. Henzinger, M.: The state of the art in dynamic graph algorithms. In: Tjoa, A.M., Bellatreche, L., Biffl, S., van Leeuwen, J., Wiedermann, J. (eds.) SOFSEM 2018. LNCS, vol. 10706, pp. 40–44. Springer, Cham (2018). https://doi.org/10.1007/978-3-319-73117-9_3
31. Holroyd, A.E., Schramm, O., Wilson, D.B.: Finitary coloring. Ann. Probab. **45**(5), 2867–2898 (2017). https://doi.org/10.1214/16-AOP1127
32. König, M., Wattenhofer, R.: On local fixing. In: Baldoni, R., Nisse, N., van Steen, M. (eds.) OPODIS 2013. LNCS, vol. 8304, pp. 191–205. Springer, Cham (2013). https://doi.org/10.1007/978-3-319-03850-6_14
33. Korman, A., Kutten, S.: On distributed verification. In: Chaudhuri, S., Das, S.R., Paul, H.S., Tirthapura, S. (eds.) ICDCN 2006. LNCS, vol. 4308, pp. 100–114. Springer, Heidelberg (2006). https://doi.org/10.1007/11947950_12
34. Korman, A., Kutten, S.: Distributed verification of minimum spanning trees. Distrib. Comput. **20**(4), 253–266 (2007). https://doi.org/10.1007/s00446-007-0025-1
35. Korman, A., Kutten, S., Peleg, D.: Proof labeling schemes. Distrib. Comput. **22**(4), 215–233 (2010). https://doi.org/10.1007/s00446-010-0095-3
36. Korman, A., Peleg, D., Rodeh, Y.: Constructing labeling schemes through universal matrices. Algorithmica **57**(4), 641–652 (2010). https://doi.org/10.1007/s00453-008-9226-7
37. Kutten, S., Peleg, D.: Fault-local distributed mending. J. Algorithms **30**(1), 144–165 (1999). https://doi.org/10.1006/jagm.1998.0972
38. Kutten, S., Peleg, D.: Tight fault locality. SIAM J. Comput. **30**(1), 247–268 (2000). https://doi.org/10.1137/S0097539797319109
39. Lenzen, C., Suomela, J., Wattenhofer, R.: Local algorithms: self-stabilization on speed. In: Guerraoui, R., Petit, F. (eds.) SSS 2009. LNCS, vol. 5873, pp. 17–34. Springer, Heidelberg (2009). https://doi.org/10.1007/978-3-642-05118-0_2
40. Linial, N.: Locality in distributed graph algorithms. SIAM J. Comput. **21**(1), 193–201 (1992). https://doi.org/10.1137/0221015
41. Naor, M., Stockmeyer, L.: What can be computed locally? SIAM J. Comput. **24**(6), 1259–1277 (1995). https://doi.org/10.1137/S0097539793254571
42. Panconesi, A., Srinivasan, A.: The local nature of Δ-coloring and its algorithmic applications. Combinatorica **15**(2), 255–280 (1995). https://doi.org/10.1007/BF01200759
43. Peleg, D.: Distributed Computing: A Locality-Sensitive Approach. SIAM (2000). https://doi.org/10.1137/1.9780898719772

Proof Labeling Schemes for Reachability-Related Problems in Directed Graphs

Yoav Ben Shimon(✉), Orr Fischer, and Rotem Oshman

Blavatnik School of Computer Science, Tel-Aviv University, Tel Aviv, Israel
{benshimon2,orrfischer,roshman}@mail.tau.ac.il

Abstract. We study proof labeling schemes in directed networks, and ask what assumptions are necessary to be able to certify reachability-related problems such as strong connectivity, or the existence of a node from which all nodes are reachable. In contrast to undirected networks, in directed networks, having unique identifiers alone does not suffice to be able to certify all graph properties; thus, we study the effect of knowing the size of the graph, and of each node knowing its out-degree. We formalize the notion of giving the nodes initial knowledge about the network, and give tight characterizations of the types of knowledge that are necessary and sufficient to certify several reachability-related properties, or to be able to certify any graph property. For example, we show that in order to certify that the network contains a node that is reachable from all nodes, it is necessary and sufficient to have any two of the assumptions we study (unique identifiers, knowing the size, or knowing the out-degree); and to certify strong connectivity, it is necessary and sufficient to have any single assumption.

1 Introduction

Proof labeling schemes (PLS) are a mechanism for certifying that a network has some desired property, by storing an efficiently-verifiable certificate at every node of the network. In the classical formalism introduced in [9], each node v is assigned a label $\ell(v) \in \{0,1\}^*$, and the property is verified by having each node examine its label and its neighbors' labels, and then decide whether to accept or reject. We require that if the network satisfies the property, then there must exist a labeling that causes all nodes to accept, whereas if the network does not satisfy the property, then for any labeling, some node must reject. The main complexity measure of proof labeling schemes is the length of the label assigned to each node. Subsequent work has considered many variations on this model, allowing nodes to send messages other than their label, use randomness, communicate further across the graph, and more.

Research funded by the Israel Science Foundation, Grant No. 2801/20, and also supported by Len Blavatnik and the Blavatnik Family foundation.

M. Parter (Ed.): SIROCCO 2022, LNCS 13298, pp. 21–41, 2022.
https://doi.org/10.1007/978-3-031-09993-9_2

Most work on proof labeling schemes considers undirected network graphs, where every edge of the graph represents bidirectional communication between its two endpoints. However, distributed networks in real world settings can sometimes have asymmetric behavior. Thus, in the current paper, we consider weakly-connected directed communication networks, and ask whether basic reachability-related properties can be certified by a proof labeling scheme in such networks.

The notion of a proof labeling scheme readily generalizes to directed networks, except that now, since communication is asymmetric, each node receives only the labels of its in-neighbors, not its out-neighbors. As a result, the ideas underlying some well-known proof labeling schemes for undirected networks may no longer work. For example, in an undirected network with unique identifiers, every graph property can be certified by a PLS where the prover gives all nodes the entire graph [8,9]; each node verifies that its neighbors received the same graph it did, and that its edges are represented accurately. In directed graphs, however, this scheme no longer works: even if we have unique identifiers and each node verifies that its in-neighbors are described by the prover accurately, we cannot verify that the prover did not add "fake" nodes to its description of the network graph, along with "fake" outgoing edges to those nodes (this will not necessarily be detected by any "real" node). Other well-known schemes such as the spanning-tree verification scheme from [9] also fail. This motivates us to ask: what is the minimal knowledge about the network that is required to restore our ability to decide any graph property? And what knowledge is required to certify specific important properties, such as strong connectivity, or the existence of a node that can reach or be reached from all other nodes in the network?

To study these questions, we formalize a notion of *initial knowledge*—reliable information that nodes initially have about the network: for example, nodes may initially know the size of the network, or they may have identifiers that are guaranteed to be unique. This notion separates information that is present prior to the verification process from information that can be gathered via communication.[1] Our main results are the following. For the problem of certifying the existence of a node that can be reached from all nodes, we show:

Theorem 1.1 (Informal). *To decide whether the network contains a* global sink*—a node that can be reached from all other nodes—it is necessary and sufficient to have two of the following assumptions: (1) the nodes have unique identifiers; (2) the size of the network is known to all nodes; (3) each node knows its out-degree.*

As for the existence of a node that can reach all other nodes, we show:

Theorem 1.2 (Informal). *To decide whether the network contains a* global source*—a node from which all nodes are reachable—it suffices to have unique identifiers, but without unique identifiers, even knowing the size of the network and the out-degrees does not suffice.*

[1] The latter type of knowledge, which is not given in advance but rather computed by the nodes during runtime, is addressed in [1].

For strong connectivity, we show:

Theorem 1.3 (Informal). *To decide whether the network is strongly connected, it is necessary and sufficient to either have unique identifiers, know the size of the network, or know the out-degrees.*

These results demonstrates the striking usefulness of knowing the out-degree of each node: surprisingly, even by itself, it already suffices to decide strong connectivity.

Finally, we show that in fact, if we have unique identifiers and know either the size of the graph or the out-degrees of the nodes, then any graph property can be certified, and we also study the usefulness of having a *marked global sink* in the graph, and show that while this can replace one of the assumptions above, by itself it does not suffice to decide all graph properties.

Our Techniques. To prove that a certain problem is undecidable under specific assumptions, we give simple examples of a pair of graphs that are indistinguishable to all nodes, but one graph has the property and the other does not. To show that a problem *is* decidable we construct a PLS.

Our constructions are based on several building blocks, which we introduce in Sect. 4. Several are known from prior work—e.g., using a DFS traversal—but their use in the context of directed networks, and with only partial knowledge of the network parameters, is not immediate. The most novel building block is an "infinite-round" protocol for anonymous networks that identifies all the nodes belonging to strongly-connected components that have no outgoing edges. In essense, the protocol calculates the distribution of an infinite random walk from a uniformly random vertex, and marks all the nodes whose visiting probability does not converge to zero. Interestingly, we show that despite this protocol having "infinite running time", it implies the existence of an $O(n^2)$-bit PLS (and in fact, we show in Sect. 4 that the existence of any proof labeling scheme—even one that uses infinitely-long labels—implies the existence of a $O(n^2)$-bit PLS). This allows us to show that knowledge of the out-degrees of the nodes is sufficient to certify strong connectivity, and it is also useful when certifying the existence of a global sink.

2 Related Work

Proof labeling schemes and related formalisms have seen a large amount of work since their introduction in [9]; we refer to the surveys of Feuilloley and Fraigniaud [2] and of Suomela [11] for a comprehensive overview of the field. Most prior work, however, is in *undirected* network graphs. Several papers study the effect of initial knowledge on decidability in proof labeling schemes: e.g., in [9] it is shown that in the absence of unique identifiers or any other knowledge about the network graph, some problems are undecidable (e.g., whether the graph is a tree). In [5], the size of the graph is considered, and in [8] it is shown that having unique identifiers, knowing the graph size, or having a unique leader are all equivalent up to an additive $O(\log n)$ term in the label size. We show that in directed networks the picture is more complicated (see Sect. 6).

It is proven in [8,9] that any graph property can be certified using $O(n^2)$-bit labels, provided we have unique identifiers. In the current paper we extend this result and show that for any combination of initial knowledge (which may or may not include UIDs), any *decidable* graph property can be decided using $O(n^2)$ bits per node. Our proof is similar in spirit to a proof from [4], where it is shown that when the prover is required to be UID-oblivious (the labels may not depend on the UIDs), any property that is decidable using UIDs can also be decided without UIDs. We note, however, that in the current paper we impose no restrictions on the prover, and incorporate arbitrary knowledge fields, so the correctness of the scheme in our case is quite different from [4].

The only prior work on proof labeling schemes in directed networks of which we are aware is [3,8]. In [8], an $O(\log \Delta)$-bit PLS is given for s-t-reachability in directed networks, and this is shown to be tight in [3]. Upper and lower bounds for cyclicity, acyclicity and certifying a spanning tree are also given in [3]; only anonymous networks are considered in [3], and no initial knowledge about the network graph is assumed. However, [3] observes that the PLS for certifying a spanning tree from [9] continues to work in directed networks, assuming we have a marked root node r, and the spanning tree is rooted at r and oriented away from it. In Sect. 5.1 we rely on a very similar observation, except that we do not require the existence of such a node r, but rather use a similar idea to certify that the network contains a global source.

Several of our protocols rely on a scheme presented in Sect. 4, where we essentially compute the distribution of a random walk initiated at a random node of the graph. A somewhat related idea was used in [10] to solve the broadcast problem in directed networks, or to assign unique identifiers in an anonymous networks, given a unique marked source node; however, the usage in [10] is quite different from ours.

3 Preliminaries

Notation and Graph Terminology. Given a set $S \subseteq X$ and a mapping $f : X \to Y$, let $f(S) = \{f(x) : x \in S\}$ denote image of S under f, as a *multiset* (i.e., if $x_1 \neq x_2 \in S$ have $f(x_1) = f(x_2)$, we will have two copies of $f(x_1)$ in $f(X)$).

A directed graph is denoted by $G = (V(G), E(G))$, where $V(G) \subseteq \mathbb{N}$ is the set of vertices and $E(G) \subseteq \mathbb{N} \times \mathbb{N}$ is the set of edges. A *graph language* is a family of graphs (this notion is closed under isomorphism, as the vertex names are immaterial). We typically use $n = |V(G)|$ to denote the size of the graph G that we are working with.

We let $u \rightsquigarrow v$ denote the existence of a directed path from node u to node v in the graph. The *underlying graph* of G is the undirected graph derived from G by considering the edges in $E(G)$ as undirected edges. We always assume G is *weakly connected*, that is, its underlying graph is a connected graph. Given a family $\mathcal{G}$ of directed graphs, let $\mathcal{G}_n$ be the restriction of $\mathcal{G}$ to graphs of size n.

For a node $v \in G$, let $N_{in}^G(v) = \{u \in V : (u,v) \in E\}$ be the *incoming neighbors* of v, and let $N_{out}^G(v) = \{u \in V : (v,u) \in E\}$ be the *outgoing neighbors* of V. We denote $\deg_{in}^G(v) = |N_{in}^G(v)|$ and $\deg_{out}^G(v) = |N_{out}^G(v)|$. When G is clear from context, we omit it from all of our notation, and write, e.g., V instead of $V(G)$, $N_{in}(v)$ instead of $N_{in}^G(v)$, and so on.

Initial Knowledge. Optionally, a node v may have some *initial knowledge* about the network graph, such as the size of the graph or a unique identifier assigned to v. One of our results (the universal prover that we present in Sect. 4) is quite general; thus, we introduce the notion of *initial knowledge fields* to capture any type of initial knowledge. Intuitively, a *field* $\mathcal{F}$ for some graph family $\mathcal{G}$ assigns to each vertex $v \in V(G)$ of a graph $G \in \mathcal{G}$ a value, which we think of as the initial knowledge corresponding to that field. In general, knowledge fields are not necessarily *functions* of the graph, e.g., unique identifiers, or a unique leader. Thus, $\mathcal{F}$ is a *relation* specifying what values are permissible for a given graph.

Initial knowledge fields are formally defined as follows: a field is a relation $\mathcal{F} \subseteq \bigcup_{n\in\mathbb{N}} \left(\mathcal{G}_n \times \left(\mathcal{X}^{\{1,\dots,n\}}\right)\right)$, specifying which values for this type of knowledge are acceptable for each graph; here, $\mathcal{X}$ is the domain from which the values are drawn, and $\mathcal{X}^{\{1,\dots,n\}}$ is the set of functions mapping $\{1,\dots,n\}$ to $\mathcal{X}$. We require that $\mathcal{F}$ be a *total relation*, that is, for each $G \in \mathcal{G}$, there exists some $F : \{1,\dots,n\} \to \mathcal{X}$ such that $(G,F) \in \mathcal{F}$. For simplicity we typically assume that $\mathcal{X} = \mathbb{N}$, that is, the field value is represented by an integer.

In the current paper we are mainly interested in the following fields:

- The UID field assigns to each vertex a unique identifier: $(G,U) \in \mathsf{UID}$ iff $U(v_1) \neq U(v_2)$ for every $v_1 \neq v_2 \in V(G)$.
- The size field assigns to each vertex the size of the network graph: $(G,S) \in \mathsf{size}$ iff $S(v) = |V(G)|$ for every $v \in V(G)$.
- The $\mathsf{out\text{-}deg}$ field assigns to each vertex its out-degree in the network graph: $(G,D) \in \mathsf{out\text{-}deg}$ iff $D(v) = \deg_{out}(v)$ for every $v \in V$.

Given fields $\mathcal{F}_1,\dots,\mathcal{F}_k$, an $(\mathcal{F}_1,\dots,\mathcal{F}_k)$-*instance* is a tuple $\hat{G} = (G, F_1,\dots,F_k)$, where $F_1,\dots,F_k$ assign to the nodes of G values corresponding to the fields $\mathcal{F}_1,\dots,\mathcal{F}_k$ (respectively), such that $(G,F_i) \in \mathcal{F}_i$ for each i. For example, a $(\mathsf{UID},\mathsf{out\text{-}deg})$-instance is given by (G,U,D), where G is a directed graph, $U : V(G) \to \mathbb{N}$ assigns a unique identifier $U(v)$ to each node $v \in V(G)$, and $D : V(G) \to \mathbb{N}$ assigns $D(v) = \deg_{out}(v)$ to each node $v \in V(G)$. For convenience, we often use the same symbol for a field $\mathcal{F}$ ("the knowledge type") and the mapping F that specifies its values in a given graph (e.g., in the sequel we use $\mathsf{UID}(v)$ to denote the UID assigned to a node v). In the case of the size field we also omit the vertex v and write simply size to represent the value given to all nodes.

Throughout the paper, we assume that the $\mathsf{UID}, \mathsf{size}$ and $\mathsf{out\text{-}deg}$ fields are encoded using $O(\log n)$ bits at each node.

Proof Labeling Schemes. We generalize the definitions of the classes D_1, D_2 from [3], which represent proof labeling schemes with unidirectional or bidirectional communication (resp.), to incorporate initial knowledge. (We are mostly interested in unidirectional communication, D_1, but on one occasion we will show that under some assumptions, unidirectional simulation can *simulate* bidirectional communication.)

A *labeling* for a graph G is a mapping $\ell : V \to \{0,1\}^*$ that assigns to each node $v \in V$ a label $\ell(v) \in \{0,1\}^*$. Given fields $\mathcal{F}_1, \ldots, \mathcal{F}_n$ and a value $i \in \{1,2\}$, a $D_i\{\mathcal{F}_1, \ldots, \mathcal{F}_n\}$*-proof labeling scheme (PLS)* w.r.t. a graph family $\mathcal{G}$ is a prover-verifier pair $(\mathbf{Prv}, \mathbf{Ver})$, where

- **Prv**, the prover, takes an $(\mathcal{F}_1, \ldots, \mathcal{F}_k)$-instance $\hat{G} = (G, F_1, \ldots, F_k)$ (where $G \in \mathcal{G}$) and produces a labeling $\ell : V(G) \to \{0,1\}^*$.
- **Ver**, the verifier, is a *decision function* at each node v, which outputs a Boolean value when given (a) the initial knowledge $F_1(v), \ldots, F_k(v)$ available to v; (b) the label $\ell(v)$ of v, (c) the labels $\ell(N_{in}(v))$ of v's in-neighbors, and (d) if $i = 2$ (bidirectional communication), also the labels $\ell(N_{out}(v))$ of v's out-neighbors. The pair $(\hat{G}, \ell)$ is *accepted* by **Ver** if all nodes of G output 1.

An $(\mathcal{F}_1, \ldots, \mathcal{F}_k)$-PLS $(\mathbf{Prv}, \mathbf{Ver})$ is said to *decide* a graph language $\mathcal{L} \subseteq \mathcal{G}$ if it satisfies:

- Completeness: for every $(\mathcal{F}_1, \ldots, \mathcal{F}_k)$-instance $\hat{G} = (G, F_1, \ldots, F_k)$ such that $G \in \mathcal{L}$, the verifier **Ver** accepts $(\hat{G}, \mathbf{Prv}(\hat{G}))$.
- Soundness: for every $(\mathcal{F}_1, \ldots, \mathcal{F}_k)$-instance $\hat{G} = (G, F_1, \ldots, F_k)$ such that $G \in \mathcal{G} \setminus \mathcal{L}$, and for every labeling $\ell : V(G) \to \{0,1\}^*$, the verifier **Ver** rejects $(\hat{G}, \ell)$.

If there exists a $D_i\{\mathcal{F}_1, \ldots, \mathcal{F}_k\}$-PLS w.r.t. graph family $\mathcal{G}$ deciding the graph language $\mathcal{L}$, we say that $\mathcal{L}$ is $D_i\{\mathcal{F}_1, \ldots, \mathcal{F}_k\}$*-decidable* w.r.t. graphs in $\mathcal{G}$. The *proof size* of a PLS $(\mathbf{Prv}, \mathbf{Ver})$ is the maximum length of the proof assigned by **Prv** to any node, as a function of the size of the graph. Throughout most of the paper we take $\mathcal{G}$ to be the family of all weakly-connected directed graphs, and omit it unless stated otherwise.

In some of our constructions, it is convenient to first construct a labeling scheme with *infinite-length labels*, and then convert it to a PLS with finite-length labels. We refer to a labeling scheme with infinite-length labels as an $D_i\{\mathcal{F}_1, \ldots, \mathcal{F}_k\}$-$\infty$-*PLS*. It satisfies the definitions above, except that the labeling is of the form $\ell : V \to \{0,1\}^{\mathbb{N}}$ (i.e., each node's label is an infinite string indexed by the natural numbers).

Problem Statements. We consider the following graph languages:

- Graphs with a global source: GLOBALSOURCE $= \{G \in \mathcal{G} : \exists v \in V \forall u \in V. v \leadsto u\}$.
- Graphs with a global sink: GLOBALSINK $= \{G \in \mathcal{G} : \exists v \in V \forall u \in V. u \leadsto v\}$.
- Strongly-connected graphs: STRONGCON $= \{G \in \mathcal{G} : \forall u, v \in V. u \leadsto v\}$.

Inputs. In the current paper we mostly discuss problems where the nodes have no inputs, but on occasion we will need to introduce an input assignment. The definitions above are extended to handle inputs in the natural way.

4 Building Blocks

We begin by presenting several schemes and results that will be used as building blocks in our constructions. Some of these are present in some related form in prior work, but their application to directed graphs, and the addition of initial knowledge fields, make their correctness non-trivial.

4.1 DFS-Based Schemes

Some schemes we show later will require proving that some specific nodes, specified by other means, are each a global sink. For this purpose we define the predicate MARKEDISSINK, where each node is given as input a bit $mark(v) \in \{0, 1\}$, and our goal is to verify that every marked node (i.e., every node v that has $mark(v) = 1$) is a global sink:

$$\text{MARKEDISSINK} = \{G \in \mathcal{G} : (mark(v) = 1) \rightarrow (\forall u \in V. u \rightsquigarrow v)\} .$$

If we had bidirectional communication, we could verify that all nodes have a path to a given node v by specifying the distance of each node to v; however, with unidirectional communication, nodes cannot check that they have an out-neighbor whose distance is smaller than their own. Instead, the MARKEDISSINK predicate can be certified by asking the prover to provide, for each marked node v, a DFS tree rooted at v: each node u is given the DFS interval $I_v(u) \subseteq [1, 2n]$, specifying the entering and leaving time of node u in a DFS initiated at v. This idea serves to construct an *ancestor labeling scheme* in [7], and it is used in [8] to argue that in undirected graphs, having a unique leader, knowing the size of the graph, or having unique identifiers are all equivalent assumptions up to an additive $O(\log n)$ term in the proof size. In our case, we can use it to verify MARKEDISSINK, requiring only knowledge of the graph size:

Claim 4.1. MARKEDISSINK *is* $D_1\{\mathsf{size}\}$*-decidable with proof size* $O(k \log n)$*, where* k *is the number of marked nodes in the graph.*

In the scheme from Claim 4.1, if there is at least one marked global sink, then we can use the (claimed) DFS intervals as unique identifiers (as in [8]):

Claim 4.2. *If there is at least one marked global sink* s *in the graph, and all nodes accept the proof given in Claim 4.1, then* $I_s(v) \neq I_s(u)$ *for each* $u \neq v$.

The details of the construction, and proofs of these claims, will appear in the full version of the paper.[2]

[2] A correctness proof was not given in [8], but more importantly, we must prove that the DFS can be verified even using only unidirectional communication.

4.2 Canonical Proof Labeling Scheme

A well-known result for undirected graphs (see [8,9]) is that when we have unique identifiers, every predicate can be decided by a PLS with proof size $O(n^2)$, by having the prover give each node a complete description of the graph. When unique identifiers are not available, not every predicate is decidable [9]; however, it is shown in [4] that every language that is closed under lifts is decidable without unique identifiers. We use a similar construction to show that even in directed graphs and with arbitrary initial knowledge, every decidable language can be decided in $O(n^2 + L)$ bits, where L is the representation length of all the initial knowledge fields.

In the construction below, the prover will need to give each node a complete description of the instance, including the initial knowledge. The *representation length* $R_{\mathcal{F}} : \mathbb{N} \to \mathbb{N}$ of a field $\mathcal{F}$ is the number of bits required to specify the values assigned by $\mathcal{F}$ in graphs of size n.

Theorem 4.3 (The canonical prover). *Given fields* $\mathcal{F}_1, \ldots, \mathcal{F}_k$ *and* $i \in \{1, 2\}$, *there exists a* $D_i\{\mathcal{F}_1, \ldots, \mathcal{F}_k\}$*-prover* $\mathbf{Prv}^{\mathrm{canon}}$ *such that*

- *For any graph language* $\mathcal{L}$ *that is recognized by some* $D_i\{\mathcal{F}_1, \ldots, \mathcal{F}_k\}$*-*$\infty$*-PLS, there is a verifier* $\mathbf{Ver}^{\mathcal{L}}$ *such that* $(\mathbf{Prv}^{\mathrm{canon}}, \mathbf{Ver}^{\mathcal{L}})$ *is a* $D_i\{\mathcal{F}_1, \ldots, \mathcal{F}_k\}$*-PLS for* $\mathcal{L}$.
- *The length of the labels assigned by* $\mathbf{Prv}^{\mathrm{canon}}$ *in graphs of size* n *is bounded by* $O\left(n^2 + \sum_{j=1}^{k} R_{\mathcal{F}_j}(n)\right)$. *However, if each field* $\mathcal{F}_j$ *is a* function *(i.e., for each graph* G *there is only one value* F *such that* $(G, F) \in \mathcal{F}_j$*), then labels of length* $O\left(n^2\right)$ *suffice.*

If in addition the predicates $G \in \mathcal{L}$ *and* $(G, F_j) \in \mathcal{F}_j$ *for all* $j = 1, \ldots, k$ *are computable in time* $O(t(n))$ *in graphs of size* n, *then* $\mathbf{Ver}$ *is a computable function, with time complexity* $O\left(n^3 + n \cdot \sum_{j=1}^{k} R_{\mathcal{F}_j}(n) + t(n)\right)$.

Proof (sketch). We ask the prover to give each node v a description of the network graph G, along with the values $F_1, \ldots, F_k$ of the initial knowledge fields; however, if all the fields are functions of G, we can omit them and have the nodes compute them by themselves. The prover also tells node v which node it corresponds to in the prover's claimed network graph. To decide a language $\mathcal{L}$ that has a $D_i\{\mathcal{F}_1, \ldots, \mathcal{F}_k\}$-$\infty$-PLS $(\mathbf{Prv}, \mathbf{Ver})$, each node v checks that it received the same description as its in-neighbors, and that the prover's claimed instance is consistent with the local view of v; then, node v verifies that the graph described by the prover is in $\mathcal{L}$.

The soundness of the scheme is proven as follows: we cannot rely on the prover to accurately describe the instance $\hat{G} = (G, F_1, \ldots, F_k)$ that we are working with, but we can use the prover's description to construct another instance $\hat{G}'$ that would be accepted by the PLS $(\mathbf{Prv}, \mathbf{Ver})$. We take the labeling $\ell' = \mathbf{Prv}(\hat{G}')$ produced by $\mathbf{Prv}$ on the "fake instance" $\hat{G}'$, and construct from it a labeling ℓ for the "real instance" $\hat{G}$, by assigning to each node the label of the node onto which it is mapped in $\hat{G}'$. We prove that the resulting labeling is also accepted by **Ver**, meaning that $G \in \mathcal{L}$.[3]

4.3 The Charge-Distribution Protocol

We now describe a PLS that can be used to identify global sinks when the out-degree of each node is known. The PLS is obtained by asking the prover to specify the execution of a distributed protocol; we begin by describing the protocol, and then show how it is used to obtain a PLS.

The Charge-Distribution Protocol. Recall that a subset $C \subseteq V$ is called a *strongly-connected component (SCC)* of a directed graph G if C is a maximal subset of V such that $u \rightsquigarrow v$ for all $u, v \in C$. We say that an SCC C is *recurrent* if there are no edges $(u, v) \in E$ such that $u \in C$ and $v \notin C$; otherwise C is called *transient*. If C is the only recurrent SCC, then we say that C is a *sink SCC*. Finally, we say that node v is *recurrent* if the SCC to which v belongs is recurrent, and otherwise v is *transient*.

The charge-distribution protocol is a synchronous protocol aimed at determining whether the SCC to which a node belongs is recurrent or transient. Each node v stores a *charge* $C(v) \in \mathbb{R}$, initially set to 1. In each round, a node v with $\deg_{out}(v) > 0$ sends the value $C(v)/\deg_{out}(v)$ to each of its neighbors, and subtracts the values sent from its own charge; if $\deg_{out}(v) = 0$, node v keeps the charge. Then, node v adds to its charge the values received from its neighbors. All together, if we denote by $C_t(v)$ the charge at node v after $t \geq 0$ rounds, then

$$C_t(v) = \begin{cases} 1, & \text{if } t = 0, \\ \sum_{u \in N_{in}(v)} \frac{C_{t-1}(u)}{\deg_{out}(u)}, & \text{if } t > 0 \text{ and } \deg_{out}(v) > 0, \\ C_{t-1}(v) + \sum_{u \in N_{in}(v)} \frac{C_{t-1}(u)}{\deg_{out}(u)}, & \text{if } t > 0 \text{ and } \deg_{out}(v) = 0. \end{cases} \quad (1)$$

The charge-distribution protocol simulates the behavior of a Markov chain over the graph G, with the addition of self-loops on nodes with no out-neighbors. Indeed, let G' be G with those added self-loops, and define on it a Markov chain M_G with a uniform initial distribution, and the following transition probabilities:

$$p(u, v) = \begin{cases} \frac{1}{\deg^{G'}_{out}(u)}, & \text{if } v \in N^{G'}_{out}(u), \\ 0, & \text{otherwise.} \end{cases}$$

Note that our definitions of recurrent and transient nodes in the graph G coincide with the standard definition of recurrent and transient states in the Markov chain M_G (e.g., Definition 4.2.5 in [6]).

Let $p_t(v)$ denote the probability that the Markov chain is at node v after t steps. Then by induction on t we have $p_t(v) = C_t(v)/n$. This observation allows

[3] This is in some sense an extension of the argument in [4], where it is shown that *if the prover cannot use the unique identifiers when choosing its proof*, then the languages that can be recognized are exactly those that are closed under lifts.

us to apply some basic results about Markov chains to analyze the behavior of the charge-distribution protocol:

Claim 4.4 ([6]). *Let M be a Markov chain starting at some node u_0 with probability 1. Let s_t denote the state M visits in the t-th step. Let v be some node such that $u_0 \rightsquigarrow v$. Then we have:*

- *If v is transient, then* $\lim_{t\to\infty} \Pr[s_t = v] = 0$.
- *If v is recurrent, then either* $\lim_{i\to\infty} \Pr[s_t = v] > 0$ *or no limit exists.*

Our Markov chain M_G starts from a uniform distribution over the nodes, rather than starting at one specific node, but it is easy to see that a similar claim holds:

Claim 4.5. *Let v be some node. Then for the charge distribution protocol described above,*

- *If v is transient, then* $\lim_{t\to\infty} C_t(v) = 0$
- *If v is recurrent, then either* $\lim_{t\to\infty} C_t(v) > 0$ *or no limit exists.*

Another consequence we get is that when there is a global sink with no outgoing edges, the charge at that node tends to the size of the graph in the limit, as all the charge in the graph flows to the sink:

Claim 4.6. *If G is a graph with a global sink v such that* $\deg_{out}(v) = 0$, *then for the charge distribution protocol described above it holds that* $\lim_{t\to\infty} C_t(v) = |V(G)|$.

A Proof-Labeling Scheme for Recurrent States. Using the charge-distribution protocol and the characterization from Claim 4.5, we construct a PLS that allows us to prove which nodes are recurrent and which are transient. Since Claim 4.5 characterizes the behavior of the charge *in the limit*, we first give an ∞-PLS, and then convert it into a finite PLS using Theorem 4.3.[4]

Consider the following graph language, with inputs drawn from $\{0, 1\}$:

$$\textsc{RecurrentNodes} = \{(G, \mathit{mark}) : \mathit{mark}(v) = 1 \text{ iff } v \text{ is a recurrent node}\}.$$

Theorem 4.7. *There exists a D_1* {out-deg}-∞-*PLS for* RecurrentNodes.

Proof. The prover specifies the execution of the charge-distribution protocol in G: the label of a node $v \in V$ is the infinite string

$$\ell(v) = \deg_{out}(v), C_0(v), C_1(v), C_2(v), \ldots$$

[4] An alternative approach might be to specify the execution of the charge-distribution protocol up to some sufficiently large number of rounds, and deduce the asymptotic behavior, but this is problematic for two reasons: first, a Markov chain might take an exponential number of steps to approach its limiting behavior; and second, the convergence time depends on the size of the chain, but we would like to use our PLS in settings where the size of the graph is not necessarily known.

where each value $C_t(v)$ is represented in binary using some prefix-free encoding of the non-negative rationals. To verify the proof, each node v examines its in-neighbors' labels and verifies that (1) is respected at v for all $t \geq 0$. Then, node v checks whether $\lim_{t\to\infty} C_t(v)$ exists and is equal to 0: if so, node v accepts iff $mark(v) = 0$, and if the limit does not exist, or exists but is not equal to 0, node v accepts iff $mark(v) = 1$. The completeness and soundness of the scheme follow immediately from Claim 4.5. □

Corollary 4.1. RECURRENTNODES *is* D_1 {out-deg}*-decidable with proof size* $O(n^2)$.

Proof. Given the ∞-PLS for RECURRENTNODES, we can apply Theorem 4.3 to get a PLS. The proof size is $O(n^2)$, since the out-deg field is a function of the graph. □

5 Decidability of Reachability-Related Problems

We are now ready to characterize the decidability of reachability-related problems, such as the existence of a global source or sink, under various combinations of initial knowledge. We consider only unidirectional communication (D_1). To prove that a problem is decidable, we give a PLS; to prove that a problem $\mathcal{L}$ is $D_1\{\mathcal{F}_1,\ldots,\mathcal{F}_k\}$-undecidable, we exhibit two instances $\hat{G}^{\text{yes}} = (G^{\text{yes}}, F_1^{\text{yes}},\ldots,F_k^{\text{yes}}), \hat{G}^{\text{no}} = (G^{\text{no}}, F_1^{\text{no}},\ldots,F_k^{\text{no}})$ and a mapping $\rho : V(G^{\text{no}}) \to V(G^{\text{yes}})$, such that

- $G^{\text{yes}} \in \mathcal{L}$ and $G^{\text{no}} \notin \mathcal{L}$,
- For every node $v \in V(G^{\text{no}})$ we have $F_1^{\text{no}}(v) = F_1^{\text{yes}}(\rho(v))$, ..., $F_k^{\text{no}}(v) = F_k^{\text{yes}}(\rho(v))$.
- For every node $v \in V(G^{\text{no}})$ we have $\rho(N_{in}^{G^{\text{no}}}(v)) = N_{in}^{G^{\text{yes}}}(\rho(v))$.

We call this type of mapping a *view-preserving map*.[5]

The existence of the mapping ρ proves that there does not exist a PLS for the language $\mathcal{L}$: if there were some labeling ℓ of $\hat{G}^{\text{yes}}$ that makes all nodes accept, then using ρ we can obtain a labeling ℓ' of $\hat{G}^{\text{no}}$, by setting $\ell'(v) = \ell(\rho(v))$ for each $v \in V(G^{\text{no}})$. Node $v \in V(G^{\text{no}})$ has the same "view" as node $\rho(v) \in V(G^{\text{yes}})$: the two nodes have the same initial knowledge, $F_1^{\text{no}}(v) = F_1^{\text{yes}}(\rho(v)),\ldots,F_k^{\text{no}}(v) = F_k^{\text{yes}}(\rho(v))$, they have the same label, $\ell'(v) = \ell(\rho(v))$, and they also see the same labels in their in-neighborhood, $\ell'(N_{in}^{G^{\text{no}}}(v)) = \ell(\rho(N_{in}^{G^{\text{no}}}(v))) = \ell(N_{in}^{G^{\text{yes}}}(\rho(v)))$. Therefore node v behaves in $\hat{G}^{\text{no}}$ the same way that $\rho(v)$ behaves in $\hat{G}^{\text{yes}}$, i.e., it accepts. This is true for every $v \in V(G^{\text{no}})$, and therefore all nodes accept; but since $G^{\text{no}} \notin \mathcal{L}$, this shows that the PLS is not sound.

[5] This notion is closely related to *covering maps* and *lifts* [4], but those classical notions are not appropriate for directed graphs, and they also do not handle initial knowledge in the way we require here.

5.1 Global Source

We begin with the problem of verifying whether the graph has a *global source*, a node from which every other node is reachable. We show that this is undecidable given the graph size and the nodes' out-degrees:

Theorem 5.1. GLOBALSOURCE *is* D_1 {size, out-deg}*-undecidable.*

Proof. Consider the YES-instance shown in Fig. 1a, where node 1 is a global source, and the NO-instance shown in Fig. 1b, where there is no global source. The numbers in Fig. 1b indicate the YES-instance node onto which each NO-instance node is mapped.

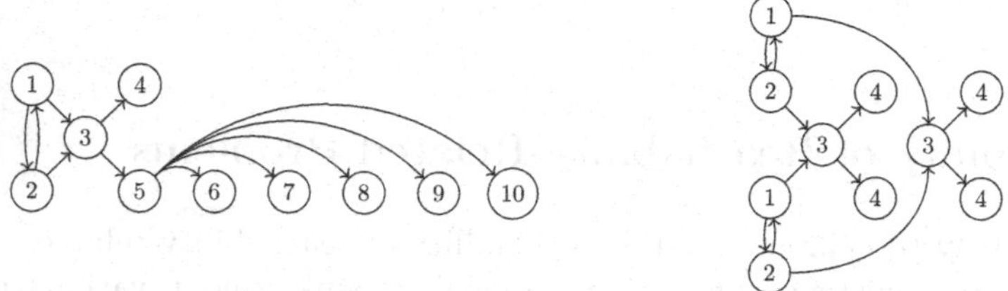

(a) A YES-instance for GLOBALSOURCE (b) A NO-instance for GLOBALSOURCE

Fig. 1. YES-instance (left) and NO-instance (right) for GLOBALSOURCE

Observe that for every node v in Fig. 1b, the node onto which v is mapped in Fig. 1a has the same view: both graphs have the same size (10), each node of Fig. 1b has the same out-degree as the node onto which it is mapped in Fig. 1a, and the in-neighborhood of each node v in Fig. 1b is mapped onto the in-neighborhood of the node onto which v is mapped in Fig. 1b. Together, the two instances show that size and out-deg do not suffice to decide GLOBALSOURCE. □

On the other hand, unique identifiers do suffice to decide GLOBALSOURCE: it is easy to verify that the classical spanning-tree PLS from [9] still works, even in directed graphs—we ask the prover to specify the ID of the root of the tree, and give each node v its distance $d(v)$ from the root; then, each node v verifies that it has an in-neighbor $u \in N_{in}(v)$ with $d(u) = d(v) - 1$, unless v itself is the root.

Claim 5.2. GLOBALSOURCE *is* D_1 {UID}*-decidable with proof size* $O(\log n)$*.*

5.2 Global Sink

We now turn to the question of whether the graph contains a *global sink*, a node that is reachable from all nodes. It turns out that the GLOBALSINK language is undecidable given any single field size, out-deg or UID, but it is decidable given any *two* of those fields.

Theorem 5.3. GLOBALSINK *is* $D_1\{\mathcal{F}\}$*-undecidable for any* $\mathcal{F} \in \{\mathsf{UID}, \mathsf{out\text{-}deg}, \mathsf{size}\}$.

Proof. For $D_1\{\mathsf{UID}\}$, consider the following YES-instance and NO-instance, with letters indicating UIDs (Fig. 2):

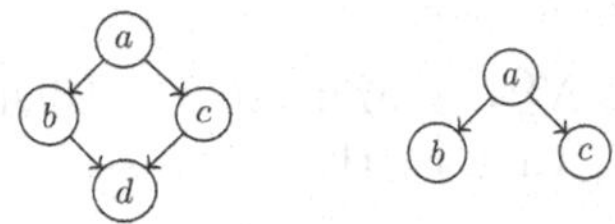

Fig. 2. YES-instance (left) and NO-instance (right) for GLOBALSINK

In the YES-instance node d is a global sink, while the NO-instance has no global sink. However, each node of the NO-instance can be mapped onto the node that has the same UID in the YES-instance, and it will then have the same local view as that node.

For both $D_1\{\mathsf{out\text{-}deg}\}$ and $D_1\{\mathsf{size}\}$ we consider the YES-instance in Fig. 3a, with the difference being the initial information given (not shown in the figure). It is clearly a YES-instance, as node 3 is a global sink. Next, consider the following NO-instances for $D_1\{\mathsf{out\text{-}deg}\}$ and $D_1\{\mathsf{size}\}$:

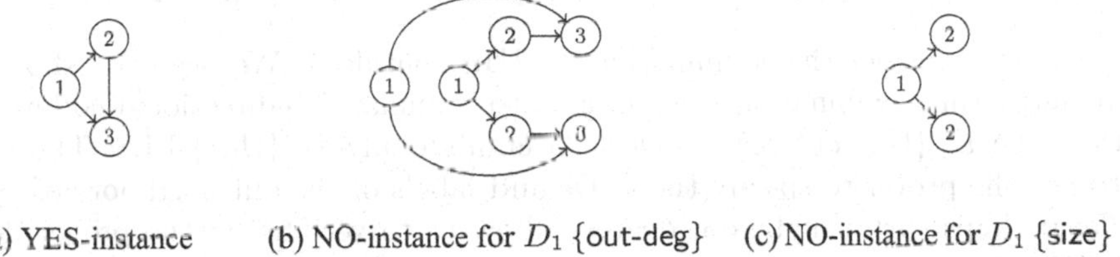

(a) YES-instance (b) NO-instance for $D_1\{\mathsf{out\text{-}deg}\}$ (c) NO-instance for $D_1\{\mathsf{size}\}$

Fig. 3. YES-instance and two NO-instances for GLOBALSINK

Figure 3b shows a NO-instance that agrees with Fig. 3a on the out-degrees but not the size, and Fig. 3c shows a NO-instance that agrees with Fig. 3a on the size but not the out-degrees. This shows that out-deg or size alone do not suffice to decide GLOBALSINK. □

We now turn to positive results, and show that any combination of two of the fields UID, out-deg, size is sufficient to decide GLOBALSINK.

Theorem 5.4. GLOBALSINK *is* $D_1\{\mathsf{UID}, \mathsf{size}\}$*-decidable with proof size* $O(n \log n)$.

Proof. We ask the prover to construct a spanning tree T, oriented upwards and rooted at some global sink r. The prover uses the UIDs of the nodes as the vertices of T, and gives $(T, \mathsf{UID}(v))$ to each node v. [6]

Let $\ell(v) = (T(v), id(v))$ be the label of each node v. The verifier at node v accepts iff:

- $T(v)$ is a rooted tree over size nodes, oriented upwards, and $\mathsf{UID}(v)$ appears as a vertex in $T(v)$.
- $T(v) = T(u)$ for each $u \in N_{in}^G(v)$, and $id(v) = \mathsf{UID}(v)$.
- $id(N_{in}^G(v)) = N_{in}^T(id(v))$, that is, the IDs of v's neighbors in T match v's in-neighbors in G.

Completeness is immediate. As for soundness, suppose that all nodes accept the labeling ℓ. Then all nodes receive the same tree, otherwise from weak connectivity some node v has a node $u \in N_{in}^G(v)$ with $T(u) \neq T(v)$, causing node v to reject. Let T be the tree given to all nodes; the nodes verify that T is a rooted tree, oriented upwards, and containing size nodes.

To show that the prover cannot "invent" or omit any nodes, let $S = \{id(v) : v \in T\}$ be the set of UIDs that appear in T, and let $R = \{\mathsf{UID}(v) : v \in G\}$ be the set of UIDs that appear in G. Since each node v verifies that $\mathsf{UID}(v)$ appears in T, we have $R \subseteq S$. But T is a tree of the same size as G (as verified by the nodes), and therefore $R = S$. Since each node v verifies that its in-neighbors in T match the UIDs of its in-neighbors in G, it must be that T is a subgraph of G. All together, T is a spanning tree of G, oriented upwards and rooted at some node $r \in V$, which implies that r is a global sink. □

We now consider the combination $D_1\{\mathsf{UID}, \mathsf{out\text{-}deg}\}$. We begin by showing that under this combination it is possible to "simulate" bidirectional communication: the $D_1\{\mathsf{UID}, \mathsf{out\text{-}deg}\}$ model can emulate any $D_2\{\mathsf{UID}\}$-PLS. The idea is to ask the prover to specify the UIDs and labels of the out-neighbors of each node; perhaps surprisingly, out-degrees allow us to verify that the prover does this faithfully.

Lemma 5.1. *If a graph language $\mathcal{L}$ has an $D_2\{\mathsf{UID}\}$-PLS of size $f(n)$, then it also has a $D_1\{\mathsf{UID}, \mathsf{out\text{-}deg}\}$-PLS with of size $O(n(\log(n) + f(n)))$.*

Proof. Given a $D_2\{\mathsf{UID}\}$-PLS $(\mathbf{Ver}, \mathbf{Prv})$ for $\mathcal{L}$, we construct a $D_1\{\mathsf{UID}, \mathsf{out\text{-}deg}\}$-PLS $(\mathbf{Ver}', \mathbf{Prv}')$ as follows: if ℓ is the labeling provided by $\mathbf{Prv}$, then the prover $\mathbf{Prv}'$ gives to each node v the label $\ell'(v) = (\mathsf{UID}(v), \ell(v), L(v))$, where $L(v) = \{(\mathsf{UID}(u), \ell(u)) : u \in N_{out}(v)\}$.

Given labels $\{(id(v), \ell(v), L(v))\}_{v \in V}$, the verifier at node v checks that the following conditions hold:

- $id(v) = \mathsf{UID}(v)$ and $|L(v)| = \mathsf{out\text{-}deg}(v)$.
- For each $u \in N_{in}(v)$ we have $(\mathsf{UID}(v), \ell(v)) \in L(u)$.

[6] Node v of course knows its own UID, but in order for v's out-neighbors to learn $\mathsf{UID}(v)$, it must be part of v's label.

– Let $A(v) = \{a : \exists u.(u, a) \in L(v)\}$ be the multiset of labels that appear in $L(v)$ (i.e., the labels the prover claims it has given to v's out-neighbors). Then **Ver** must accept at node v when given $\ell(v), \ell(N_{in}(v)), A(v)$.

Completeness is straightforward. As for soundness, let $\{(id(v), \ell(v), L(v))\}_{v \in V}$ be a labeling that is accepted by all nodes under **Ver**$'$. We show that the labeling $\{\ell(v)\}_{v \in V}$ is accepted by **Ver**, which implies that $G \in \mathcal{L}$. Since each node v checks that **Ver** accepts $\ell(v), \ell(N_{in}(v)), A(v)$, it suffices to show that for each node $v \in V$ we have $A(v) = \ell(N_{out}(v))$.

Fix $v \in V$, and let $R(v) = \{(\mathsf{UID}(u), \ell(u)) : u \in N_{out}(v)\}$ be the real list of v's out-neighbors together with their labels. First, observe that $R(v) \subseteq L(v)$, because if there were some out-neighbor $w \in N_{out}(v)$ such that $(\mathsf{UID}(w), \ell(w)) \notin L(v)$, node w would reject. Since $|L(v)| = |R(v)| = \mathsf{out\text{-}deg}(v)$, this implies that $L(v) = R(v)$, which completes the proof of soundness.

The size of the proof is dominated by the number of bits required to represent $L(v)$, which is $O(n(\log(n) + f(n)))$ in the worst case. □

Next, we observe that the $D_1\,\{\mathsf{UID}\}$-PLS for GLOBALSOURCE from Claim 5.2 can actually serve as a $D_2\,\{\mathsf{UID}\}$-PLS for GLOBALSINK:

Lemma 5.2. GLOBALSINK *is* $D_2\,\{\mathsf{UID}\}$*-decidable with proof size* $O(\log(n))$.

Proof. We apply the PLS from Claim 5.2, replacing the in-neighborhood $N_{in}(v)$ with the out-neighborhood $N_{out}(v)$ everywhere, which has the effect of reversing all edge directions. The resulting PLS certifies that the graph G' obtained by reversing all the edges of G has a global source; of course, this is true iff G itself has a global sink. □

By combining Lemma 5.1 and Lemma 5.2, we obtain:

Corollary 5.1. GLOBALSINK *is* $D_1\,\{\mathsf{UID}, \mathsf{out\text{-}deg}\}$*-decidable using* $O(n \log n)$*-bit labels.*

Remark. *In Theorem 5.4 and Corollary 5.1, unique identifiers were used to "mark" a global sink* r*, by asking the prover to describe a spanning tree rooted at* r*. This provided us with the guarantee that the graph contains at least one node that* knows *that it should be a global sink, and can then verify that indeed it is. In some sense, this is the only manner in which UIDs are useful: in Sect. 6 we show that we can replace UIDs with the assumption that we have at least one marked sink in the graph, and together with either* size *or* out-deg*, this suffices to decide every graph language.*

Finally, we turn to what is perhaps the most challenging combination: size and out-deg. Without unique identifiers, we cannot ask the prover to "commit" to a node v that it claims is a global sink, and then prove that node v is indeed a sink (as we did in Theorem 5.4 and Corollary 5.1). Instead, we use the charge-distribution PLS from Sect. 4.3 to identify all the recurrent nodes in the graph, and then ask the prover to "convince" each recurrent node that it is a global sink, using the DFS scheme from Sect. 4.1. The correctness of this idea relies on two crucial observations: first, every graph contains *at least one* recurrent node, so we are guaranteed that at least one node will verify that it is a global sink. Second, if a graph contains a global sink, then all its recurrent nodes are global sinks (see the proof below).

Theorem 5.5. GLOBALSINK *is* D_1 {size, out-deg}*-decidable with proof size* $O(n^2)$.

Proof. We compose the PLS for identifying recurrent nodes from Corollary 4.1 with the PLS for verifying that every marked node is a global sink, from Claim 4.1. The label of node v is given by $\ell(v) = (mark(v), \ell_1(v), \ell_2(v))$, where $mark(v) \in \{0, 1\}$ indicates whether the prover claims that v is a global sink, ℓ_1 is the proof for RECURRENTNODES from Corollary 4.1, and ℓ_2 is the proof for MARKEDISSINK from Claim 4.1. The nodes verify that

- The verifier from Corollary 4.1 accepts $\{(mark(v), \ell_1(v))\}_{v \in V}$ (using the out-deg field), and
- The verifier from Claim 4.1 accepts $\{(mark(v), \ell_2(v))\}_{v \in V}$ (using the size field).

To prove that the scheme is complete we must prove that if G contains a global sink, then all recurrent nodes in G are global sinks. Let G be a graph with a global sink $s \in V(G)$, and let v be a recurrent node of G. The SCC C to which v belongs is, by definition, recurrent: there are no outgoing edges (u, w) such that $u \in C$ and $w \notin C$. However, we know that $v \rightsquigarrow s$, as s is a global sink. Therefore v must be in the same SCC as s, which implies that v is also a global sink.

As for soundness, if both verifiers accept, we know that $mark(v) = 1$ iff v is recurrent, and that if $mark(v) = 1$ then v is a global sink. To complete the proof, we observe that the prover *must* mark at least one node, as every graph G contains some recurrent node: if we take a topological sort of the SCC-graph of G, then the last SCC in the topological sort is recurrent, and all its nodes are recurrent nodes. Thus, if the prover fails to mark any node, the nodes of the last SCC will reject.

The proof size of the PLS constructed above is dominated by the size of the labeling for RECURRENTNODES, which is $O(n^2)$. □

5.3 Strong Connectivity

The final problem we consider is *strong connectivity.* Counter-intuitively, even though it is a stronger property, strong connectivity is "easier" to verify than the existence of a global source or sink: the difficulty in verifying GLOBALSOURCE and GLOBALSINK is that we must check whether there *exists* a node satisfying some condition, which allows nodes to "shirk their responsibility" and accept, believing that some other node satisfies the condition. In contrast, strong connectivity requires *all* nodes to satisfy the condition (e.g., all must be global sinks and global sources). Strong connectivity is not decidable without any initial knowledge, but it is decidable given any of the fields UID, size, out-deg.

Theorem 5.6. STRONGCON *is* $D_1\{\}$*-undecidable.*

Proof. Consider the following simple YES-instance and NO-instance (Fig. 4):

Fig. 4. YES-instance (left) and NO-instance (right) for STRONGCON

In the absence of any initial knowledge, each node knows only its own label and the label of its in-neighbors. Thus, in the YES-instance, node 1 receives the label of node 2, and vice-versa; the same happens in the NO-instance. □

The example above does not work if we provide any of the fields UID, size or out-deg. Indeed, we prove that with any one of these fields the problem becomes decidable. We rely on the fact that a graph G is strongly connected if and only if one of the following conditions holds: (1) all nodes of G are global sources; (2) all nodes of G are recurrent; or (3) G contains a node that is both a global source and a global sink.

If we are given UIDs, we can use them to certify that each node in the graph is a global source, by asking the prover to provide the distance from each node to every node in the graph. The distances are verified by checking that for each node $u \neq v$, node v has some in-neighbor whose claimed distance from u is one less than v's.

Theorem 5.7. STRONGCON *is* $D_1\{\text{UID}\}$*-decidable with proof size* $O(n \log n)$.

Now suppose that we are given the size of the graph instead of UIDs. We give a PLS that is similar to the one that uses UIDs, except that each node is represented by an index in the range $\{1, \ldots, n\}$, where n is the size of the graph. Of course, the prover may try to cheat by giving two nodes the same index, but we prove that this will be detected.

Theorem 5.8. STRONGCON *is* $D_1\{\text{size}\}$*-decidable with proof size* $O(n \log n)$.

Proof. The prover orders the nodes of the graph arbitrarily, and for each $i = 1, \ldots, n$, it provides the i-th node with the label $\ell(i) = (i, d(i))$, where $d(i) : \{1, \ldots, n\} \to \{1, \ldots, n\}$ is a mapping such that $[d(i)](j)$ is the distance from the j-th node to the i-th node.

For convenience, let $d(j, v) = [d(v)](j)$ be the claimed distance from the node with index j to node v. Given labels $\{(i(v), d(v))\}_{v \in V}$ and size $= n$, the verifier at each node v checks that:

- $i(v) \in \{1, \ldots, n\}$, and $d(i(v), v) = 0$.
- For each $j \in \{1, \ldots, n\} \setminus \{i\}$ we have $d(j, v) > 0$, and there is some in-neighbor $u \in N_{in}(v)$ such that $d(j, u) = d(j, v) - 1$.

Completeness is straightforward to prove. As for soundness, suppose that the labeling $\{(i(v), d(v))\}_{v \in V}$ is accepted at all nodes. If the prover assigned unique indices, $i(u) \neq i(v)$ whenever $u \neq v$, then the proof from Theorem 5.7 shows that every node is a global source. Thus, let us prove that if the prover does not assign unique indices to all nodes, the proof is rejected.

For each index $i \in \{1, \ldots, n\}$, let $V(i) = \{v \in V : i(v) = i\}$. Note that, since each node v appears in exactly one set $V(i)$, we have $\sum_{i=1}^{n} |V(i)| = n$. Our goal is to prove that $|V(i)| = 1$ for each i. Suppose not, and let i be an index with $|V(i)| \geq 2$. Then there must exist some $j \neq i$ such that $|V(j)| = 0$ (because $\sum_{i=1}^{n} |V(i)| = n$), in other words, the prover did not assign index j to any node. But this will be detected, because every node verifies that it has a path back to some node indexed j: let v be a node such that $d(j, v) = \min_{u \in V} d(j, u)$. Since $|V(j)| = 0$, we know that $v \notin V(j)$, that is, $i(v) \neq j$. Node v therefore verifies that $d(j, v) > 0$, and it also verifies that some in-neighbor $u \in N_{in}(v)$ has $d(j, u) = d(j, v) - 1$. But this contradicts the minimality of $d(j, v)$. □

Finally, suppose we are given only the out-degrees of the nodes. We use the charge-distribution PLS from Sect. 4.3 to certify that all nodes are recurrent, as this is equivalent to strong connectivity:

Theorem 5.9. StrongCon *is* D_1 {out-deg}*-decidable with proof size* $O(n^2)$.

To conclude this section, we observe that while any single one of the fields UID, size,out-deg suffices to decide strong connectivity, if we are given both UID and size, we can give a very short PLS:

Theorem 5.10. StrongCon *is* D_1 {UID, size}*-decidable with proof size* $O(\log n)$.

To do this, we ask the prover to choose some node $v \in V$, and prove that v is both a global source and a global sink, composing the D_1 {UID}-PLS for GlobalSource from Claim 5.2 and the D_1 {size}-PLS for MarkedIsSink from Claim 4.1. Both require $O(\log n)$ bits.

6 Sufficient and Necessary Conditions for Universal Decidability

In this section we ask what initial knowledge is needed to guarantee that *every* graph language can be decided by a proof labeling scheme. It is well-known that in undirected graphs with unique identifiers, every graph language is decidable [8, 9], but we have already seen that in directed graphs this is not the case. It turns out that adding either the size or out-deg fields to UID restores our ability to decide every graph language:

Theorem 6.1. *For every graph language* $\mathcal{L}$*, there exist a* D_1 {UID, size}*-PLS and a* D_1 {UID, out-deg}*-PLS that recognize* $\mathcal{L}$.

Proof (sketch). We show that the canonical prover-verifier pair ($\mathbf{Prv}^{\text{canon}}, \mathbf{Ver}^{\mathcal{L}}$) from Theorem 4.3 decides $\mathcal{L}$, given either UID and size or UID and out-deg. Recall that $\mathbf{Prv}^{\text{canon}}$ gives each node v a complete description $\hat{G}' = (G', F_1', \ldots, F_k')$ of what it *claims* is the input instance $\hat{G} = (G, F_1, \ldots, F_k)$ (the graph and the initial knowledge), and also specifies a mapping $\rho : V(G) \to V(G')$ that tells each node $v \in V(G)$ what node it is mapped onto in $V(G')$. We prove that if we are given unique identifiers, together with either the size of the graph or the out-degrees, then the prover cannot lie without being detected, and thus, if all nodes accept, we must have $G' = G$.

Let $R = \{\mathsf{UID}(v) : v \in V(G)\}$ be the real UIDs of the nodes of G, and let $S = \{\mathsf{UID}(v') : v' \in V(G')\}$ be the UIDs in the prover's instance. To prove that $G' = G$, it suffices to show that $R = S$: each $v \in V(G)$ verifies that the in-neighbors of $\rho(v)$ in G' match those of v in G, and if this is true for all nodes and $S = R$, then $G' = G$ (up to isomorphism). Thus, let us prove that $S = R$, otherwise the proof is rejected.

Each $v \in V(G)$ verifies that $\mathsf{UID}(\rho(v)) = \mathsf{UID}(v)$, and thus we have $R \subseteq S$. If we know the size of the graph, then under $\mathbf{Ver}^{\mathcal{L}}$, the nodes verify that G has the same size as G'. This implies that $S = R$. If we have out-degrees instead of the size of the graph, then we can use an argument similar to Lemma 5.1 to show that the prover must accurately describe the out-neighbors of each node v: we must have $\rho(N_{out}^{G}(v)) = N_{out}^{G'}(v)$. This suffices to show that the prover cannot "invent" any new nodes, i.e., $S = R$: in the cut between the "real nodes" $\rho(V)$ in G' and any "fake nodes" $V(G') \setminus \rho(V)$ invented by the prover, some real node $v \in V$ will notice a "fake edge" e' incident to $\rho(v)$ (either incoming or outgoing), such that no real edge e incident to v is mapped onto e'. This will cause node v to reject.

Note that many of the results from previous sections (e.g., Theorem 5.4 and Corollary 5.1) can be viewed as special cases of the result above, but in those cases we are able to give shorter proofs, while here the proof size is $O(n^2)$.

The Effect of Having a Marked Global Sink. In undirected graphs, having a unique marked leader suffices to decide all graph languages, even in the absence of unique identifiers [8]. Next we ask whether a similar scenario exists in directed graphs. It is clear that merely having a unique leader is not useful, so we strengthen the requirement, and ask for a *marked global sink*. For the remainder of the section, we consider proof labeling schemes with respect to the family $\mathcal{G}$ = GLOBALSINK of graphs that contain a global sink.

We show that even if the marked global sink is not unique (i.e., more than one node can be marked), together with any of the fields UID, size, we can decide any graph language. Formally, let marked-sink be an initial knowledge field where each node v is given a bit $mark(v) \in \{0, 1\}$, such that (a) at least one node $v \in V$ has $mark(v) = 1$, and (b) any node $v \in V$ with $mark(v) = 1$ is a global sink. We show that the combination of marked-sink together with any one of the fields suffices to guarantee that all graph languages are decidable:

Theorem 6.2. *For any graph language $\mathcal{L}$ and field $\mathcal{F} \in \{\mathsf{UID}, \mathsf{size}\}$, there is a $D_1\{\mathsf{marked\text{-}sink}, \mathcal{F}\}$-PLS which recognizes $\mathcal{L}$ with regards to graphs in* GLOBALSINK.

Proof (sketch). We show that if we have marked-sink and UID, then as in Theorem 6.1, the prover cannot lie about the graph: suppose the prover's claimed graph G' contains at least one "fake node" $f \in V(G') \setminus \rho(V(G))$, where ρ is the mapping the prover gave from the input graph to the graph G'. Since we have unique identifiers, we conflate the nodes of G, G' with their UIDs. Let $s \in V(G)$ be some marked sink. Node s verifies that $\rho(s) \in V(G')$ and that $\rho(s)$ is marked in G', and all nodes verify that in G', every marked node is a global sink. Thus, there is some path from the "fake node" f to $\rho(s)$ in G'. Since $s \in \rho(V(G))$, the path must at some point cross an edge $(u, w) \in E(G')$ such that $u \notin \rho(V(G))$ but $w \in \rho(V(G))$. Node w notices that it has no in-neighbor in G whose UID matches u, causing it to reject.

As for the other two knowledge fields, if we have size and marked-sink, we can use the size to assign UIDs as in Claim 4.2, and then use the scheme above.

It can also be shown that even a *unique* marked global sink by itself does not suffice to decide all graph languages, and neither does a unique marked global source, or even, in strongly connected graphs, a unique marked node which is both a sink and a source.

7 Conclusion

We conclude by discussing several open problems related to the complexity of proof-labeling schemes in directed networks.

Many of the proofs we constructed in this paper are based on asking the prover to specify a tree, or a collection of trees; for example, the distance-based PLS from Theorem 5.7, where the prover gives each node its distance from every other node, essentially amounts to specifying a collection of BFS trees, one rooted at each node. This type of PLS typically has proof length at most $O(n \log n)$, but we see at least three different reasons for this cost:

- In Theorem 5.7 (strong connectivity) the prover essentially specifies n different trees, and since each tree requires $O(\log n)$ bits per node, the total is $O(n \log n)$ bits per node.
- In Theorem 5.4 (global sink using UID, size), the prover only specifies one tree, but it gives the entire tree to all the nodes, to allow them to verify it. This again amounts to $O(n \log n)$ bits per node.
- In Corollary 5.1 (global sink using UID, out-deg), the prover also specifies one tree, but this time the tree is encoded using $O(\log n)$ bits per node. However, the prover needs to give each node the labels of all of its out-neighbors, to allow the verification to go through. This once again leads to the cost of $O(n \log n)$ bits per node.

It is interesting to ask which of these is inherent, and which can perhaps be eliminated to obtain a shorter proof, given that the information-theoretic cost of encoding a tree is only $O(n \log n)$ bits in total across all the nodes together (rather than $O(n \log n)$ per node).

More curious still is the usefulness of out-degrees: in Theorems 5.5 and 5.9 we used them to apply the charge-distribution PLS, yielding to an ∞-PLS that we compressed down to $O(n^2)$ bits per node—essentially, we asked the prover to describe the entire graph. However, the problems we studied in this paper are not sensitive to *all* edges in a dense graph: there is always a subset of $O(n)$ edges such that even if we remove all the other edges, the reachability property will be preserved. Thus, it would be interesting (and surprising) if the true complexity of strong connectivity using only out-degrees is $\Omega(n^2)$. One potential direction to reduce the length of the proof is to apply the charge-distribution protocol over a spanning, strongly-connected subgraph containing only $O(n)$ edges, rather than the entire graph. However, it is not clear how we can verify that the prover described the execution correctly, or even how to specify a sparse subgraph that can be verified to be spanning and preserve the reachability property we are interested in, without using UIDs. Resolving these questions—or alternatively, developing lower bound techniques that can exploit our uncertainty about all but the out-degrees of the nodes—remain fascinating open problems.

References

1. Bick, A., Kol, G., Oshman, R.: Distributed zero-knowledge proofs over networks. In: SODA, pp. 2426–2458. SIAM (2022)
2. Feuilloley, L., Fraigniaud, P.: Survey of distributed decision. Bull. EATCS **119** (2016). http://bulletin.eatcs.org/index.php/beatcs/issue/view/21
3. Foerster, K.T., Luedi, T., Seidel, J., Wattenhofer, R.: Local checkability, no strings attached:(a) cyclicity, reachability, loop free updates in SDNs. Theoret. Comput. Sci. **709**, 48–63 (2018)
4. Fraigniaud, P., Halldórsson, M.M., Korman, A.: On the impact of identifiers on local decision. In: PODC, pp. 224–238 (2012)
5. Fraigniaud, P., Korman, A., Peleg, D.: Local distributed decision. In: FOCS, pp. 708–717 (2011)
6. Gallager, R.G.: Stochastic Processes: Theory for Applications. Cambridge University Press, Cambridge (2013)
7. Gavoille, C., Peleg, D.: Compact and localized distributed data structures. Distrib. Comput. **16**(2), 111–120 (2003)
8. Göös, M., Suomela, J.: Locally checkable proofs in distributed computing. Theory Comput. **12**(1), 1–33 (2016)
9. Korman, A., Kutten, S., Peleg, D.: Proof labeling schemes. Distrib. Comput. **22**(4), 215–233 (2010)
10. Langberg, M., Schwartz, M., Bruck, J.: Distributed broadcasting and mapping protocols in directed anonymous networks. In: PODC, pp. 382–383 (2007)
11. Suomela, J.: Survey of local algorithms. ACM Comput. Surv. **45**(2), 24:1–24:40 (2013)

On the Computational Power of Energy-Constrained Mobile Robots: Algorithms and Cross-Model Analysis

Kevin Buchin[1], Paola Flocchini[2], Irina Kostitsyna[3], Tom Peters[3], Nicola Santoro[4], and Koichi Wada[5(✉)]

[1] TU Dortmund, Dortmund, Germany
kevin.buchin@tu-dortmund.de
[2] University of Ottawa, Ottawa, Canada
pflocchi@uottawa.ca
[3] TU Eindhoven, Eindhoven, The Netherlands
{i.kostitsyna,t.peters1}@tue.nl
[4] Carleton University, Ottawa, Canada
santoro@scs.carleton.ca
[5] Hosei University, Tokyo, Japan
wada@hosei.ac.jp

Abstract. We consider distributed systems of identical autonomous computational entities, called *robots*, moving and operating in the plane in synchronous *Look-Compute-Move* (*LCM*) cycles. The algorithmic capabilities of these systems have been extensively investigated in the literature under four distinct models ($\mathcal{OBLOT}$, $\mathcal{FSTA}$, $\mathcal{FCOM}$, $\mathcal{LUMI}$), each identifying different levels of memory persistence and communication capabilities of the robots. Despite their differences, they all always assume that robots have unlimited amounts of energy.

In this paper, we remove this assumption and start the study of the computational capabilities of robots whose energy is limited, albeit renewable. We first study the impact that memory persistence and communication capabilities have on the computational power of such energy-constrained systems of robots; we do so by analyzing the computational relationship between the four models under this energy constraint. We provide a complete characterization of this relationship.

We then study the difference in computational power caused by the energy restriction and provide a complete characterization of the relationship between energy-constrained and unrestricted robots in each model. We prove that within $\mathcal{LUMI}$ there is no difference; an integral part of the proof is the design and analysis of an algorithm that in $\mathcal{LUMI}$ allows energy-constrained robots to execute correctly any protocol for robots with unlimited energy. We then show the (apparently

This work was supported in part by JSPS KAKENHI No. 20K11685 and 21K11748, Israel & Japan Science and Technology Agency (JST) SICORP (Grant#JPMJSC1806), and by the Natural Sciences and Engineering Research Council of Canada (NSERC) under Discovery Grants A2415 and 203254.

M. Parter (Ed.): SIROCCO 2022, LNCS 13298, pp. 42–61, 2022.
https://doi.org/10.1007/978-3-031-09993-9_3

counterintuitive) result that in all other models, the energy constraint actually provides the robots with a computational advantage.

Keywords: Oblivious robots · Luminous robots · Energy-constrained robots · Comparison of models

1 Introduction

In this paper, we consider distributed systems composed of identical autonomous computational entities, viewed as points and called *robots*, moving and operating in the Euclidean plane in synchronous *Look-Compute-Move* (*LCM*) cycles. In each synchronous round, a non-empty set of (possibly all) robots is activated, each performs its *LCM* cycle simultaneously and terminates by the end of the round. Each cycle is composed of three phases: in the *Look* phase, an entity obtains a snapshot of the robots' configuration showing the positions of all the other robots; in the *Compute* phase, the robot executes its algorithm (the same for all robots) and computes a destination point using the snapshot as input; in the *Move* phase, the robot moves towards the computed destination. Repeating these cycles, the robots can collectively perform some tasks and solve some pattern formation problems.

The activation of robots is controlled by an adversarial scheduler, who selects which robots are activated in each round. This general setting is usually called *semi-synchronous* (SSYNCH); the special restricted setting where every robot is activated in every round is called *fully-synchronous* (FSYNCH).

These distributed robot systems have been extensively investigated within distributed computing. The research aim has been to understand the nature and the extent of the impact that factors, such as *memory persistence* and *communication capability*, have on the solvability of problems and thus on the computational power of the system. To this end, four models have been identified and investigated: $\mathcal{OBLOT}$, $\mathcal{FSTA}$, $\mathcal{FCOM}$, and $\mathcal{LUMI}$.

In the most common (and weakest) model, $\mathcal{OBLOT}$, in addition to the standard assumptions of anonymity and uniformity (robots have no IDs and run identical algorithms), the robots are *oblivious* (they have no persistent memory to record information of previous cycles) and they are *silent* (without explicit means of communication). The restrictions imposed by the absence of persistent memory and the incapacity of explicit communication severely limit what the robots can do. Computability in this model has been the object of intensive research since its introduction in [25] (e.g., see [1,3–6,12,13,16,17,19,25,25,28,29], as well as the recent book [10]).

In the stronger $\mathcal{LUMI}$ model, formally introduced and defined in [7], robots are provided with some (albeit limited) persistent memory and means for communication. In this model each robot is equipped with a constant-sized memory (called *light*), whose value (*color*) can be set during the *Compute* phase. The light is visible to all the robots and is persistent between the robot activations. Hence, these luminous robots are capable of both remembering and communicating a constant number of bits. Design of algorithms and feasibility of solving problems for luminous robots have been extensively studied [7,8,14,18,20–23,26,27]; for a recent survey, see [9]. The availability of both persistent memory and communication, however limited, clearly renders luminous robots more powerful than oblivious robots (see e.g., [7]).

To better understand the computational power of persistent memory and communication individually, models $\mathcal{FSTA}$ and $\mathcal{FCOM}$ (which fall in between $\mathcal{OBLOT}$ and $\mathcal{LUMI}$) were introduced in [14] (and studied in [20,26]). In the first model, $\mathcal{FSTA}$, the light of a robot is "internal", i.e., visible only to that robot, while in the second model, $\mathcal{FCOM}$, the light of a robot is visible only to the other robots but not to the robot itself. Thus in $\mathcal{FSTA}$, the color merely encodes an internal state, and the robots are *finite-state* and *silent*. On the contrary, in $\mathcal{FCOM}$, a robot can communicate to the other robots by means of the light but forgets the content of its transmission by the next cycle; that is, robots are *oblivious* and *finite-communication*.

To understand the computational power of these distributed systems, one needs to explore and determine the computational power of the robots within each of these models, as well as (and more importantly) with respect to each other. This type of cross-model investigation has been taking place but is rather limited in scope (e.g., [14]). Recently, a substantial step has been taken in [15] where, by integrating existing bounds and establishing new results, a comprehensive map of the computational relationship between the four models, $\mathcal{OBLOT}$, $\mathcal{FSTA}$, $\mathcal{FCOM}$, and $\mathcal{LUMI}$, has been drawn (and hence the computational impact of the presence/absence of persistent memory and/or communication capabilities has been established) for the two fundamental synchronous settings: fully-synchronous and semi-synchronous.

The Energy Problem. In the vast existing literature, surprisingly, no consideration has been made so far on the energy required for the robots to be able to operate. Existing works share the same implicit assumption that the robots have an unlimited amount of energy enabling them to perform their activities in every activation round. In this paper, we remove this assumption and initiate the study of the robots whose energy is limited, albeit renewable. More precisely, we consider systems where an activated entity uses all its energy to execute an *LCM* cycle, and once this happens, the robot is not operational and cannot be activated in the next round; the energy, however, can be restored through a period of inactivity. This would be the case if, for example, the robot's power is provided by a battery rechargeable by energy harvesting (as it is done in conceptually related systems such as wireless mobile sensors [24]).

The immediate natural questions are: *what is the computational power of these energy-constrained robots*? and, in particular, *what is the impact of the crucial factors (memory and communication) in this case*? In this paper, we start investigating these questions.

Contributions. We consider systems where the energy of a robot is sufficient to execute exactly one *LCM* cycle, and the depleted energy is restored after one round of inactivity. We investigate the computational power of the distributed robot systems described by the four models when the robots are subject to such energy constraint.

We establish an equivalence between a system of energy-constrained robots under the semi-synchronous (SSYNCH) scheduler and a system of classic robots with unlimited energy under a new scheduler, which we call RSYNCH. By our definition of RSYNCH, the sets of robots activated in any two consecutive rounds are required to be disjoint. This direct correspondence enables us to reduce the cross-model investigation of the energy-constrained robots to the cross-model investigation of energy-unbounded robots under the RSYNCH scheduler. Furthermore, it allows us to determine the change (if any) in computational power due to the energy restriction, by determining the relationship between RSYNCH and the general unrestricted SSYNCH scheduler.

Let M^S and M^{RS} denote the systems of unlimited-energy robots defined by model $M \in \{\mathcal{LUMI}, \mathcal{FCOM}, \mathcal{FSTA}, \mathcal{OBLOT}\}$ under SSYNCH and RSYNCH, respectively (the latter being equivalent to the systems of energy-constrained robots under SSYNCH). We first study the impact that memory persistence and communication capabilities have on the computational power of energy-constrained systems of robots; we do so by analyzing the computational relationship between the four models under RSYNCH scheduler. We provide a complete characterization:

$$\mathcal{LUMI}^{RS} \equiv \mathcal{FCOM}^{RS} > \mathcal{FSTA}^{RS} > \mathcal{OBLOT}^{RS},$$

where (as formally defined in Sect. 2) $\mathcal{X} > \mathcal{Y}$ denotes that model $\mathcal{X}$ is strictly more powerful than $\mathcal{Y}$, and $\mathcal{X} \equiv \mathcal{Y}$ denotes that $\mathcal{X}$ and $\mathcal{Y}$ are computationally equivalent. An integral part of the proof that $\mathcal{FCOM}^{RS}$ is more powerful than $\mathcal{FSTA}^{RS}$ (that is, it is better to communicate than to remember), is the design and analysis of an algorithm that allows robots in $\mathcal{FCOM}^{RS}$ to correctly execute any protocol for the more powerful $\mathcal{LUMI}^{RS}$.

We then study what impact on computational power is created by restricting the energy of the robots, by comparing the computational difference between energy-constrained and unrestricted robots in each of the four models (i.e., between M^{RS} and M^S for each $M \in \{\mathcal{LUMI}, \mathcal{FCOM}, \mathcal{FSTA}, \mathcal{OBLOT}\}$). We provide a complete characterization. In particular, we prove that for $\mathcal{LUMI}$ robots, the strongest model, there is no difference between energy-constrained and unlimited-energy robots; i.e., $\mathcal{LUMI}^{RS} \equiv \mathcal{LUMI}^S$. An integral part of the proof is the design and analysis of an algorithm that allows energy-constrained robots in $\mathcal{LUMI}^{RS}$ to correctly execute any protocol for robots with unlimited energy in $\mathcal{LUMI}^S$. In all other models, we prove that restricting energy actually provides the robots with a definite computational advantage; this apparently counterintuitive result is due to the fact that the energy restriction reduces the adversarial power of the activation scheduler. Let us stress that the established characterization covers all the cross-model and cross-scheduler relationships.

Finally, we complete the study of systems of energy-constrained robots by analyzing the relationship between their computational power and that of robots with unlimited energy under the most benign synchronous activation scheduler FSYNCH (i.e., fully synchronous). In this case, perhaps not surprising, we prove that, in each robot model, energy-constrained robots are strictly less powerful than fully-synchronous ones with unbounded energy. We again cover all the cross-model and cross-scheduler relationships.

The details of omitted parts and proofs are shown in a full paper [2].

2 Models and Preliminaries

The Basics. The system consists of a set $R = \{r_0, \ldots, r_{n-1}\}$ of computational entities, called robots, modeled as geometric points, that live in $\mathbb{R}^2$, where they can move freely and continuously. The robots are autonomous without a central control. They are indistinguishable by their appearance, do not have internal identifiers, and execute the same algorithm. Each robot has its own local coordinate system, which may be inconsistent with the coordinate systems of the other robots. A robot perceives itself at the origin of its coordinate system, and is capable of observing the positions of the other robots in it.

The robots operate in *Look-Compute-Move* (*LCM*) cycles. When activated, a robot executes a cycle by performing the following three operations:

Look. The robot obtains a snapshot of the positions occupied by robots expressed with respect to its own coordinate system, and their colors. This operation is assumed to be instantaneous.

Compute. The robot executes the algorithm using the snapshot as input; the result of the computation is a destination point.

Move. The robot moves towards the computed destination. If the destination is the current location, the robot stays still.

The system is *synchronous.* That is, time is divided into discrete intervals, called *rounds.* In each round a robot is either active or inactive. The robots active in a round perform their *LCM* cycle in perfect synchronization; if not active, the robot is idle in that round. All robots are initially idle. In the following, we use round and time interchangeably.

Each robot has a bounded amount of *energy*, which is totally consumed whenever it performs a cycle; its energy however is restored after being idle for a round. A robot with depleted energy cannot be active.

Movements are said to be *rigid* if the robots always reach their destination. They are said to be *non-rigid* if they may be unpredictably stopped by an adversary whose only limitation is the existence of $\delta > 0$, unknown to the robots, such that if the destination is at distance at most δ the robot will reach it, else it will move at least δ towards the destination.

There might not be consistency between the local coordinate systems and their unit of distance. The absence of any a priori assumption on consistency of the local coordinate systems is called *disorientation.* The type of disorientation can range from *fixed*, where each local coordinate system remains the same through all the rounds, to *variable* where the direction, the orientation, and the unit distance of a robot may vary between successive rounds. In this paper we consider only fixed disorientation.

Let $x_i(t)$ denote the location of robot r_i at time t in a global coordinate system (unknown to the robots), and let $X(t) = \{x_i(t) : 0 \leq i \leq n-1\} = \{x_0(t), x_1(t), \ldots, x_{m-1}(t)\}$; note that $|X(t)| = m \leq n$ since several robots might be at the same location at time t. A *configuration* $C(t)$ at time t is the multi-set of the n pairs of the $(x_i(t), c_i(t))$, where $c_i(t)$ is the color (formally defined below) of robot r_i at time t.

The robots are said to have *chirality* if they share the same circular orientation of the plane (i.e., they agree on "clockwise" direction). Notice that, in presence of chirality, at any time t, there would exist a unique circular ordering of the locations $X(t)$ occupied by the robots at that time; let `suc` and `pred` be the functions denoting the ordering and, without loss of generality, let $\texttt{suc}(x_i(t)) = x_{i+1 \bmod m}(t)$ and $\texttt{pred}(x_i(t)) = x_{i-1 \bmod m}(t)$ for $i \in \{0, 1, \ldots, m-1\}$.

The Computational Models. In $\mathcal{OBLOT}$ model the robots are *silent*: they have no explicit means of communication; furthermore they are *oblivious*: at the start of a cycle, a robot has no memory of observations and computations performed in previous cycles.

In $\mathcal{LUMI}$ model each robot r is equipped with a persistent visible state variable *Light*$[r]$, called *light*, whose values are taken from a finite set C of states called *colors* (including the color that represents the initial state when the light is off). The color of the light is set by r at the end of its *Compute* operation. The lights are *persistent* from one computational cycle to the next: the color is not automatically reset at the end of a cycle; the robot is otherwise oblivious, forgetting all other information from previous cycles. In $\mathcal{LUMI}$ the *Look* operation produces a colored snapshot; i.e., it returns the set of pairs (*position*, *color*) of the other robots. Note that if $|C| = 1$, then the light is not used, and the model is equivalent to $\mathcal{OBLOT}$.

As mentioned above, the lights provide simultaneously persistent memory and direct means of communication, although both limited to a constant number of bits per cycle. The two sub-models $\mathcal{FSTA}$ and $\mathcal{FCOM}$ of $\mathcal{LUMI}$ each offers only one of these two capabilities. In $\mathcal{FSTA}$ model a robot can only see the color of its own light; that is, the light is *internal* and its color merely encodes an internal state. Hence the robots are *silent* as in $\mathcal{OBLOT}$; but are *finite-state* as in $\mathcal{LUMI}$. Observe that a snapshot in $\mathcal{FSTA}$ is the same as in $\mathcal{OBLOT}$.

In $\mathcal{FCOM}$ the lights are *external* and visible only to the other robots: a robot can communicate to the others by setting its color, but forgets it by the next cycle; that is, robots are *finite-communication* but *oblivious*. A snapshot that robot r perceives in $\mathcal{FCOM}$, as in $\mathcal{LUMI}$, contains the information about robots' colors, except that the color *Light*$[r]$ is omitted from the set of colors associated with the position of r.

Activation Schedulers and Energy Restriction. In each synchronous round, some robots become active and they execute their *LCM* cycle in complete synchrony. The choice of which non-empty subset of the robots is activated in a specific round is under the control of an adversarial *activation scheduler* constrained to be fair; that is, every robot will become active infinitely often. Given a synchronous scheduler $\mathcal{S}$ and a set of robots R, an *activation sequence* of R under $\mathcal{S}$ is an infinite sequence $E = \langle e_1, e_2, \ldots, e_i, \ldots \rangle$, where $e_i \subseteq R$ denotes the set of robots activated in round i, satisfying the *fairness constraint*:

$$\forall r \in R, i \geq 1 \, \exists j > i : r \in e_j \, .$$

Let $\mathcal{E}(\mathcal{S}, R)$ denote the set of all activation sequences of R by $\mathcal{S}$. In the standard synchronous scheduler (SSYNCH), first studied in [25] and often called *semi-synchronous*, each sequence $E = \langle e_1, e_2, \ldots, e_i, \ldots \rangle \in \mathcal{E}(\text{SSYNCH}, R)$ satisfies the *basic condition*

$$\forall i \geq 1, \varnothing \neq e_i \subseteq R \, .$$

The special *fully-synchronous* (FSYNCH) setting, where every robot is activated in every round, corresponds to further restricting the activation sequences by imposing $\forall i \geq 1, e_i = R$. Notice that, in this setting, the activation scheduler has no adversarial power. Another special setting is defined by the well-known *round robin* scheduler (e.g., see [26]), whose generalized definition corresponds to adding the restriction:

$$[\exists p > 1 : (\bigcup_{1 \leq i \leq p} e_i = R) \ \textbf{and} \ (\forall 1 \leq i \neq j \leq p, [e_i \cap e_j = \varnothing]) \ \textbf{and} \ (\forall i \geq 1, [e_i = e_{i+p}])] \, .$$

We study systems of *energy-constrained robots* under the standard synchronous activation scheduler. More precisely, we study systems where a robot (i) has just enough energy to execute a cycle, (ii) it cannot be activated in a round unless it has full energy, and (iii) its depleted energy is regenerated after one round. These three conditions clearly have an impact on the possible activation sequences of the robots. In particular, since a robot with depleted energy cannot be activated, the basic condition on e_i becomes

$$\forall i \geq 1, e_i \subseteq R^*[i] \ \textbf{and} \ R^*[i] \neq \varnothing \Rightarrow e_i \neq \varnothing \, , \tag{1}$$

where $R^*[i] \subseteq R$ denotes the set of robots with full energy in round i. Furthermore, since it takes a round to regenerate depleted energy, e_i must also satisfy

$$\forall i \geq 1, (e_i \cap e_{i+1} = \varnothing) \, .$$

Notice that, since $e_i \subseteq R^*[i]$, it is possible that $e_i = R$ when $R^*[i] = R$. Should this be the case, since a robot has just enough energy to execute a cycle, then $R^*[i+1] = \varnothing$ and thus $e_{i+1} = \varnothing$. Furthermore, due Equation (1),

$$\exists i, (\varnothing \neq e_i \neq R) \Rightarrow \forall j \geq i, (\varnothing \neq e_j \neq R) \, .$$

That is, if fewer than $|R|$ but a positive number of robots are activated in any round i, then the set of robots with full energy $R^*[j] \neq R$, and thus $\varnothing \neq e_j \neq R$, for all $j \geq i$. Thus, the activation sequences of the energy-constrained robots are infinite sequences where the prefix is a (possibly empty) alternating sequence of R and $\varnothing$, and, if the prefix is finite, the rest are non-empty sets satisfying the constraint $(e_i \cap e_{i+1} = \varnothing)$. Notice that this set of sequences, denoted by $\mathcal{E}(\text{SSYNCH}_{res}, R)$, is not a proper subset of $\mathcal{E}(\text{SSYNCH}, R)$ since some sequences might have empty sets in their prefix.

Consider now the synchronous scheduler, we shall call RSYNCH, obtained from SSYNCH by adding the following *restricted-repetition condition* to its activation sequences:

$$\left[\forall i \geq 1, e_i = R\right] \textbf{ or}$$

$$\left[\exists p \geq 0 : \left(\left[\forall i \leq p, (e_i = R)\right] \textbf{ and } \left[\forall i > p, (\varnothing \neq e_i \neq R \textbf{ and } e_i \cap e_{i+1} = \varnothing)\right]\right)\right],$$

that is, $\mathcal{E}(\text{RSYNCH}, R)$ is composed of sequences where the prefix is a (possibly empty) sequence of R and, if the prefix is finite, the rest are non-empty sets satisfying the constraint $(e_i \cap e_{i+1} = \varnothing)$.

There is an obvious bijection ϕ between $\mathcal{E}(\text{SSYNCH}_{res}, R)$ and $\mathcal{E}(\text{RSYNCH}, R)$, where $\phi(E)$ corresponds to removing all empty sets from $E \in \mathcal{E}(\text{SSYNCH}_{res}, R)$. Thus computation performed by R under E is equivalent to one performed under $\phi(E)$. Informally, under SSYNCH_{res}, if all robots are activated in the same round i, they will be all idle in round $i+1$, and they will all be with full energy in round $i+2$. Since no activity takes place in round $i+1$, we can ignore all such empty rounds and assume that round $i+2$ occurs right after round i. Thus, the computation by a set of energy-constrained robots R under the standard synchronous scheduler SSYNCH is equivalent to the one if the robots in R were energy-unbounded but the activation was controlled by scheduler RSYNCH. This restricted-repetition setting has never been studied before; observe that it includes both fully synchronous FSYNCH and round robin as special cases.

Computational Relationships. Let $\mathcal{M} = \{\mathcal{LUMI}, \mathcal{FCOM}, \mathcal{FSTA}, \mathcal{OBLOT}\}$ be the set of models under investigation, and $\mathcal{S} = \{\text{FSYNCH}, \text{RSYNCH}, \text{SSYNCH}\}$ be the set of activation schedulers under consideration. We denote by $\mathcal{R}$ the set of all teams of robots satisfying the core assumptions (i.e., they are identical, autonomous, and operate in *LCM* cycles), and $R \in \mathcal{R}$ a team of robots having identical capabilities (e.g., common coordinate system, persistent storage, internal identity, rigid movements etc.). By $\mathcal{R}_n \subset \mathcal{R}$ we denote the set of all teams of size n.

Given a model $M \in \mathcal{M}$, a scheduler $S \in \mathcal{S}$, and a team of robots $R \in \mathcal{R}$, let $Task(M, S; R)$ denote the set of problems solvable by R in M under adversarial scheduler S. Let $M_1, M_2 \in \mathcal{M}$ and $S_1, S_2 \in \mathcal{S}$.

- We say that model M_1 under scheduler S_1 is *computationally not less powerful than* model M_2 under S_2, denoted by $M_1^{S_1} \geq M_2^{S_2}$ if $\forall R \in \mathcal{R}$ we have $Task(M_1, S_1; R) \supseteq Task(M_2, S_2; R)$.
- We say that M_1 under S_1 is *computationally more powerful than* M_2 under S_2, denoted by $M_1^{S_1} > M_2^{S_2}$, if $M_1^{S_1} \geq M_2^{S_2}$ and $\exists R \in \mathcal{R}$ such that $Task(M_1, S_1; R) \setminus Task(M_2, S_2; R) \neq \varnothing$.

- We say that M_1 under S_1 and M_2 under S_2 are *computationally equivalent*, denoted by $M_1^{S_1} \equiv M_2^{S_2}$, if $M_1^{S_1} \geq M_2^{S_2}$ and $M_2^{S_2} \geq M_1^{S_1}$.
- Finally, we say that M_1 under S_1 and M_2 under S_2 are *computationally orthogonal* (or *incomparable*), denoted by $M_1^{S_1} \bot M_2^{S_2}$, if $\exists R_1, R_2 \in \mathcal{R}$ such that $\mathit{Task}(M_1, S_1; R_1) \setminus \mathit{Task}(M_2, S_2; R_1) \neq \varnothing$ and $\mathit{Task}(M_2, S_2; R_2) \setminus \mathit{Task}(M_1, S_1; R_2) \neq \varnothing$.

For brevity, for a model $M \in \mathcal{M}$, let M^F, M^{RS}, and M^S, denote M^{FSYNCH}, M^{RSYNCH}, and M^{SSYNCH}, respectively. Furthermore, with a slight abuse of notation, let $M^F(R)$, $M^{RS}(R)$, and $M^S(R)$, denote $\mathit{Task}(M, \text{FSYNCH}; R)$, $\mathit{Task}(M, \text{RSYNCH}; R)$, and $\mathit{Task}(M, \text{SSYNCH}; R)$, respectively. Trivially, for all $M \in \mathcal{M}$,

$$M^F \geq M^{RS} \geq M^S,$$

and, for all $P \in \mathcal{S}$,

$$\mathcal{LUMI}^P \geq \mathcal{FSTA}^P \geq \mathcal{OBLOT}^P \text{ and } \mathcal{LUMI}^P \geq \mathcal{FCOM}^P \geq \mathcal{OBLOT}^P.$$

3 Computational Relationship Between RSYNCH and SSYNCH

We begin by studying the impact that constraining the energy has on the computational capability of the robots. We do so by analyzing the computational relationship between RSYNCH and SSYNCH in each of the four models.

3.1 Power of RSYNCH in $\mathcal{FCOM}, \mathcal{FSTA}$ and $\mathcal{OBLOT}$

First, we show that RENDEZVOUS problem (RDV) [11], where two robots a and b must gather in the same location not known in advance, cannot be solved in RSYNCH. Recall that $\mathcal{R}_2 \subset \mathcal{R}$ is the set of all teams of robots of size 2.

Lemma 1. *$\exists R \in \mathcal{R}_2$, RDV $\notin \mathcal{OBLOT}^{RS}(R)$. This result holds even in presence of chirality and rigidity of movement.*

The problem SHRINKING ROTATION (SRO) was introduced to show that $\mathcal{OBLOT}^F > \mathcal{OBLOT}^S$, and that models $\mathcal{OBLOT}^F$ and $\mathcal{FCOM}^S$ (or $\mathcal{FSTA}^S$) are incomparable [15].

Lemma 2 ([15]). *$\exists R \in \mathcal{R}_2$, SRO $\notin \mathcal{FCOM}^S(R) \cup \mathcal{FSTA}^S(R)$. This result holds even in presence of chirality and rigidity of movement.*

This problem can also play a role in showing $\mathcal{OBLOT}^{RS} > \mathcal{OBLOT}^S$ and orthogonality of $\mathcal{OBLOT}^{RS}$ and $\mathcal{FCOM}^S$ (or $\mathcal{FSTA}^S$).

Lemma 3. *$\forall R \in \mathcal{R}_2$, SRO $\in \mathcal{OBLOT}^{RS}(R)$, assuming common chirality and rigid movement.*

Lemma 1 and the fact that RDV can be solved by $\mathcal{FCOM}$ and $\mathcal{FSTA}$ in SSYNCH [14], and Lemmas 2 and 3 imply:

Theorem 1. *The following relations hold:* $\mathcal{OBLOT}^{RS} \perp \mathcal{FCOM}^S$, $\mathcal{OBLOT}^{RS} \perp \mathcal{FSTA}^S$, $\mathcal{FCOM}^{RS} > \mathcal{FCOM}^S$, $\mathcal{FSTA}^{RS} > \mathcal{FSTA}^S$, *and* $\mathcal{OBLOT}^{RS} > \mathcal{OBLOT}^S$.

We now show the dominance of $\mathcal{LUMI}^S$ over $\mathcal{FSTA}^{RS}$ and the orthogonality of $\mathcal{FSTA}^{RS}$ with $\mathcal{FCOM}^S$. In order to show these results, we use the following problem.

Definition 1 (CYCLIC CIRCLES (CYC)). *Let* $n \geq 3$, $k = 2^{n-1}$, *and* $d : \mathbb{N} \to \mathbb{R}$ *is a non-invertible function. The problem is to form a cyclic sequence of patterns* $C, C_0, C, C_1, C, C_2, \ldots, C, C_{k-1}$, *where* C *is a pattern of* $n-1$ *robots occupying vertices of a regular* n*-gon with the* n*th robot in its center (see Fig. 1 (a)), and* C_i *(for* $0 \leq i \leq k-1$*) is a configuration with the* $n-1$ n*-gon robots in the same position, but the center robot occupying a point at distance* $d(i)$ *from the center in the direction of the empty vertex (see Fig. 1 (b)). In other words, the central robot moves to the designated position at distance* $d(i)$ *and comes back to the center. The process repeats after all* 2^{n-1} *configurations* C_i *have been formed.*

Lemma 4. *Let* $n \geq 3$. $\exists R \in \mathcal{R}_n$, CYC $\notin \mathcal{FSTA}^F(R)$. *This result holds even in presence of chirality and rigid movements.*

Next, we show that $\mathcal{FCOM}$ robots can solve CYC under SSYNCH. Intuitively, the $n-1$ robots on the circle act as a distributed counter using their lights to display the binary representation of the index i of the next configuration to be formed (see Fig. 1 (c)). The increment of the counter is performed by changing the bits accordingly and maintaining the carry, as in a full adder. Whenever activated, the central robot "reads" the information and understands when it is time to move and to which destination.

Lemma 5. *Let* $n \geq 3$. $\forall R \in \mathcal{R}_n$, CYC $\in \mathcal{FCOM}^S(R)$, *assuming chirality.*

The orthogonality of $\mathcal{FSTA}^{RS}$ and $\mathcal{FCOM}^S$ follows from Lemmas 3–5, and the dominance of $\mathcal{LUMI}^S$ over $\mathcal{FSTA}^{RS}$ follows from Lemmas 4 and 5.

Theorem 2. $\mathcal{FSTA}^{RS} < \mathcal{LUMI}^S$ *and* $\mathcal{FSTA}^{RS} \perp \mathcal{FCOM}^S$.

3.2 Power of RSYNCH in $\mathcal{LUMI}$

We now show that, in spite being more powerful in $\mathcal{FCOM}$ and $\mathcal{FSTA}$, RSYNCH robots with full lights ($\mathcal{LUMI}$) are *not* more powerful than SSYNCH robots with full lights. That is, $\mathcal{LUMI}^S$ is computationally equivalent to $\mathcal{LUMI}^{RS}$. To do so we prove the following theorem.

Theorem 3. $\forall R \in \mathcal{R}$, $\mathcal{LUMI}^{RS}(R) \leq \mathcal{LUMI}^S(R)$.

The approach is to show that $\mathcal{LUMI}$ robots under SSYNCH can simulate any algorithm designed for $\mathcal{LUMI}$ robots under RSYNCH (see algorithm `sim-RS-by-S(`A`)` in [2]). Let algorithm A for $\mathcal{LUMI}$ robots in RSYNCH use light with ℓ colors: $C = \{c_0, c_1, \ldots, c_{\ell-1}\}$. SSYNCH robots run algorithm `sim-RS-by-S(`A`)` simulating the execution of A as follows. Each robot r has the following sets of colors:

r.color $\in C$ indicating its own light used in algorithm A, initially set to *r.color* $= c_0$;

r.step $\in \{1, 2, 3, 4, 5, m\}$ indicating the step of the simulation currently under execution. Initially *r.step* is set to 1 for all robots, and thus initially the simulation is in Step 1;

r.executed $\in$ {*True*, *False*} indicating whether r has executed algorithm A in the current mega-cycle (see below). Initially *r.executed* is set to *False* for all robots;

r.charged $\in \{C, E, M\}$, where C, E, and M stand for "charged", "empty", and "just moved" respectively. The flag is used to ensure the validity of the simulated RSYNCH activation sequence, it indicates whether r is charged and can execute the algorithm A. Initially *r.charged* is set to C for all robots.

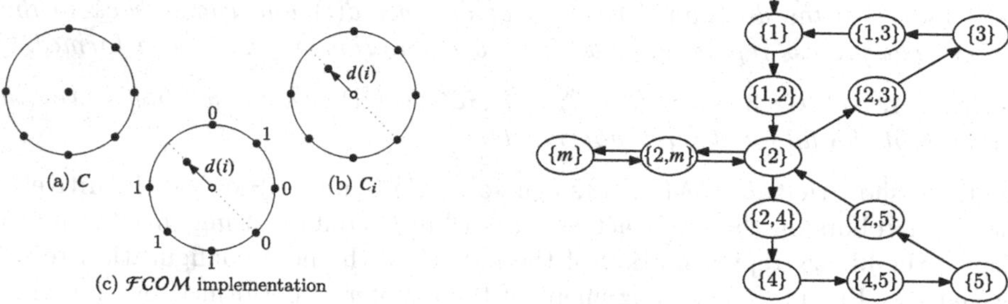

Fig. 1. The configurations of CYC.

Fig. 2. Transition diagram `sim-RS-by-S(A)`.

The simulation proceeds in two phases. The first phase of the simulation corresponds to the first $p \geq 0$ (where p can be ∞) activations in a simulated RSYNCH activation schedule where all robots in R are activated at each round. The second phase corresponds to the remaining activation cycles where a strict subset of R is activated at each round.

The state of the robot system is defined by the set of colors *step* currently set by the robots (see Fig. 2). By construction, there can be at most two different *step* colors in the system at each moment. If all robots have the same *step* color i (for $i \in \{1, 2, 3, 4, 5, m\}$), we say the robots execute Step i of the simulation. If not all robots have the same *step*, this corresponds to the system transitioning from one step to another.

The two phases of the simulation consist of executing so called *mega-cycles*. The mega-cycle of the first phase corresponds to one cycle $\{1\} \to \{2\} \to \{3\} \to \{1\}$. The mega-cycle of the second phase consists of several cycles $\{2\} \to \{4\} \to \{5\} \to \{2\}$ (until all robots have executed it) with one cycle $\{2\} \to \{m\} \to \{2\}$ to reset the flags and start the new mega-cycle. The robots execute A in Step 1 in the first phase, and in Step 2 in the second phase. The remaining steps serve for bookkeeping of flags *executed* and *charged*.

Specifically, the states of the simulation are (refer to Fig. 2):

- State $\{i, j\}$ are the transition states between Step i and Step j. Activated robots do not execute A nor change the values of the *charged* and *executed* flags, but only set their $r.step$ to either i or to j (depending on the specific case).
- Step 1: activated robot executes one cycle of the simulated algorithm A. Note that when Step 1 begins execution, it holds that $\forall \rho \in R\,(\rho.executed = False$ and $\rho.charged = C)$.
- Step 2: activated robot checks if all robots have their *charged* flag set to E, that is, if all robots were activated in Step 1. If so, we are in the first phase, and the simulation proceeds to Step 3 resetting all the *charged* and *executed* flags to prepare for the next move in Step 1. Otherwise, we are in the second phase, and there are two cases:
(1) If all robots have their *executed* flag set to *True*, the mega-cycle has finished. The robot proceeds to Step m.
(2) If some robots have their *executed* flag set to *False*, the mega-cycle has not finished yet. The simulation proceeds executing A. As soon as among the activated robots there is a non-empty subset R' with $charged = C$ and $executed = False$, they execute algorithm A, set *charged* to M and *executed* to *True*, and proceed to Step 4. Afterwards, the remaining robots proceed to Step 4 without executing A or changing their *charged* and *executed* flags.
- Step 3: robot resets the *charged* and *executed* flags to prepare the next move in Step 1. Note that as long as Step 3 is executed, full activation continues.
- Step 4: robot updates the *charged* flag if it did not execute A in preceding Step 2 from E to C (the robots that did not execute are recharged).
- Step 5: robot updates the *charged* flag if it executed A in preceding Step 2 from M to E (the robots that executed are discharged).
- Step m executes when one mega-cycle of phase two is completed. Robots reset their *executed* flags and transition back to Step 2. Note that Step m does not affect the *charged* flag, thus the robots which executed A in the last activation cycle of the preceding mega-cycle are discharged in the first activation cycle of the new mega-cycle and cannot be activated.

The initial configuration of `sim-RS-by-S(`A`)` satisfies $r.step = 1$, $r.charged = C$ and $r.executed = False$ for any robot r. For a configuration K, if the set of values of $r.step$ appearing in K is V, we say that the *step configuration* of K is V.

Observe that at any moment in time of `sim-RS-by-S(`A`)` the step configuration $V \in \{\{1\}, \{2\}, \{3\}, \{4\}, \{5\}, \{m\}, \{1,2\}, \{2,3\}, \{3,1\}, \{2,4\}, \{4,5\}, \{2,5\}, \{2,m\}\}$. Indeed, the case when $|V| = 1$ corresponds to a (sub)set of activated robots performing the same action (executing A, or modifying the flags *charged* and *executed*). These robots may update their *step* value leading to a new step configuration V' of a size at most two. In the case when $|V| = 2$ no actions are performed by the robots except of updating their *step* to some value $i \in V$ (remaining robots transition to Step i), eventually leading to the next step configuration $V' = \{i\}$.

The correctness of `sim-RS-by-S(`A`)` is shown by proving properties of the step configurations. The details and their proofs are shown in [2]. Note that, if algorithm A uses ℓ colors, the simulating algorithm `sim-RS-by-S(A)` uses $O(\ell)$ colors.

Lemma 6. *Algorithm* `sim-RS-by-S(A)` *correctly simulates execution of algorithm A run on* $\mathcal{LUMI}^R$ *robots by* $\mathcal{LUMI}^S$ *robots.*

Theorem 4. $\mathcal{LUMI}^S \equiv \mathcal{LUMI}^{RS}$.

4 Computational Relationship Between FSYNCH and RSYNCH

We have seen that RSYNCH is more powerful than SSYNCH in $\mathcal{FCOM}$, $\mathcal{FSTA}$, $\mathcal{OBLOT}$, and it has the same computational power in $\mathcal{LUMI}$. To better understand the power of RSYNCH among the classical synchronous schedulers, we now turn our attention to the relationship between RSYNCH and FSYNCH.

4.1 Dominance of FSYNCH over RSYNCH

The problem CENTER OF GRAVITY EXPANSION (CGE) was used to show dominance of FSYNCH over SSYNCH, that is, CGE is solvable in $\mathcal{FCOM}^F$ and $\mathcal{FSTA}^F$ but is not solvable in $\mathcal{LUMI}^S$ [15]. This problem can also be used to show dominance of FSYNCH over RSYNCH. In fact, we can obtain a stronger result showing that CGE is not solvable in $\mathcal{LUMI}^{F'}$, where F' is any scheduler such that the first activation does not contain all robots.

Lemma 7. *Let* $n \geq 2$. $\exists R \in \mathcal{R}_n$: CGE $\notin \mathcal{LUMI}^{F'}(R)$, *where* F' *is any scheduler such that the first activation does not contain all robots.*

Since RSYNCH contains patterns in F', we have the following.

Corollary 1. *Let* $n \geq 2$. $\exists R \in \mathcal{R}_n$: CGE $\notin \mathcal{LUMI}^{RS}(R)$.

As a consequence, we obtain dominance of FSYNCH over RSYNCH.

Theorem 5. $\forall \mathcal{X} \in \{\mathcal{OBLOT}, \mathcal{FCOM}, \mathcal{FSTA}, \mathcal{LUMI}\}$ $[\mathcal{X}^F > \mathcal{X}^{RS}]$.

4.2 Orthogonality of FSYNCH with RSYNCH

We now proceed to show incomparability of $\mathcal{FSTA}^F$ with $\mathcal{X}^{RS}$ for $\mathcal{X} \in \{\mathcal{LUMI}, \mathcal{FCOM}\}$, and of $\mathcal{OBLOT}^F$ with $\mathcal{X}^{RS}$ for $\mathcal{X} \in \{\mathcal{LUMI}, \mathcal{FCOM}, \mathcal{FSTA}\}$. The former can be obtained by observing that (1) CYC is not in $\mathcal{FSTA}^F$ (Lemma 4) and is in $\mathcal{FCOM}^S$ (Lemma 5), and thus CYC is in $\mathcal{FCOM}^{RS}$ and $\mathcal{LUMI}^{RS}$, and (2) CGE is in $\mathcal{FSTA}^F$ ([15]) and is not in $\mathcal{LUMI}^{RS}$ (Corollary 1), and thus CGE is not in $\mathcal{FCOM}^{RS}$.

Theorem 6. $\mathcal{FSTA}^F \perp \mathcal{LUMI}^{RS}$ *and* $\mathcal{FSTA}^F \perp \mathcal{FCOM}^{RS}$.

The problems OSP and CGE*[1] were used to show $\mathcal{OBLOT}^F \perp \mathcal{LUMI}^S$ [15]. That is, problem OSP can be solved in $\mathcal{LUMI}^S$, but not in $\mathcal{OBLOT}^F$, and problem CGE* can be trivially solved in $\mathcal{OBLOT}^F$, but not in $\mathcal{LUMI}^S$. These problems can also be used to show $\mathcal{OBLOT}^F \perp \mathcal{X}^{RS}$ (for $\mathcal{X} \in \{\mathcal{LUMI}, \mathcal{FCOM}, \mathcal{FSTA}\}$) by using the below Lemmas 8–9, Corollary 1, and the fact that CGE* can be solved in $\mathcal{FSTA}^F$ [15].

Lemma 8. ([7]). $\exists R \in \mathcal{R}_2$, OSP $\notin \mathcal{OBLOT}^F(R)$.

Lemma 9. $\forall R \in \mathcal{R}_2$, OSP $\in \mathcal{FSTA}^{RS}(R) \cap \mathcal{FCOM}^{RS}(R)$.

Theorem 7. *$\mathcal{OBLOT}^F \perp \mathcal{LUMI}^{RS}$, $\mathcal{OBLOT}^F \perp \mathcal{FCOM}^{RS}$, and $\mathcal{OBLOT}^F \perp \mathcal{FSTA}^{RS}$.*

5 Analysis Within RSYNCH

In this section, we study the impact that memory persistence and communication capabilities have on the computational power of energy-constrained systems of robots.

We start by establishing the following theorem. Note that due to Theorem 4, this would imply that $\mathcal{LUMI}^{RS} \equiv \mathcal{FCOM}^{RS}$.

Theorem 8. $\forall R \in \mathcal{R}$, $\mathcal{LUMI}^S(R) \subseteq \mathcal{FCOM}^{RS}(R)$.

We do so, once again, by developing an algorithm `sim-LUMI-by-FCOM(A)` for $\mathcal{FCOM}$ robots that simulates a given algorithm A for $\mathcal{LUMI}$ robots (see [2] for details). Let A be an algorithm for $\mathcal{LUMI}$ robots with disorientation and non-rigid movement under SSYNCH, and let A use light with ℓ colors: $C = \{c_0, c_1, \dots c_{\ell-1}\}$. We now extend the simulation algorithm of $\mathcal{LUMI}$ robots by $\mathcal{FCOM}$ robots in FSYNCH described in [15], designing a more complex simulation algorithm `sim-LUMI-by-FCOM(A)` by $\mathcal{FCOM}$ robots in RSYNCH.

The main ideas of the simulation remain the same. Robots first copy the lights of their neighbors. This in turn allows them to look at their neighbors to gain information about the color of their own light. Next, some robots activate and have enough information to execute a step in A. Lastly, the robots reset their states such that they can start copying the lights of their neighbors again. This cycle repeats itself and every time some robots execute algorithm A.

Contrary to the simulation algorithm in FSYNCH, we need to take extra care here to ensure that the resulting schedule is fair and every robot executes a step in A infinitely often. After all, the same subset of robots might execute algorithm A each time the robots perform this cycle. As with `sim-RS-by-S(A)`, we consider the concept of *mega-cycles*, in which every robot executes a step in A exactly once. To facilitate this, each robot gets a flag indicating if the robot has already executed A in this mega-cycle. To give a robot information about its own execute status, we let the robots copy this flag in the same way as they copy the other lights of their neighbors. As soon as all robots have executed A in one mega-cycle, these flags are reset and the next mega-cycle begins.

[1] OSP is OSCILLATING POINTS and CGE* is PERPETUAL CENTER OF GRAVITY EXPANSION.

Recall that **suc** and **pred** are the successor and predecessor functions defined on a unique circular order on robots' positions $X(t)$. Figure 3 shows the transition diagram as the robots run the simulation. Each robot r has the following colors of $\mathcal{FCOM}$:

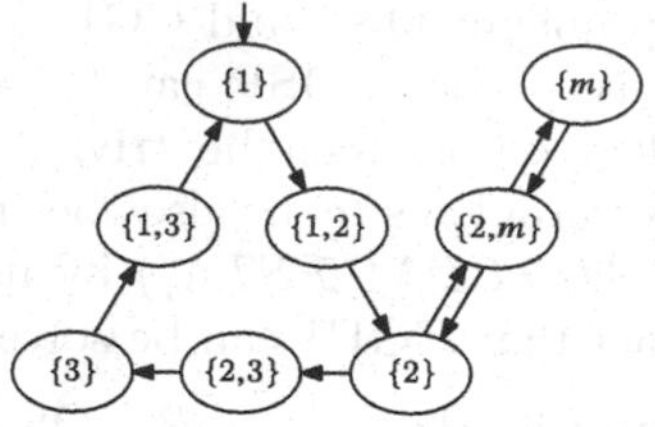

Fig. 3. Transition diagram of `sim-LUMI-by-FCOM(A)`.

$r.color \in C$ indicating its own light used in algorithm A, initially set to $r.color = c_0$;
$r.\texttt{suc}.color$ indicating a set of colors from 2^C at $\texttt{suc}(x)$, initially set to $r.\texttt{suc}.color = \emptyset$;
$r.step \in \{1, 2, 3, m\}$ indicating the step of the algorithm. See below for more details on each step. Initially $r.step$ is set to 1;
$r.executed \in \{True, False\}$ indicating whether r has executed the algorithm A in the current mega-cycle, initially set to $r.executed = False$;
$r.\texttt{suc}.executed$ indicating a set from $2^{\{True, False\}}$ at $\texttt{suc}(x)$ and having the same role of $r.\texttt{suc}.color$, initially set to $r.\texttt{suc}.executed = \emptyset$;
$r.me.checked$, $r.\texttt{suc}.checked \in \{True, False\}$ indicating whether all *color* and *executed* flags are correctly set, initially set to $r.me.checked, r.\texttt{suc}.checked = False$.

The algorithm is a sequence of mega-cycles, each of which lasts until all robots execute the simulated algorithm exactly once, and the end of a mega-cycle is checked at the beginning of Step 2. During each mega-cycle, this algorithm is composed of the following three steps:

- In Step 1, every robot obtains the neighbor's information so that robots can recognize their own lights.
- In Step 2, activated robots execute algorithm A using the color of their light obtained.
- In Step 3, every robot resets check flags to prepare the next cycle of execution of algorithm A.

Repeating the three steps, the end of the current mega-cycle is checked at the beginning of Step 2 and if the mega-cycle is finished, each robot r proceeds to Step m to reset.

- In Step m, every robot resets flag $r.executed$ indicating whether r has executed in this mega-cycle or not to *False* and the control returns to Step 2.

The details of these steps are as follows:

Step 1: Copy colors and activation ($\forall \rho \neq r\,(\rho.step = 1)$). In the *Look* phase, r checks if it is in Step 1 by detecting $\rho.step = 1$ for all other robots $\rho \neq r$. Every robot r stores (and displays) in variable $r.\texttt{suc}.color$ and $r.\texttt{suc}.executed$, respectively, its successors' sets of colors and also their *executed* flags indicating whether the successor robots have been activated in this mega-cycle. None of the robots move in this step. Note that every robot can recognize its own color of light and its flags in the configuration after Step 1 is completed. This step ends when all robots store their neighbors' sets of *color* and *executed* flags. To be able to detect this, every robot sets $r.checked$ to *True*, indicating that it has executed Step 1. Moreover, whenever all successors of r have set their *checked* flags to *True*, r also sets $r.\texttt{suc}.checked$ to *True*. Now every robot can determine the end of Step 1 when the two checked flags ($\rho.me.checked$, $\rho.\texttt{suc}.checked$) of every robot ρ become *True*. When this condition is satisfied, r changes $r.step$ to 2.

Step 2: Perform simulation ($\forall \rho \neq r\,(\rho.step = 2)$). In the *Look* phase, r recognizes to be in Step 2 by observing $\rho.step = 2$ of all other robots ρ. Robot r can derive $r.color$ and $r.executed$ by using its predecessor's $\texttt{suc}.color$ and $\texttt{suc}.executed$. The rest of Step 2 consists of two phases: checking the end of a mega-cycle, and executing the simulated algorithm.

***Checking end of a mega-cycle*:** After calculating $r.color$ and $r.executed$, r checks if the current mega-cycle has ended. If all robots have executed algorithm A in this mega-cycle (all robots have $\rho.executed$ set to *True*), then r moves to Step m (resetting all the executed flags) without changing colors of lights.

***Execution of simulation*:** If the mega-cycle has not ended yet, the robots activated in this round will perform a step in A. The activated robots (indicated by set S_2) at this time, actually perform their cycle according to algorithm A if they have not performed algorithm A in this mega-cycle yet ($r.executed$ is *False*). The robots that executed algorithm A update $r.executed$ and furthermore update $r.color$ and move according to algorithm A. After the actual execution of algorithm A, each robot r performing the simulation sets $r.step = 3$ to reset all checking flags (Step 3) to prepare for the next Step 1. If all robots in S_2 have performed algorithm A so far in this mega-cycle, no light is changed, and Step 2 is performed again at the next round. This continues until some robots not having executed algorithm A are activated. This is guaranteed by the fairness of the scheduler.

Step 3: Reset check flags ($\forall \rho \neq r(\rho.step = 3)$). In the *Look* phase, r checks if it is in Step 3 by observing $\rho.step = 3$ for all other robots ρ. In Step 3, all robots reset $r.checked$, $r.\texttt{suc}.checked$ to *False*, and $r.\texttt{suc}.executed$ to $\{False\}$, enabling the robots to perform another round in this cycle. Each robot resetting its flags changes $r.step$ to 1.

Step m: Reset mega-cycle ($\forall \rho \neq r(\rho.step = m)$). In the *Look* phase, r checks if it is in Step m by observing $\rho.step = m$ for all other robots ρ. In Step m, each robot r sets $r.executed = False$ and $r.\texttt{suc}.executed = \{False\}$ and the step returns to 2 to begin a new mega-cycle after all robots reset their activated flags.

We can show that this simulation algorithm works correctly for $\mathcal{FCOM}$ robots in RSYNCH. Considering the specifics of $\mathcal{FCOM}$, the following two

Algorithm 1. `Scheme-for-modification-flags` for robot r at location x.

Assumptions

Let configurations be $same(step = \alpha)$ or $except1(step = \alpha; \gamma)$

State Compute

if $\forall \rho \neq r\,(\rho.step = \alpha)$ **then** // step α
  if not $\forall \rho \neq r\,(\rho.flag = True)$ **then**
    $r.step \leftarrow \alpha$, $r.flag \leftarrow True$
  else if $\forall \rho \neq r\,(\rho.flag = True)$ **then**
    $r.flag \leftarrow True$, $r.step \leftarrow \beta$
else if $\forall \rho \neq r\,(\rho.step = \beta)$ **then** $r.step \leftarrow \beta$ // step β
else if $\forall \rho \neq r\,((\rho.step = \alpha)$ **or** $(\rho.step = \beta))$ **then** $r.step \leftarrow \beta$

schemes are essential for the algorithm to work correctly: (1) transition scheme between step configurations, and (2) recognition of own light color.

(1) Transition Scheme Between Step Configurations: Since we consider $\mathcal{FCOM}$ robots, when a robot r checks a predicate, for example $\forall \rho\,(\rho.step = \alpha)$, r cannot see its own value $r.step$. Therefore, r only checks $\forall \rho \neq r\,(\rho.step = \alpha)$ and observes that the predicate may be satisfied although $r.step$ is not α. Thus, predicates appearing in the simulation must be of the form $\forall \rho \neq r\,(\cdots)$, and we must consider configurations in which only one robot r observes that some predicate holds while any of the other robots would observe that the predicate does not hold.

To provide some intuition, consider a simple example where every robot ρ has two lights $\rho.step \in \{\alpha, \beta, \gamma\}$ and $\rho.flag \in \{True, False\}$ (Algorithm 1). Let us consider a configuration, denoted $same(step = \alpha)$, where $\rho.step = \alpha$ for any robot ρ, and a configuration, denoted $except1(step = \alpha; \gamma)$, where there exists just one robot r_e with $r_e.step = \gamma$ while $\rho.step = \alpha$ for any other robot ρ. Lemma 10 below shows the following: if a step configuration is $same(step = \alpha)$ or $except1(step = \alpha; \gamma)$ with $\rho.flag = False$ for every robot ρ, Algorithm 1 transforms the step configuration into one of type $same(step = \beta)$ or $except1(step = \beta; \alpha)$ with $\rho.flag$ changed to $True$ for every robot ρ. The idea is that, as long as a robot still observes at least one other robot with $flag = False$, it sets its $flag$ to $True$, while maintaining $step = \alpha$. As soon all robots have their $flag = True$, the simulation needs to progress to $step = \beta$. However, the activated robot r may see that all other robots have their $flag = True$, but r itself may not have been active before and therefore still needs to set its $flag$ to $True$. Now any robot active that sees another robot with $step = \beta$ knows that all values $flag$ are $True$, including its own. The formal proof is shown in [2].

Lemma 10. *Let $C(t_a)$ be a step configuration either of type $same(step = \alpha)$ or $except1(step = \alpha, \gamma \neq \beta)$ at time t_a, and let $\forall \rho\,(\rho.flag = False)$ hold at t_a. If Algorithm 1 is executed on $C(t_a)$, then there is a time $t_b > t_a$ when $C(t_b)$ satisfies the following conditions:*

(a) The step configuration $C(t_b)$ *is either* $same(step = \beta)$ *or* $except1(step = \beta; \alpha)$.
(b) $\forall \rho\, (\rho.flag = True)$ *holds at* t_b.

This scheme of transitions between step configurations is used in our simulation algorithm. Refer to [2] for the correctness of transitions between the steps and of the scheme as a whole.

(2) Recognition of Own Light Color: The color $r.color$ used in the execution of algorithm A is determined as follows: Assuming all robots have correctly set their $\texttt{suc}.color$, $\rho.\texttt{suc}.color$ for any $\rho \in \texttt{pred}(x)$ contains all colors at location x. Let $r.color.here$ be the set of colors seen at x. Note that, by definition, this does not include $r.color$, since r is on location x and it cannot see its own light. Now r can calculate $r.color$ in the following way:

$$r.color = \texttt{pred}(x).\texttt{suc}.color - x.color.here\,.$$

Since the executed flag of robot r contains two colors, $r.executed$ can be determined similarly to the case of determining $r.color$ by using $\texttt{pred}(x).\texttt{suc}.executed$ instead of $\texttt{pred}(x).\texttt{suc}.color$.

Note, that we can remove the assumption of chirality by using the method of [15]. The robots copy colors of both predecessors and successors, and deduce their colors by looking at the colors mirrored by the robots in the two-hop neighborhood.

It is fairly straightforward that this algorithm works well in the first phase of RSYNCH (equivalent to FSYNCH). Once a non-full activation occurs, full activation will never happen and thereafter activations of disjoint sets of robots happen alternatingly (see [2]).

Algorithm `sim-LUMI-by-FCOM(`A`)` executes Steps 1–3 and Step m in infinite rounds in RSYNCH and the execution of A obeys SSYNCH. Let E be the sequence of the set of activated robots that execute steps of algorithm A in simulation `sim-LUMI-by-FCOM(`A`)`. Since, we can show that any mega-cycle is completed and every robot executes exactly once in every mega-cycle, E is fair. Then we have obtained Theorem 8. Note that, if algorithm A uses ℓ colors, the simulating algorithm `sim-LUMI-by-FCOM(`A`)` uses $O(\ell 2^\ell)$ colors.

Therefore it holds that $\mathcal{FCOM}^{RS} \geq \mathcal{LUMI}^S$. Since $\mathcal{LUMI}^S \equiv \mathcal{LUMI}^{RS}$ by Theorem 4, we have that: $\mathcal{FCOM}^{RS} \geq \mathcal{LUMI}^{RS}$. Moreover, since the reverse relation is trivial and $\mathcal{FSTA}^{RS} < \mathcal{LUMI}^S$ (Theorem 2), the next theorem follows. The proof is shown in [2].

Theorem 9. $\mathcal{FCOM}^{RS} \equiv \mathcal{LUMI}^{RS} \equiv \mathcal{LUMI}^S$, *and* $\mathcal{FCOM}^{RS} > \mathcal{FSTA}^{RS} > \mathcal{OBLOT}^{RS}$.

6 Concluding Remarks

In this paper, we have started the investigation of the algorithmic and computational issues arising in distributed systems of autonomous mobile entities in the Euclidean plane where their energy is limited, albeit renewable.

We have studied the difference in computational power caused by the energy restriction and provided a complete characterization of the computational difference between energy-constrained and unrestricted robots in all four models considered in the literature: $\mathcal{OBLOT}, \mathcal{FSTA}, \mathcal{FCOM}, \mathcal{LUMI}$. We have also examined the difference with robots with unlimited energy, operating under a fully-synchronous scheduler. Furthermore, we have studied the impact of memory persistence and communication capabilities on the computational power of such energy-constrained systems of robots. Some of these results have been obtained through the design and analysis of novel *simulators*: algorithms that allow a set of robots with a given set of capabilities to execute correctly any protocol designed for robots with more powerful capabilities.

References

1. Agmon, N., Peleg, D.: Fault-tolerant gathering algorithms for autonomous mobile robots. SIAM J. Comput. **36**(1), 56–82 (2006)
2. Buchin, K., Flocchini, P., Kostitsyna, I., Peters, T., Santoro, N., Wada, K.: On the computational power of energy-constrained mobile robots: Algorithms and cross-model analysis. arXiv.org cs(ArXiv:2203.06546) (2022)
3. Canepa, D., Potop-Butucaru, M.: Stabilizing flocking via leader election in robot networks. In: Proceedings of the 10th International Symposium on Stabilization, Safety, and Security of Distributed Systems (SSS), pp. 52–66 (2007)
4. Cicerone, S., Stefano, D., Navarra, A.: Gathering of robots on meeting-points. Distrib. Comput. **31**(1), 1–50 (2018)
5. Cieliebak, M., Flocchini, P., Prencipe, G., Santoro, N.: Distributed computing by mobile robots: Gathering. SIAM J. Comput. **41**(4), 829–879 (2012)
6. Cohen, R., Peleg, D.: Convergence properties of the gravitational algorithms in asynchronous robot systems. SIAM J. Comput. **34**(15), 1516–1528 (2005)
7. Das, S., Flocchini, P., Prencipe, G., Santoro, N., Yamashita, M.: Autonomous mobile robots with lights. Theor. Comput. Sci. **609**, 171–184 (2016)
8. Di Luna, G., Flocchini, P., Chaudhuri, S., Poloni, F., Santoro, N., Viglietta, G.: Mutual visibility by luminous robots without collisions. Inf. Comput. **254**(3), 392–418 (2017)
9. Di Luna, G., Viglietta, G.: Robots with lights. Chapter 11 of [10], pp. 252–277 (2019)
10. Flocchini, P., Prencipe, G., Santoro, N.: Distributed Computing by Mobile Entities. Springer (2019)
11. Flocchini, P., Prencipe, G., Santoro, N.: Distributed Computing by Oblivious Mobile Robots. Morgan & Claypool (2012)
12. Flocchini, P., Prencipe, G., Santoro, N., Widmayer, P.: Gathering of asynchronous robots with limited visibility. Theor. Comput. Sci. **337**(1–3), 147–169 (2005)
13. Flocchini, P., Prencipe, G., Santoro, N., Widmayer, P.: Arbitrary pattern formation by asynchronous oblivious robots. Theor. Comput. Sci. **407**, 412–447 (2008)
14. Flocchini, P., Santoro, N., Viglietta, G., Yamashita, M.: Rendezvous with constant memory. Theor. Comput. Sci. **621**, 57–72 (2016)
15. Flocchini, P., Santoro, N., Wada, K.: On memory, communication, and synchronous schedulers when moving and computing. In: Proceedings of the 23rd International Conference on Principles of Distributed Systems (OPODIS), pp. 25:1–25:17 (2019)
16. Fujinaga, N., Yamauchi, Y., Ono, H., Kijima, S., Yamashita, M.: Pattern formation by oblivious asynchronous mobile robots. SIAM J. Comput. **44**(3), 740–785 (2015)

17. Gervasi, V., Prencipe, G.: Coordination without communication: the case of the flocking problem. Disc. Appl. Math. **144**(3), 324–344 (2004)
18. Hériban, A., Défago, X., Tixeuil, S.: Optimally gathering two robots. In: Proceedings of 19th Int. Conference on Distributed Computing and Networking (ICDCN), pp. 1–10 (2018)
19. Izumi, T., Souissi, S., Katayama, Y., Inuzuka, N., Défago, X., Wada, K., Yamashita, M.: The gathering problem for two oblivious robots with unreliable compasses. SIAM J. Comput. **41**(1), 26–46 (2012)
20. Okumura, T., Wada, K., Défago, X.: Optimal rendezvous $\mathcal{L}$-algorithms for asynchronous mobile robots with external-lights. In: Proceedings of the 22nd International Conference on Principles of Distributed Systems (OPODIS), pp. 24:1–24:16 (2018)
21. Okumura, T., Wada, K., Katayama, Y.: Brief announcement: Optimal asynchronous rendezvous for mobile robots with lights. In: Proceedings of the 19th International Symposium on Stabilization, Safety, and Security of Distributed Systems (SSS), pp. 484–488 (2017)
22. Sharma, G., Alsaedi, R., Bush, C., Mukhopadyay, S.: The complete visibility problem for fat robots with lights. In: Proceedings of the 19th International Conference on Distributed Computing and Networking (ICDCN), pp. 21:1–21:4 (2018)
23. Sharma, G., Vaidyanathan, R., Bush, C., Rai, S., Borzoo, B.: Complete visibility for robots with lights in $O(1)$ time. In: Proceedings of the 18th International Symposium on Stabilization, Safety, and Security of Distributed Systems (SSS), pp. 327–345 (2016)
24. Sharma, H., Ahteshamul, H., Jaffery, Z.A.: Solar energy harvesting wireless sensor network nodes: a survey. J. Renew. Sustain. Energy **10**(2), 023704 (2018)
25. Suzuki, I., Yamashita, M.: Distributed anonymous mobile robots: Formation of geometric patterns. SIAM J. Comput. **28**, 1347–1363 (1999)
26. Terai, S., Wada, K., Katayama, Y.: Gathering problems for autonomous mobile robots with lights. arXiv.org cs(ArXiv:1811.12068) (2018)
27. Viglietta, G.: Rendezvous of two robots with visible bits. In: 10th International Symposium on Algorithms and Experiments for Sensor Systems, Wireless Networks and Distributed Robotics (ALGOSENSORS), pp. 291–306 (2013)
28. Yamashita, M., Suzuki, I.: Characterizing geometric patterns formable by oblivious anonymous mobile robots. Theor. Comput. Sci. **411**(26–28), 2433–2453 (2010)
29. Yamauchi, Y., Uehara, T., Kijima, S., Yamashita, M.: Plane formation by synchronous mobile robots in the three-dimensional Euclidean space. J. ACM **64:3**(16), 16:1-16:43 (2017)

Randomized Strategies for Non-additive 3-Slope Ski Rental

Toni Böhnlein[1(✉)], Sapir Erlich[2], Zvi Lotker[2], and Dror Rawitz[2]

[1] Weizmann Institute of Science, Rehovot, Israel
toni.bohnlein@weizmann.ac.il
[2] Bar Ilan University, Ramat-Gan, Israel
{zvi.lotker,dror.rawitz}@biu.ac.il

Abstract. The Ski Rental problem captures the dilemma of choosing between renting and buying, and it is one of the fundamental problems in online computation. In many realistic scenarios there may be several intermediate *lease* options which are modelled by the Multi-Slope Ski Rental problem. An instance consists of k *states* where each state $i \in \{0, \ldots, k-1\}$ is characterized by a *buying cost* b_i and a *rental rate* r_i. Previous work on instance-dependent strategies for Multi-Slope Ski Rental dealt with deterministic strategies or strategies for *additive* instances (instance where going from state i to state j costs $b_j - b_i$). However, obtaining instance-dependent randomized strategies for non-additive instances remains open.

In this paper, we advance towards answering this open question by characterizing optimal randomized strategies in the non-additive case, and providing an algorithm that computes near-optimal instance-dependent randomized strategies for Multi-Slope Ski Rental with three slopes ($k = 2$). The algorithm uses parametric search and a decision algorithm which is based on the characterization of optimal randomized strategies.

Keywords: Online algorithms · Competitive analysis · Randomized algorithm · Ski-rental

1 Introduction

The Ski Rental problem (also called the Leasing problem [7]) is one of the fundamental problems in the area of online computation. In its original formulation [14], one needs to use a resource (skis) for an unknown amount of time (ski vacation of unknown duration), and there are two ways to do it: either pay a price and use the resource from that point onward for any amount of time (the *buy* option), or pay proportionally to the usage time (the *rent* option). An optimal deterministic 2-competitive strategy was given in [14] (see also [6]). An optimal randomized $\frac{e}{e-1}$-competitive strategy was given in [13].

In many realistic scenarios there may be several intermediate *lease* options between the two extreme alternatives of pure buy and pure rent. The problem

M. Parter (Ed.): SIROCCO 2022, LNCS 13298, pp. 62–78, 2022.
https://doi.org/10.1007/978-3-031-09993-9_4

where there are more than two options is called MULTI-SLOPE SKI RENTAL (MULTI-SLOPE SR), and it is defined as follows. An instance consists of k *states* (or *slopes*), where each state $i \in \{0, \ldots, k-1\}$ is characterized by a *buying cost* b_i and a *rental rate* r_i. Using the resource under state i for t time units implies costs of $b_i + r_i \cdot t$ (see Fig. 1). MULTI-SLOPE SR introduces entirely new difficulties when compared to SKI RENTAL. Intuitively, whereas the only question in the classical version is when to buy, in the multislope version we also need to answer the question of what to buy. Another way to see the difficulty is that the number of potential transitions from one slope to another in a strategy is one less than the number of slopes, i.e., $k-1$, and finding a single point of transition is qualitatively easier than finding more than one such point. In addition, the possibility of multiple transitions forces us to define the relation between multiple "buys". More formally, we need to consider transition costs between any two states i and j, where $i < j$. Hence a MULTI-SLOPE SR instance contains a transition cost $b_{i,j}$ from i to j, for every $i < j$, where $b_{0,j} = b_j$, for every $j > 0$. Following [2], we distinguish between two types of transition costs. In the *additive* case, buying costs are cumulative, namely to move from state i to state j we only need to pay the difference in buying prices $b_{i,j} = b_j - b_i$. This means that the total buying cost at state j is always b_j. In the *non-additive* case, the transition cost satisfies $b_{i,j} > b_j - b_i$, for each pair of states $i < j$. The extreme non-additive case is when $b_{i,j} = b_j$, for any i, i.e., it costs b_j to enter state j no matter where we are coming from. We refer to such transition costs as *entry fees*.

The optimal offline strategy remains the same regardless of whether the instance is additive, since it only depends on $\{b_1, \ldots, b_{k-1}\}$ (see Fig. 2). Hence, without loss of generality, we assume that $b_i < b_{i+1}$ and $r_i > r_{i+1}$, for all i. (If $b_i \leq b_j$ and $r_i \leq r_j$, then slope i dominates slope j, and the optimal offline strategy therefore ignores slope j.) Observe that a c-competitive strategy with respect to entry fees is also c-competitive with respect to any non-additive transition costs, since transition costs do not affect the optimum.

An extension of non-additive MULTI-SLOPE SR, where slopes become available over time, was considered in [3]. They obtained an online strategy whose competitive ratio is $4 + 2\sqrt{2} \approx 6.83$ for the *convex case* in which a more expensive machine comes with a lower production cost. They gave an almost matching upper and lower bounds for the *non-convex case*. Motivated by re-routing in ATM networks, Bejerano, Cidon, and Naor [5] studied non-additive MULTI-SLOPE SR. In this case, the buying costs of a slope is the setup cost of a virtual channel, while the rental cost is the cost of holding the links along the channel. They gave a deterministic 4-competitive strategy and showed that the factor of 4 holds even when the slopes are concave, i.e., when the rent of a slope may decrease with time. Damaschke [8] also considered non-additive MULTI-SLOPE SR. For deterministic strategies, he gave an upper bound of 4 and a lower bound of $\frac{5+\sqrt{5}}{2} \approx 3.618$. He also presented a randomized strategy whose competitive ratio is $\frac{2}{\ln 2} \approx 2.88$. We note that [3,5,8] assume that transition costs are entry fees.

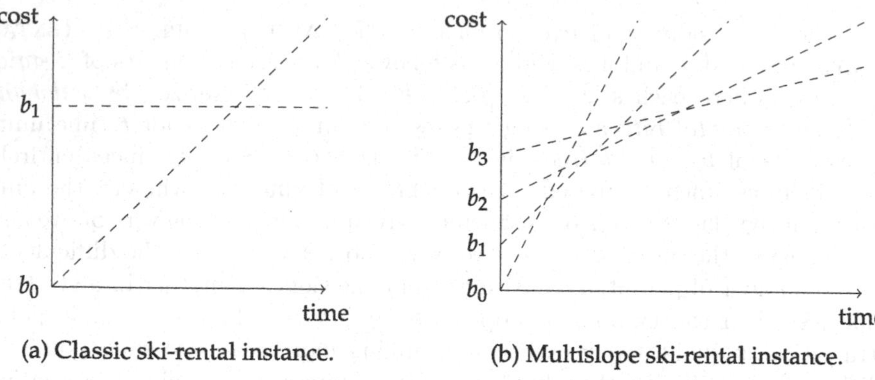

(a) Classic ski-rental instance.

(b) Multislope ski-rental instance.

Fig. 1. Visual representation of different ski-rental instances. The lines represent the different options. The classic problem has only pure rent and pure buy, while the MULTI-SLOPE SR problem might not have a pure buy option.

Irani, Swamy, and Gupta [12] studied additive MULTI-SLOPE SR with pure rent and pure buy and motivated their work by an energy saving application: each slope corresponds to some partial *hibernation* mode of the system. They showed that the lower envelope strategy, which buys a state when the optimal strategy moves to it, is 2-competitive. The competitive ratio of 2 is tight, for every k, since there are instances with pure rent, pure buy, and $k-2$ leasing slopes that are infinitesimally close to either option. Augustine, Irani, and Swami [2] considered non-additive MULTI-SLOPE SR. They devised an algorithm that, given an instance, computes a near optimal deterministic strategy. More specifically, the algorithm produces a strategy whose competitive ratio is within an additive ε of the optimal *instance-dependent* deterministic competitive ratio, for every $\varepsilon > 0$. The algorithm is based on a parametric search using a decision algorithm based on dynamic programming, and its running time is $O(k^2 \log k \cdot \log \frac{1}{\varepsilon})$. Note that the best instance-dependent competitive ratio may be lower than the upper bound of 4 [5] that applies to all instances. The case where the length of the game is a random variable with known distribution was also considered in both [2,12].

Randomized strategies for MULTI-SLOPE SR were studied by Lotker, Patt-Shamir, and Rawitz [16]. For the additive problem, they defined the notion of *profile* which describes a strategy by specifying, for every time $t \geq 0$, a probability distribution over the set of states. They gave a characterization of optimal profiles and used it to design an algorithm which computes a strategy whose competitive ratio is within an additive ε of the optimal randomized instance-dependent competitive ratio, for every $\varepsilon > 0$. The running time of the algorithm is $O(k \log \frac{1}{\varepsilon})$. This algorithm is also based on a parametric search, but in this case the decision algorithm is based on solving $O(k)$ differential equations. They also presented a randomized strategy whose competitive ratio is $\frac{e-r_k/r_0}{e}$ for any additive instance. For the non-additive case, they gave an e-competitive randomized strategy which improved the upper bound from [8]. Azar, Cohen, and Roytman [4] used LP-duality to provide a matching lower bound of e for non-additive MULTI-SLOPE SR. Their approach is based on formulating an infinite sequence of linear programs, each describing a finite prefix of the discrete version of the problem and where the objective function is the competitive ratio. They obtained a lower bound on the competitive ratio using LP-duality and ad-hoc arguments. It is claimed without details in [4] that their techniques can be used to obtain instance-dependent lower bounds.

Our Contribution. Previous work on instance-dependent strategies for MULTI-SLOPE SR dealt with deterministic strategies or strategies for additive instances. However, obtaining instance-dependent randomized strategies for non-additive instances remains an open problem. In this paper, we make an advancement towards closing this open question by providing an algorithm that computes near-optimal instance-dependent randomized strategies for MULTI-SLOPE SR with three slopes ($k = 2$).

The notion of *profile* was defined in [16] in order to describe randomized strategies for additive instances, by specifying, for every time $t \geq 0$, the probability distribution over the set of slopes. Profiles can capture strategies for additive instances, since the total spending in buying depends only on the current state. However, this does not suffice to describe strategies for non-additive instances since previous states have an impact on the total buying costs. We cope with this issue in Sect. 2 by introducing the notion of *transition function* that can describe a randomized strategy for non-additive instances.

In Sect. 3, we explore the structure of randomized transition function. We show that an optimal transition function buys the slopes in a very specific order. More specifically, time is partitioned into three intervals: (i) probability is shifted from slope 0 to slope 1 during $[0, \tau_1)$; (ii) probability is shifted from slope 0 to slope 2 during $[\tau_1, \tau_2)$; and (iii) probability is shifted from slope 1 to slope 2 during $[\tau_2, \infty)$. where $0 \leq \tau_1 \leq \tau_2$. We call such transition functions *progressive*. We note that this characterization is quite different than the one that was shown for additive instances in [16]. In the additive setting, an optimal strategy always goes through all states, and at any given time, at most two consecutive states may have positive probabilities. On the other hand, we show that in the non-additive case, it is viable to jump from state 0 to state 2 without visiting state 1. Moreover, there are times in which all states are associated with non-zero probabilities. Next, as was done in [2,16], we show that it may be assumed that an optimal strategy is *tight*, namely that it spends all available funds in buying. Given this characterization we show that a tight strategy is fully determined by its competitive ratio c and the time in which the strategy moves from buying state 1 to buying state 2 (i.e., τ_1).

In Sect. 4, we provide an algorithm that produces a near-optimal instance-dependent randomized strategy for 3-SLOPE SR. The algorithm uses parametric search and a decision algorithm which is based on the above characterization. Given $c \in [1, e]$, the algorithm computes a feasible interval for τ_1 for any possible ordering of events (τ_1, τ_2, and intersections between slopes). If all intervals are empty, then c is not attainable, and otherwise it is. The computation of the end-points of the intervals relies on the solution of several differential equations that correspond to a specific ordering of events. Finally, it is shown how to transform the computed transition function into a randomized strategy.

Additional Related Work. Fujiwara, Kitano, and Fujito [9] focused on the best and the worst instance-dependent deterministic competitive ratios for Multi-Slope SR. They proved that the best ratio is achieved for a specific additive instance where all slopes intersect at a single point. For three and four slopes ($k = 2, 3$) they showed that in the worst instances the transitions costs are entry costs and the intermediate rents are close to zero. Fujiwara, Konno, and Fujito [10] continued this line of work by presenting lower bounds on the optimal deterministic competitive ratios for $k \in \{4, \ldots, 9\}$. Levi and Patt-Shamir [15] studied Ski Rental with two lease options with an additional twist – at time 0 the strategy must decide whether it buys slope 0 or slope 1. We model this variant as a special case of non-additive Multi-Slope SR with $k = 2$, where $r_0 = \infty$ and $b_0 = 0$. Since $r_0 = \infty$, the strategy must leave state 0 when the game starts either to state 1 or to state 2. Levi and Patt-Shamir [15] provided upper bounds and matching lower bounds for the deterministic and randomized competitive ratios. Fujiwara, Satou, and Fujito [11] also studied Multi-Slope SR with $k = 2$. They analysed the optimal deterministic competitive ratio as a function of the discount rate $d \triangleq \frac{b_{02}}{b_{01}+b_{12}}$.

Patt-Shamir and Yadai [18] studied a non-linear extension of Ski Rental, where the two options of rent and buy are described using functions that are continuous and satisfy certain monotonicity conditions and computational requirements. They gave an algorithm that computes the best deterministic strategy and an algorithm that computes a near-optimal randomized strategy.

Another multi-slope extension of Ski Rental, called Multi-Shop SR was defined in [1]. Suppose there are n shops, each having its own rent and buying cost. A shop must be selected upon arrival, and then skis can be rented or bought at the selected shop. The optimal offline strategy is to rent at cost r_n, if the game ends before $\frac{b_1}{r_n}$, and otherwise to buy at cost b_1. Using similar techniques to those used in [16], a characterization of an instance-dependent optimal randomized strategy and an algorithm to compute such a strategy are given in [1]. They also considered several variants. In the first, the shops have entry fees. They showed that the optimal strategy can be partitioned into segments following the same ordering as in the standard case. In the second variant, there are shop-switching costs. In Multi-Slope SR terms, these are instances composed of n pure rent slopes and a single pure buy slope. However, as opposed to [2], the transition costs do not satisfy $b_{i,j} \leq b_j = 0$, for $i < j$, where j is one of the pure rent states. For this case they showed that an optimal strategy switches only before it buys, and this led to a reduction to standard Multi-Shop SR. In a third variant, there are entry fees and switching is allowed, where the switching costs are the entry fees. They provided partial characterization of an optimal strategy by showing that this variant can be reduced to MULTI-SHOP SR with entry fees with $O(n^2)$ shops.

2 Preliminaries: Profiles and Transition Functions

The MULTI-SLOPE SR problem is defined as follows: An instance consists of k *states* (or *slopes*), where each state $i \in \{0, \ldots, k-1\}$ is characterized by two numbers: a *buying cost* b_i and a *rental rate* r_i. Each state can be represented by a line: the ith state corresponds to the line $y = b_i + r_i x$. Such a line stands for the cost of buying the state at $t = 0$, and never leaving it. Figure 2 gives a geometrical interpretation of a multislope ski rental instance with five states. In this work we consider the case where costs are non-additive and assume an "entry fee" cost model, i.e., the cost of entering state i from any state $j < i$ is b_i. Recall that we may assume that for all i, $b_i < b_{i+1}$ and $r_i > r_{i+1}$.

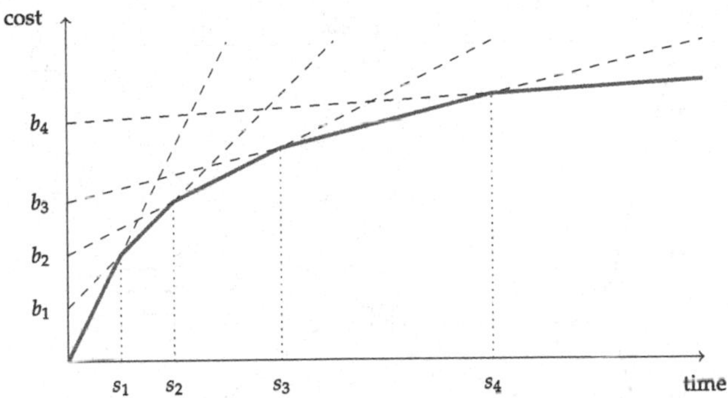

Fig. 2. A MULTI-SLOPE SR instance with 5 slopes ($k = 4$). The dashed lines are the slopes and the thick line is the optimal cost as a function of time.

Given a MULTI-SLOPE SR instance, if the game ends at time t, the optimal solution is to select the slope with the least cost at time t (the thick line in Fig. 2 denotes the optimal cost for any given t). More formally, the optimal offline cost at time t is $\text{OPT}(t) = \min_i \{b_i + r_i \cdot t\}$. Define $s_0 \triangleq 0$. For $i > 0$, let s_i denote the time where slopes $i-1$ and i intersect, i.e., s_i is the time t satisfying $b_{i-1} + r_{i-1}t = b_i + r_i t$, or

$$s_i \triangleq \frac{b_i - b_{i-1}}{r_{i-1} - r_i}. \tag{1}$$

Henceforth, we assume that $s_i < s_{i+1}$, since otherwise slope i is dominated by slopes $i-1$ and $i+1$. See the full version of the paper for a proof for the three slope case. Observe that the optimal slope for a game ending at time t is the slope i for which $t \in [s_i, s_{i+1}]$ (if $t = s_i$ for some i then both slopes $i-1$ and i are optimal).

Given a MULTI-SLOPE SR instance, the requirement is to choose a state for all times t. We assume that state transitions can be only forward. Hence, a *deterministic strategy* for a MULTI-SLOPE SR instance is a non-decreasing sequence of transition times $(t_1, \ldots, t_{k-1})$, where $t_i \in [0, \infty) \cup \{\varnothing\}$ corresponds to the transition time into state i, and $t_i = \varnothing$ means that state i was skipped. A transition $i \to j$ at time t is described by $t_\ell = \varnothing$, for $i < \ell < j$, and $t_j = t$. Notice that in this case, we must have $t_i = t' \leq t$. A *randomized strategy* can be described using a probability distribution over the deterministic strategies. However, following [16] we consider, for all times t, a probability distribution over the set of k slopes. The intuition is that this distribution determines the rent paid by a strategy at time t. Formally, a *randomized profile* (or simply a *profile*) is specified by a vector $p(t) = (p_0(t), \ldots, p_{k-1}(t))$ of k functions, where $p_i(t)$ is the probability to be in state i at time t. The correctness requirement of a profile is to always be at some slope, i.e., it must be that $\sum_{i=0}^{k-1} p_i(t) = 1$, for every time $t \geq 0$. Also, $p(0) = (1, 0, \ldots, 0)$. Figure 3 presents a 3-slope profile. Clearly, any strategy induces a profile.

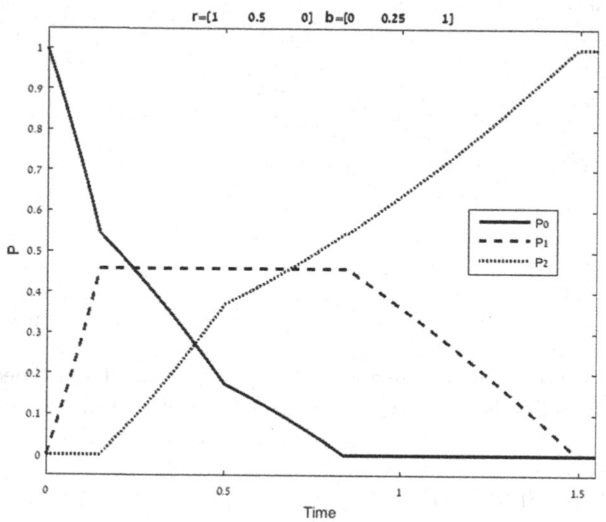

Fig. 3. A profile for a 3-slope instance.

Notice that we may consider only strategies that are monotone over time with respect to majorization [17], i.e., strategies such that for every state j and any two times $t \leq t'$ we have

$$\sum_{i=0}^{j} p_i(t) \geq \sum_{i=0}^{j} p_i(t'). \tag{2}$$

Intuitively, there is no point in "rolling back" from an advanced slope to a previous one: if at a given time we have a certain composition of the slopes, then at any later time the composition of slopes may only improve.

Given a profile p, the *rental rate* of p at time t and the *total rental cost* up to time t are defined as:

$$\rho_p(t) \triangleq \sum_{i=0}^{k} p_i(t) \cdot r_i \qquad\qquad R_p(t) \triangleq \int_0^t \rho_p(z)dz.$$

As opposed to the additive model, the *total buying cost* at time t cannot be determined by $p(t)$ alone, since it depends on the values of p before t. For instance, the cost of reaching state i through j is not the same as reaching state i directly. It follows that we need to look at the derivative of p. Given a profile p we define $P(t) = \{j : \frac{dp_j(t)}{dt} > 0\}$ and $N(t) = \{j : \frac{dp_i(t)}{dt} < 0\}$. Observe that in the 3-slope case (i.e., $k = 2$), $N(t) \subseteq \{0, 1\}$ and $P(t) \subseteq \{1, 2\}$. Since $\sum_j \frac{dp_j(t)}{dt} = 0$, we have that $\sum_{j \in P(t)} \frac{dp_j(t)}{dt} = -\sum_{j \in N(t)} \frac{dp_j(t)}{dt}$. Moreover, Eq. (2) implies that, for every state i, we have that $\sum_{j \in P(t), j \leq i} \frac{dp_i(t)}{dt} \leq -\sum_{j \in N(t), j \leq i} \frac{dp_i(t)}{dt}$. Given a profile p, let $\pi : \{0, \ldots, k-1\} \times \{1, \ldots, k\} \to \mathbb{R}$ be a function such that

$$\frac{dp_j(t)}{dt} = \begin{cases} \sum_i \pi_{j \to i}(t) & j \in N(t), \\ \sum_i \pi_{i \to j}(t) & j \in P(t). \end{cases}$$

Such a function is called a *transition function*. It can be found from $\frac{dp(t)}{dt}$ using a Maximum Flow algorithm on the bipartite graph $G(t) = (N(t), P(t), E(t))$, where $E(t) = \{(i, j) : i \in N(t), j \in P(t), i < j\}$. Intuitively, $\pi_{i \to j}$ is the amount of probability mass that passes from state i to state j at time t.

We describe the transition function π for $k = 2$. If $\frac{dp_1(t)}{dt} \geq 0$, we have that $N(t) = \{0\}$ and $P(t) \subseteq \{1, 2\}$, and in this case $\pi_{0 \to 1}(t) = \frac{dp_1(t)}{dt}$, $\pi_{0 \to 2}(t) = \frac{dp_2(t)}{dt}$, and $\pi_{1 \to 2}(t) = 0$. Otherwise, $N(t) \subseteq \{0, 1\}$, $P(t) = \{2\}$, and $\pi_{0 \to 1}(t) = 0$, $\pi_{0 \to 2}(t) = -\frac{dp_0(t)}{dt}$, and $\pi_{1 \to 2}(t) = -\frac{dp_1(t)}{dt}$.

Now we are ready to define the buying cost of a transition function π up to time t. The *buying rate* at time t and the *total buying cost* up to time t are defined as

$$\beta_\pi(t) \triangleq \sum_i b_i \sum_{j:j<i} \pi_{j \to i}(t) \qquad\qquad B_\pi(t) \triangleq \int_0^t \beta_\pi(z)dz.$$

We also define the total cost paid by moving from i to j until time t which is given by

$$B_\pi^{i \to j}(t) = b_j \cdot \int_0^t \pi_{i \to j}(z)dz.$$

Observe that $B_\pi(t) = \sum_{i,j:i<j} B_\pi^{i \to j}(t)$.

Note that (2) implies that ρ_p is monotone non-increasing and B_π is monotone non-decreasing, i.e., over time, a strategy invests non-negative amounts in buying, resulting in non-increasing rental rates.

Finally the *total cost* for p and π at time t is defined as

$$X_p(t) \triangleq B_\pi(t) + R_p(t).$$

The goal of an online strategy is to minimize total cost up to time t for any given $t \geq 0$, with respect to the best possible strategy. Intuitively, we think of a game that may end at any time. For any possible ending time, we compare the total cost of the algorithm with the best possible (offline) cost.

3 Optimal Profile for Three Slopes

In this section, we focus on the three slope case, namely on the case where $k = 2$. We show that there exists an optimal profile and transition function that take a very specific form.

3.1 Progressive and Tight Pairs of Profile and Transition Function

We show that there exists an optimal profile in which only one entry of π and two states are active at any given moment. Furthermore, the active entries of π follow a specific order.

Definition 1 (Progressive). *Let p be a profile and let π be a transition function that corresponds to p. The pair (p, π) is called* progressive *if there are times $0 \leq \tau_1 \leq \tau_2$ such that $\pi_{0\to1}(t) = 0$, for every $t \geq \tau_1$, $\pi_{0\to2}(t) = 0$, for every $t \notin (\tau_1, \tau_2]$, and $\pi_{1\to2}(t) = 0$, for every $t \leq \tau_2$,*

The following lemma shows that it can be assumed that the transition to state 1 is done before the transition to state 2. The proof is given in the full version of the paper.

Lemma 1. *Let (b, r) be a 3-*Slope SR *instance, such that $s_1 < s_2$, and let (p, π) be a c-competitive pair, for some $c > 1$. Then there exists a c-competitive pair $(\tilde{p}, \tilde{\pi})$ such that there exists τ, where (i) $\tilde{\pi}_{0\to1}(t) = 0$, for $t \geq \tau$, and (ii) $\tilde{\pi}_{0\to2}(t) = 0$ and $\tilde{\pi}_{1\to2}(t) = 0$, for $t \leq \tau$.*

Next, we show that it can be assumed that transition from state 0 to 2 is done before the transition from 1 to 2.

Lemma 2. *Let (b, r) be a 3-*Slope SR *instance, and let (p, π) be a c-competitive pair, for some $c > 1$. Then there exists a c-competitive pair $(\tilde{p}, \tilde{\pi})$ such that there exists τ, where (i) $\tilde{\pi}_{0\to2}(t) = 0$, for $t \geq \tau$, and (ii) $\tilde{\pi}_{1\to2}(t) = 0$, for $t \leq \tau$.*

Proof. We define $\tilde{p}$ and $\tilde{\pi}$ as follows. Let τ be the time in which $B_\pi^{0\to2}(\tau) + B_\pi^{1\to2}(\tau) = B_\pi^{0\to2}(\infty)$. Also, let

$$\tilde{\pi}_{i\to j}(t) = \begin{cases} \pi_{0\to2}(t) + \pi_{1\to2}(t) & (i,j) = (0,2), t < \tau, \\ 0 & (i,j) = (0,2), t \geq \tau, \\ \pi_{0\to2}(t) + \pi_{1\to2}(t) & (i,j) = (1,2), t \geq \tau, \\ 0 & (i,j) = (1,2), t < \tau, \\ \pi_{i\to j}(t) & (i,j) = (0,1). \end{cases}$$

Observe that $B_{\tilde{\pi}}(t) = B_{\pi}(t)$, for all t, by definition.

We now prove that $\rho_{\tilde{p}}(t) \leq \rho_p(t)$ for every t. For $t < \tau$, we have that

$$p_0(t) - \tilde{p}_0(t) = \int_0^t \pi_{1\to 2}(z)dz, \qquad p_1(t) - \tilde{p}_1(t) = \int_0^t -\pi_{1\to 2}(z)dz, \text{ and}$$
$$p_2(t) - \tilde{p}_2(t) = 0.$$

Hence, for $t \leq \tau$,

$$\rho_p(t) - \rho_{\tilde{p}}(t) = (r_0 - r_1)\int_0^t \pi_{1\to 2}(z)dz \geq 0.$$

Similarly, for $t \geq \tau$, we have that

$$p_0(t) - \tilde{p}_0(t) = \int_t^\infty \pi_{0\to 2}(z)dz, \qquad p_1(t) - \tilde{p}_1(t) = \int_0^t -\pi_{0\to 2}(z)dz, \text{ and}$$
$$p_2(t) - \tilde{p}_2(t) = 0.$$

Hence, for $t \leq \tau$, we have that

$$\rho_p(t) - \rho_{\tilde{p}}(t) = (r_0 - r_1)\int_0^t \pi_{0\to 2}(z)dz \geq 0.$$

The lemma follows.

Both lemmas imply that one may assume that the profile is progressive.

Theorem 1. *Let (b, r) be a 3-*SLOPE SR *instance where $s_1 < s_2$. If there exists a c-competitive pair (p, π), for $c > 1$, then there exists a progressive c-competitive pair $(\tilde{p}, \tilde{\pi})$.*

Proof. By Lemma 1 there exists a c-competitive pair (p', π') such that there exists τ_1, where $\pi'_{0\to 1}(t) = 0$, for $t \geq \tau_1$, and $\pi'_{0\to 2}(t) = 0$ and $\pi'_{1\to 2}(t) = 0$, for $t \leq \tau_1$. Given (p', π'), by Lemma 2 we have that there exists a pair (p'', π'') such that there exists τ_2, where $\pi''_{0\to 2}(t) = 0$, for $t \geq \tau_2$, and $\pi''_{1\to 2}(t) = 0$, for $t \leq \tau_2$. Moreover, the pair (p'', π'') is constructed such that $\pi''_{0\to 2}(t) = 0$ and $\pi''_{1\to 2}(t) = 0$, for $t \leq \tau_1$.

Next, We define a progressive profile which invests all free funds in buying.

Definition 2 (Tight). *Let (p, π) be a progressive c-competitive pair of profile and transition function that corresponds to p. The pair is called* tight *if $X_p(t) = c \cdot \text{OPT}(t)$, for every t where $p_2(t) < 1$.*

The next two observations provide some insight to the behavior of a tight pair.

Observation 3. *Let (p, π) be a tight c-competitive pair. Then, $\rho_p(t) + \beta_\pi(t) = c \cdot \frac{d\text{OPT}(t)}{dt}$, for every t such that $p_2(t) < 1$.*

Observation 4. *Let (p,π) be a tight c-competitive pair. Then, $\rho_p(s_2) \le c \cdot r_2$.*

Proof. If $p_2(s_2) = 1$ then, $\rho_p(s_2) = r_2 < c \cdot r_2$. Otherwise, $p_2(s_2) < 1$. Since p is tight, due to Observation 3, we have that $\rho_p(s_2) + \beta_\pi(s_2) = c \cdot \frac{d\text{OPT}(s_2)}{dt} = c \cdot r_2$.

We show that if there exists a c-competitive pair, then there is a tight c-competitive pair. Before doing so we define an ordering on profiles.

Definition 3. *Let p and p' be profiles. If $\sum_{j=0}^{i} p_j(t) \le \sum_{j=0}^{i} p'_j(t)$, for every i, we write $p(t) \preceq p'(t)$.*

Lemma 5. *Let p and p' be profiles such that $p(t) \preceq p'(t)$. Then, $\rho_p(t) \le \rho_{p'}(t)$.*

Proof. $\rho_p(t) = \sum_i p_i(t) r_i = \sum_i (r_i - r_{i+1}) \sum_{j=0}^{i} p_i(t) \le \sum_i (r_i - r_{i+1}) \sum_{j=0}^{i} p'_i(t) = \sum_i p'_i(t) r_i = \rho_{p'}(t)$.

Theorem 2. *Assuming that $s_1 < s_2$, if there exists a progressive c-competitive pair (p,π), then there exists a tight c-competitive pair $(\tilde{p},\tilde{\pi})$.*

Proof. Let τ_1 be a time such that $B_\pi^{0\to1}(\tau_1) = B_\pi^{0\to1}(\infty)$. Construct a new pair (p',π') as follows. The pair (p',π') behaves in a tight manner until it invests $B_\pi^{0\to1}(\infty)$ in the transition from state 0 to state 1. This process terminates at $\tau'_1 \le \tau_1$. The pair (p',π') only pays rent in the interval $[\tau'_1, \tau_1]$, and afterwards it behaves like (p,π). Hence $p^1(t) \preceq p(t)$, for every $t \le \tau_1$, and Lemma 5 implies that $\rho_{p'}(t) \le \rho_p(t)$ for every $t \le \tau_1$. It follows that $X_{p'}(t) \le c \cdot \text{OPT}(t)$, for every $t \le \tau_1$, and $X_{p'}(t) \le X_p(t)$, for every $t > \tau_1$.

Next, let τ_2 be a time at which p_0 becomes 0, i.e., $\tau_2 = \min\{t : p_0(t) = 0\}$. Construct a new pair (p'',π'') as follows. The pair (p'',π'') behaves in a tight manner until p''_0 becomes 0. (Notice that funds that became available due to moving from (p,π) to (p',π') may be used.) This process terminates at $\tau'_2 \le \tau_2$. The pair (p'',π'') only pays rent in the interval $[\tau'_2, \tau_2]$, and afterwards it mimics (p,π). Using similar arguments as above, we have that $X_{p'}(t) \le c \cdot \text{OPT}(t)$, for $t \le \tau_2$, and $X_{p'}(t) \le X_p(t)$, for $t > \tau_2$.

Finally, let $(\tilde{p},\tilde{\pi})$ be a tight pair. We have that $X_{\tilde{p}}(t) \le c \cdot \text{OPT}(t)$, for every t. Furthermore, let τ_3 be the time where $B_\pi^{0\to2}(\tau_3) + B_\pi^{1\to2}(\tau_3) = B_\pi^{0\to2}(\infty) + B_\pi^{1\to2}(\infty)$. By Lemma 5 we have that $\rho_{\tilde{p}}(\tau_3) \le \rho_p(\tau_3)$ and it follows that $\rho_{\tilde{p}}(t) \le \rho_p(t)$, for $t > \tau_3$.

3.2 Reconstruction: Profile and Transition Function

In this section, we show that a tight c-competitive pair (p,π) can be reconstructed using a ratio $c > 1$ and a time $\tau_1 \ge 0$. First, we define a pair which stops buying state 1 at τ_1.

Definition 4. *Given a 3-*Slope SR *instance, a ratio $c > 1$, and a time $\tau_1 \ge 0$, we define (p,π) as the profile pair, where $\frac{d}{dt}X_p(t) = c \cdot \frac{d}{dt}\text{OPT}(t)$, for every t such that $p_2(t) < 1$, and with the following three phases:*

- $0 \to 1$ *phase: from* $t = 0$ *to* $t = \tau_1$*, buy* 1 *from* 0 *with all available funds, e.g.,* $\beta_\pi(t) = \pi_{0\to 1}(t) \cdot b_1$.
- $0 \to 2$ *phase: from* $t = \tau_1$ *and as long as* $p_0(t) > 0$*, buy* 2 *from* 0 *with all available funds, e.g.* $\beta_\pi(t) = \pi_{0\to 2}(t) \cdot b_2$.
- $1 \to 2$ *phase: from the end of the previous phase and as long as* $p_2(t) < 1$*, buy 2 from 1 with all available funds, e.g.* $\beta_\pi(t) = \pi_{1\to 2}(t) \cdot b_2$.

The pair (p, π) *is denoted by* $(p[c, \tau_1], \pi[c, \tau_1])$.

It is important to note that, given a 3-Slope SR instance, not every ratio c and time τ_1 induce a tight c-competitive pair for the following reasons:

(i) τ_1 is too late, in the sense that option 1 is fully bought at a time $t' < \tau_1$. In this case the pair $(p[c, \tau_1], \pi[c, \tau_1])$ does not describe a tight profile because, for $t' < t < \tau_1$, $\beta_{\pi[c,\tau_1]}(t) = 0$ while $\rho_{p[c,\tau_1]}(t)$ is smaller than $c \cdot \frac{d}{dt}\text{OPT}(t)$.
(ii) the total costs of the profile simply exceed the available funds. In this case, $X_p(t) > c \cdot \text{OPT}(t)$ for some t, and the profile is not c-competitive.

To handle Reason (i) we define the following notion.

Definition 5. *A ratio* $c > 1$ *and a time* $\tau_1 \geq 0$ *are called* valid*, if* $p[c, \tau_1](t) < 1$*, for every* $t < \tau_1$.

Regarding Reason (ii), we check the rental of a profile, that is c and τ_1 are feasible if the rental rate of the profile does not exceed the available funds at any time.

Definition 6. *A ratio* $c > 1$ *and a time* $\tau_1 \geq 0$ *are called* feasible *if* $\rho_{p[c,\tau_1]}(t) \leq c \cdot \frac{d\text{OPT}(t)}{dt}$*, for every* t.

Notice that if c and τ_1 are not feasible, then $p[c, \tau_1]$ does not describe a c-competitive profile, as there exists a time t where $\rho_{p[c,\tau_1]}(t) > c \cdot \frac{d\text{OPT}(t)}{dt}$, but it means that at that time $\beta_{\pi[c,\tau_1]}(t) < 0$ which is impossible.

Observation 6. *Given a* 3-Slope SR *instance, if* $c > 1$ *and* $\tau_1 \geq 0$ *are valid and feasible, then the pair* $(p[c, \tau_1], \pi[c, \tau_1])$ *is tight and c-competitive.*

The next lemma helps us to deal with Reason (ii). If $c > 1$ and $\tau_1 \geq 0$ are valid, then to check whether the pair $(p[c, \tau_1], \pi[c, \tau_1])$ is c-competitive, we only need to evaluate the rent at times s_1 and s_2. The proof is given in the full version of the paper.

Lemma 7. *The pair* $(p[c, \tau_1], \pi[c, \tau_1])$ *is c-competitive if and only if*
(i) $\rho_{p[c,\tau_1]}(s_1) < c \cdot r_1$*, and (ii)* $\rho_{p[c,\tau_1]}(s_2) \leq c \cdot r_2$.

Given a 3-Slope SR instance, a ratio c, and a time τ_1, due to Lemma 7 we can determine if the pair $(p[c, \tau_1], \pi[c, \tau_1])$ is tight and c-competitive. To do so, we try to construct the pair in a piece-wise fashion until we either fail or succeed.

First, we analyse how the profile $p[c,\tau_1]$ spends funds. The spending rate of a tight profile at time t is:

$$\beta_\pi(t) + \rho_p(t) = c \cdot \frac{d}{dt}\text{OPT}(t) = c \cdot r_j \tag{3}$$

where $r_j = r_0$, if $t \in [0, s_1)$, $r_j = r_1$, if $t \in [s_1, s_2)$, and $r_j = r_2$, if $t \geq s_s$. Hence, the spending rate changes at times s_1 and s_2.

Moreover, the pair $(p[c,\tau_1], \pi[c,\tau_1])$ is progressive and has phases as described in Definition 4. Recall that time τ_2 is defined as the earliest time when $p_0(t) = 0$ (if such a time exists). The rental rates and buying rates for the three phases are as follows:

$$\rho_p(t) = \begin{cases} p_0(t)r_0 + p_1(t)r_1 & t \in (0, \tau_1] \\ p_0(t)r_0 + p_1(t)r_1 + p_2(t)r_2 & t \in (\tau_1, \tau_2] \\ p_1(t)r_1 + p_2(t)r_2 & t > \tau_2 \end{cases}$$

$$\beta_\pi(t) = \begin{cases} \pi_{0\to1}(t)b_1 = \frac{dp_1(t)}{dt} \cdot b_1 & t \in (0, \tau_1] \\ \pi_{0\to2}(t)b_2 = \frac{dp_2(t)}{dt} \cdot b_2 & t \in (\tau_1, \tau_2] \\ \pi_{1\to2}(t)b_2 = \frac{dp_2(t)}{dt} \cdot b_2 & t > \tau_2 \end{cases}$$

Recall that $p_1(t)$ is a constant function for $t \in [\tau_1, \tau_2]$.

It follows that the pair $(p[c,\tau_1], \pi[c,\tau_1])$ can have up to five pieces separated by s_1, s_2, τ_1, and τ_2. With Eq. (3) we set up a differential equation for the pieces separated by τ_1 and τ_2.

$0 \to 1$ Phase: When $t \in (0, \tau_1]$, we have that $p_2(t) = 0$ and $p_1(t) = 1 - p_0(t)$. Hence,

$$\frac{d}{dt}p_1(t) \cdot b_1 + (1 - p_1(t))r_0 + p_1(t) \cdot r_1 = c \cdot r_j \ ,$$

and equivalently $\frac{d}{dt}p_1(t) + p_1(t) \cdot \frac{r_1 - r_0}{b_1} = \frac{c \cdot r_j - r_0}{b_1}$.

The solution to this differential equation is $p_1(t) = \frac{c \cdot r_j - r_0}{r_1 - r_0} + \Gamma \cdot \exp(-\frac{r_1 - r_0}{b_1}t)$, where Γ depends on a boundary condition.[1]

$0 \to 2$ Phase: When $t \in (\tau_1, \tau_2]$, we have that $p_1(t) = p_1(\tau_1)$ and $p_2(t) = 1 - p_1(\tau_1) - p_0(t)$. Hence,

$$\frac{d}{dt}p_2(t) \cdot b_2 + (1 - p_1(\tau_1) - p_2(t))r_0 + p_1(\tau_1) \cdot r_1 + p_2(t) \cdot r_2 = c \cdot r_j,$$

and equivalently $\frac{d}{dt}p_2(t) + p_2(t)\frac{r_2 - r_0}{b_2} = \frac{c \cdot r_j - r_0 + (r_0 - r_1)p_1(\tau_1)}{b_2}$.

$1 \to 2$ Phase:
When $t > \tau_2$ we have that $p_0(t) = 0$ and $p_2(t) = 1 - p_1(t)$ and we get

$$\frac{d}{dt}p_2(t) \cdot b_2 + (1 - p_2(t))r_1 + p_2(t) \cdot r_2 = c \cdot r_j,$$

and equivalently $\frac{d}{dt}p_2(t) + p_2(t)\frac{r_2 - r_1}{b_2} = \frac{c \cdot r_j - r_1}{b_2}$.

[1] A solution to a differential equation of the form $y'(t) + \alpha y(t) = \beta$ where α and β are constants is $y = \frac{\beta}{\alpha} + \Gamma \cdot e^{-\alpha t}$, where Γ depends on some boundary condition.

At time $t = 0$, we have the boundary condition $p(0) = (1, 0, 0)$. For each of the following pieces, there is a boundary condition determined by the previous piece ensuring that each function of the profile is continuous (cf. Fig. 3).

Theorem 3. *Given a* 3-SLOPE SR *instance and ratio c and time τ_1, it can be determined whether c and τ_1 describe a c-competitive profile. If so, the tight c-competitive pair $(p[c,\tau_1], \pi[c,\tau_1])$ can be constructed.*

Proof. We prove the theorem by presenting an algorithm that constructs the profile $p[c, \tau_1]$ or determines that c and τ_1 are either not valid or not feasible. The construction is done in a piece-wise manner. The algorithm starts by using the boundary condition $p_1(0) = 0$ to solve the first differential equation (Phase $0 \rightarrow 1$, $r_j = r_0$). A new piece starts every time a phase ends or when the optimal spending rate changes. Each piece corresponds to a new differential equation. While going through the pieces we need to verify that (i) $p_1(\tau_1) \leq 1$,(ii) $\rho_{p[c,\tau_1]}(s_1) < c \cdot r_1$, and (iii) $\rho_{p[c,\tau_1]}(s_2) \leq c \cdot r_2$. The theorem follows.

4 Computing a Near-Optimal Strategy

In this section, we present an algorithm to compute an instance-dependent near-optimal strategy for 3-SLOPE SR with entry fees. The algorithms is based on tools that were developed in the previous section.

Given an instance of 3-SLOPE SR, we want to find a pair $(p[c,\tau_1], \pi[c,\tau_1])$ where c is a near-optimal competitive ratio. We are motivated by the approach that is taken in [16], where the optimal competitive ratio for an additive multi-slope instance is approximated to arbitrary precision by performing binary search. This is possible since, in the additive case, a profile can be fully described by a ratio $c > 1$, and checked whether it is indeed c-competitive. However, in the non-additive case we need two parameters to describe a profile, namely ratio c and time τ_1. Our approach to this two parametric search is to fix the ratio c and compute all times τ_1 that yield a tight c-competitive pair $(p[c,\tau_1], \pi[c,\tau_1])$ or decide that no such time exists. Consequently, we may use binary search to approximate the optimal competitive ratio to arbitrary precision.

4.1 Monotonicity of the Competitive Ratio and Interval

Applying binary search on the competitive ratio would work only if the competitive ratio is monotone. That is, it must be that if there exists a c-competitive pair, then there exists a c'-competitive pair for every $c' > c$. We prove this property for non-additive 3-SLOPE SR with entry fees.

Lemma 8. *Given a* 3-SLOPE SR *instance with entry fees, let c and τ_1 be valid and feasible, i.e., assume that there is a tight c-competitive pair $(p[c,\tau_1], \pi[c,\tau_1])$. Then for every $c' > c$, one of the following holds:*

- *c' and τ_1 are valid and feasible.*

– *There exist a time $\tau_1' < \tau_1$, such that c' and τ_1' are valid and feasible. Moreover, $p_1[c', \tau_1'](\tau_1') = 1$.*

For the proof of this and of the following lemma we refer to the full version of the paper. From Lemma 8 we now know that, if we find a c-competitive pair $(p[c, \tau_1], \pi[c, \tau_1])$, for some τ_1, then the optimal competitive ratio is at most c, and if we cannot find a c-competitive pair $(p[c, \tau_1], \pi[c, \tau_1])$, for any τ_1, then the optimal competitive factor must be larger than c.

The following lemma shows that, for a given c, all of the possible times τ_1 where c and τ_1 are valid and feasible form an interval.

Lemma 9. *Given a 3-*Slope SR *instance with entry fees, let (p^1, π^1) and (p^2, π^2) be c-competitive pairs that are induced by c and τ_1^1 and c and τ_1^2, receptively, where $\tau_1^1 < \tau_1^2$. Then, c and $\tilde{\tau}_1$, for every $\tilde{\tau} \in [\tau_1^1, \tau_1^2]$, are valid and feasible.*

4.2 Outline of Our Algorithm

In the proof of Theorem 3, we show how to construct a piece-wise, progressive profile that buys in a tight manner based on a ratio $c > 1$ and time $\tau_1 \geq 0$. Following this approach, we fix a ratio c, construct a tight profile but keep τ_1 as a variable. Our goal is to derive expressions for the rent at times s_1 and s_2 as (simple) functions of τ_1. Then, Lemma 7 allows us to determine which values τ_1 form a competitive pair together with c.

As τ_1 is a variable, we need to distinguish between three main cases in the analysis. The cases result from the relations of s_1, τ_1 and τ_2 which separate the pieces of the profile: (i) $\tau_1 < s_1 < \tau_2$, (ii) $s_1 \leq \tau_1$, and (iii) $\tau_2 < s_1$.
For the first two cases there are two sub-cases: either $s_2 \leq \tau_2$ or $\tau_2 < s_2$. We construct a profile for each case.

We start by solving the differential equation for Phase $0 \to 1$ with $r_j = r_0$ and use the boundary condition $p_1(0) = 0$. The result are three functions $(p_0^1(t), p_1^1(t), p_2^1(t))$ that describe the first piece of each profile in the three cases. Note that this piece does not depend on τ_1.

To outline the approach further, we concentrate on Case (i): We assume that $\tau_1 < s_1$. Hence, the first piece is valid in the interval $[0, \tau_1]$. Next, we derive the second piece of the profile using the differential equation for Phase $0 \to 2$ with $r_j = r_0$. As boundary condition at time τ_1 we use $p_2(\tau_1) = 0$. We derive three functions $(p_0^2(t), p_1^2(t), p_2^2(t))$ that describe our profile in the interval $[\tau_1, s_1]$.

With the first two pieces, we are able to derive an expression for the rent $\rho_p(s_1)$ as a function of τ_1 in Case (i) as follows: $\rho_p(s_1) = r_0 \cdot p_0^2(\tau_1) + r_1 \cdot p_1^2(\tau_1) + r_2 \cdot p_2^2(\tau_1)$. In the full version, we present the detailed analysis where the expression is simplified to

$$\rho_{s_1}(t) \triangleq c \cdot r_0 - (c-1) r_0 \cdot \exp\left(\frac{r_0 - r_2}{b_2} \cdot s_1 + \left(\frac{r_0 - r_1}{b_1} - \frac{r_0 - r_2}{b_2}\right) \cdot t\right).$$

We show that the expression defines a continuous and decreasing function. It follows that all values t such that $\rho_{s_1}(t) < c \cdot r_1$, i.e., values that satisfy the first condition of Lemma 7, form an interval. Continuing in this manner, we derive an expression for $\rho_p(s_2)$ that depends on τ_1 for Case (i). We show that this expression defines a continuous and convex function, and can determine another interval of values for τ_1 that satisfy the second condition of Lemma 7. The intersection of these two intervals, contains values for τ_1 that form a competitive pair together with c. We present the approach in detail in the full version where we show how these intervals are derived for each of the three cases.

Our algorithm performs binary search for the optimal competitive ratio c in the range $[1, e]$. Recall that an upper bound of e was given in [16]. For each candidate ratio c, Algorithm FEASIBLE (presented in the full version) computes the respective intervals of τ_1 for the three cases and checks whether one of them contains a feasible value for τ_1. Each such interval is obtained by using the conditions for validity and feasibly that were given in the previous section. If all intervals are empty, then a ratio of c cannot be obtained, otherwise a time τ_1 is obtained such that c and τ_1 are both valid and feasible. It follows that there exists a tight c-competitive pair $(p[c, \tau_1], \pi[c, \tau_1])$.

We summarize the main result of the section by the following theorem.

Theorem 4. *Given a* 3-SLOPE SR *instance with entry fees, such that* $s_1 < s_2$*, and* $c > 1$*, Algorithm* FEASIBLE *computes time* τ_1 *such that* $(p[c, \tau_1], \pi[c, \tau_1])$ *is a tight and c-competitive pair or decides that no c-competitive profile exists.*

In the full version of the paper we show how to compute a c-competitive strategy, given a tight c-competitive pair (p, π).

5 Future Directions

We analyse non-additive 3-SLOPE SKI RENTAL and provide a characterization of an optimal randomized transition function. Based on the analysis, we provide an algorithm that produces a near-optimal instance-dependent randomized strategy for 3-SLOPE SR.

There are two natural future directions. The first is to extend the result to MULTI-SLOPE SKI RENTAL with k slopes, where $k > 3$. This work already provides several ideas. For example, it is not hard to extend the transition ordering that was given in Sect. 3. That is, one can prove that the transition $i \to j$ must precede the transition $i' \to j'$, if $i = i'$ and $j < j'$ or if $i' > i$ and $j = j'$. This leads to a partial order, where (i, j) must precede (i', j') if $(i, j) \leq (i', j')$. Another direction is to consider a non-additive transition cost function where b_{ij} may be smaller than b_j, but still larger than $b_j - b_i$. As mentioned in the introduction, any c-competitive strategy that works for the entry fees case would be c-competitive for any non-additive instance with the same cost vector b. However, the best strategy for the former does not need to be the best strategy for the latter.

References

1. Ai, L., Wu, X., Huang, L., Huang, L., Tang, P., Li, J.: The multi-shop ski rental problem. In: ACM International Conference on Measurement and Modeling of Computer Systems (SIGMETRICS), pp. 463–475 (2014)
2. Augustine, J., Irani, S., Swamy, C.: Optimal power-down strategies. SIAM J. Comput. **37**(5), 1499–1516 (2008)
3. Azar, Y., Bartal, Y., Feuerstein, E., Fiat, A., Leonardi, S., Rosén, A.: On capital investment. Algorithmica **25**(1), 22–36 (1999)
4. Azar, Y., Cohen, I.R., Roytman, A.: Online lower bounds via duality. In: 28th Annual ACM-SIAM Symposium on Discrete Algorithms, pp. 1038–1050 (2017)
5. Bejerano, Y., Cidon, I., Naor, J.S.: Dynamic session management for static and mobile users: a competitive on-line algorithmic approach. In: 4th International Workshop on Discrete Algorithms and Methods for Mobile Computing and Communications, pp. 65–74. ACM (2000)
6. Black, D.L., Sleator, D.D.: Competitive algorithms for replication and migration problems. Technical Report, CMU-CS-89-201, CMU, November 1989
7. Borodin, A., El-Yaniv, R.: Online Computation and Competitive Analysis. Cambridge University Press, Cambridge (1998)
8. Damaschke, P.: Nearly optimal strategies for special cases of on-line capital investment. Theoret. Comput. Sci. **302**(1–3), 35–44 (2003)
9. Fujiwara, H., Kitano, T., Fujito, T.: On the best possible competitive ratio for the multislope ski-rental problem. J. Comb. Optim. **31**(2), 463–490 (2014). https://doi.org/10.1007/s10878-014-9762-9
10. Fujiwara, H., Konno, Y., Fujito, T.: Analysis of lower bounds for the multislope ski-rental problem. IEICE Trans. Fundam. Electron. Commun. Comput. Sci. **97-A**(6), 1200–1205 (2014)
11. Fujiwara, H., Satou, S., Fujito, T.: Competitive analysis for the 3-slope ski-rental problem with the discount rate. IEICE Trans. Fundam. Electron. Commun. Comput. Sci. **99-A**(6), 1075–1083 (2016)
12. Irani, S., Shukla, S.K., Gupta, R.K.: Online strategies for dynamic power management in systems with multiple power-saving states. ACM Trans. Embedded Comput. Syst. **2**(3), 325–346 (2003)
13. Karlin, A.R., Manasse, M.S., McGeoch, L.A., Owicki, S.S.: Competitive randomized algorithms for nonuniform problems. Algorithmica **11**(6), 542–571 (1994)
14. Karlin, A.R., Manasse, M.S., Rudolph, L., Sleator, D.D.: Competitive snoopy caching. Algorithmica **3**(1), 77–119 (1988)
15. Levi, A., Patt-Shamir, B.: Non-additive two-option ski rental. Theoret. Comput. Sci. **584**, 42–52 (2015)
16. Lotker, Z., Patt-Shamir, B., Rawitz, D.: rent, lease, or buy: randomized algorithms for multislope ski rental. SIAM J. Discret. Math. **26**(2), 718–736 (2012)
17. Marshall, A.W., Olkin, I.: Inequalities: Theory of Majorization and Its Applications. Academic Press, Cambridge (1979)
18. Patt-Shamir, B., Yadai, E.: Non-linear ski rental. In: 32nd ACM Symposium on Parallelism in Algorithms and Architectures, pp. 431–440 (2020)

Accelerated Information Dissemination on Networks with Local and Global Edges

Sarel Cohen[1], Philipp Fischbeck[2(✉)], Tobias Friedrich[2], Martin S. Krejca[3], and Thomas Sauerwald[4]

[1] School of Computer Science, The Academic College of Tel Aviv-Yaffo, Tel Aviv-Yaffo, Israel
sarelco@mta.ac.il

[2] Hasso Plattner Institute, University of Potsdam, Potsdam, Germany
{philipp.fischbeck,tobias.friedrich}@hpi.de

[3] Sorbonne University, CNRS, LIP6, Paris, France
martin.krejca@lip6.fr

[4] Department of Computer Science and Technology, University of Cambridge, Cambridge, UK
thomas.sauerwald@cl.cam.ac.uk

Abstract. Bootstrap percolation is a classical model for the spread of information in a network. In the round-based version, nodes of an undirected graph become active once at least r neighbors were active in the previous round. We propose the *perturbed* percolation process: a superposition of two percolation processes on the same node set. One process acts on a local graph with activation threshold 1, the other acts on a global graph with threshold r – representing local and global edges, respectively. We consider grid-like local graphs and expanders as global graphs on n nodes.

For the extreme case $r = 1$, all nodes are active after $O(\log n)$ rounds, while the process spreads only polynomially fast for the other extreme case $r \geq n$. For a range of suitable values of r, we prove that the process exhibits both phases of the above extremes: It starts with a polynomial growth and eventually transitions from at most cn to n active nodes, for some constant $c \in (0, 1)$, in $O(\log n)$ rounds. We observe this behavior also empirically, considering additional global-graph models.

Keywords: Bootstrap percolation · Random graphs · Expanders · Rumor spreading

1 Introduction

Information spreads very fast in networks (see, e.g., [23]). Several practical and theoretical studies concern n agents (nodes) interacting within a network and exchanging information via incident edges. These works have demonstrated that if each agent, once informed, informs all its agents in the neighborhood, the entire network is typically informed in a time that is at most logarithmic in the

M. Parter (Ed.): SIROCCO 2022, LNCS 13298, pp. 79–97, 2022.
https://doi.org/10.1007/978-3-031-09993-9_5

number of agents. This behavior even holds if each agent chooses only *one random* neighbor at each iteration (and a slightly faster dissemination is possible if an agent does not choose the same agent twice in a row [22]). A similar behavior occurs in the *bootstrap percolation* model [17], in which agents are informed once the number of informed neighbors reaches a certain threshold. This model has been extensively analyzed on a range of graph models, including hypercubes [9], grids [10], Erdős–Rényi graphs [31], preferential attachment graphs [4], random regular graphs [11], random geometric graphs [14], hyperbolic random graphs [15], inhomogeneous random graphs [1,5], geometric inhomogeneous random graphs [32], Kleinberg's small world model [24,27], as well as superpositions of Erdős–Rényi graphs and other graphs [37].

In the bootstrap percolation model, the process usually either reaches almost all agents quickly or terminates without having reached most of the agents [9, 31]. This speed is often attributed to the logarithmic diameter of the network, as well as to the existence of high-degree nodes, which are both prevalent in many real-world graphs as well as in their mathematical models. However, these models assume that information spreads the same way among all edges. If this is not the case, e.g., because agents need to be convinced of some information by more than one agent, the resulting behavior can be fundamentally different [12,16,26,28,35].

Typically, the edges of a graph describe the closeness of agents, i.e., two agents connected via an edge are close, while non-edges represent separation. This is particularly true for graph models that utilize an underlying geometry for determining the edge set. However, another perspective, found in epidemics, is that every pair of agents has an activation probability defined, e.g., by splitting the agents into groups [30]. In bootstrap percolation, one can set different activation thresholds based on the groups [13]. Further, one can model the closeness via different graphs on the same agents, namely via *local* and *global* edges, which are assigned different activation probabilities [7,8]. The underlying graphs represent different interactions, e.g., contacts within and across households [6].

We aim at understanding the effect of edge types on the speed of information dissemination. To this end, we analyze graphs that have two types of edges: one representing short edges, and another one representing long edges. The graph induced by the short edges (the *local* graph) models the local neighborhood of agents. These model whether two agents are close, e.g., people an agent is exposed to more often, such as colleagues, relatives or neighbours. The graph induced by the long edges (*global* graph) models non-local (global) contacts. This represents people who the agent has not that much contact with, e.g., people who live further away or celebrities from social media who the agent may never meet personally but is influenced by.

We employ the classic *bootstrap percolation* model as a foundation for the spread of information in networks as described above. In this model, each agent either has a certain piece of information (it is *active*) or it has not (*inactive*). Given a parameter $r \in \mathbb{N}^+$ (the *activation threshold*) and a set of initially active nodes, iteratively, at each round $t \in \mathbb{N}^+$, a node becomes (and remains) active if it has at least r neighbors that were active in round $t - 1$.

Model. We propose the *perturbed percolation model*, which is the superposition of two bootstrap percolation processes on the same node set but with two different edge sets. One process acts on the *local* graph with an activation threshold of 1. The other process acts on the *global* graph with an activation threshold of r. This is similar to the above-mentioned models where nodes have local and global contacts with varying activation probabilities [8]. Note that a perturbed percolation process always percolates completely if the local graph is connected. However, the overall speed is majorly influenced by the global graph via r.

Theoretically and empirically, we analyze *how quickly* nodes become active in this model. We are interested in the two following activation rates: a *polynomial* rate, i.e., the number of active nodes in round t is a polynomial in t, and a *rapid* rate, i.e., the number of inactive nodes reduces from at least $(1 - c)n$, for some constant $c \in (0, 1)$, to none in $O(\log n)$ rounds.

Results. For our theoretical results, we analyze the activation rate of the perturbed percolation model on local graphs that we refer to as *polynomial-neighborhood graphs* (PNGs) with n nodes, characterized by having a polynomially expanding neighborhood w.r.t. the hop distance, including grid graphs, cycles, and, asymptotically almost surely (a.a.s.), random geometric graphs with expected polylogarithmic node degree. We prove the following landscape of perturbed percolation w.r.t. the activation threshold r, using PNGs as local and expanders as global graph:

- For the extreme case $r \geq n$, the process has a **polynomial** rate (Theorem 1).
- For the other extreme case $r = 1$, the process has a **rapid** rate (Corollary 1), i.e., adding global edges changes the rate immediately from polynomial to rapid.
- Our **main result** is that the process with suitable values of r between the extreme cases above, including $r = 2$, has a **polynomial-to-rapid** rate (Corollary 2), i.e., the process has a polynomial rate for a polynomial number of rounds (w.r.t. n) and then ends with a rapid rate. This result highlights that while the edges from the global graph speed up the overall process, it takes some (long) time for the process to actually switch to a rapid rate.

We complement our theoretical results by empirical analyses (Figs. 1 and 2). Next to Erdős–Rényi graphs as global graphs, we also include Barabási–Albert and hyperbolic random graphs, which are not covered by our theoretical analysis. For all cases, we observe a clear distinction between the polynomial and the rapid rate.

Framework (Informal Description). Our main result follows from our more general result (Theorem 3) based on proving the following three independent properties, assuming a graph with n nodes:

1. Any *bootstrap* percolation process on the local graph, for any initial active set of size 1, has polynomial rate.

2. For the perturbed percolation process, a.a.s. for an initial number of rounds polynomial in n, no inactive node has at least r global edges to active nodes.
3. Asymptotically a.s., any *bootstrap* percolation process on the global graph, for any initial active set of linear size, percolates completely in a number of rounds logarithmic in n.

Combining all three properties shows a polynomial-to-rapid rate. We note that Property 3 considers the classic bootstrap percolation setting but requires to *first* fix the random graph and *then* the initial set (even adversarially). Typically, this order is reversed. Thus, we believe our results proving this property (Theorem 6 and 8) to be of independent interest. In addition, in Theorem 8 we provide an improved bound of $r \cdot n / \ln n$ for the size of the initial set in Property 3 for the special case of Erdős–Rényi graphs.

Outline. In Sect. 2, we introduce our notation as well as our model and the graph classes we consider. Sections 3 and 4 contain our theoretical results. The former considers the extreme cases of the activation threshold r, the latter suitable intermediate values. Our main result of these sections is Corollary 2. In Sect. 5, we discuss our empirical results, and we provide an outlook in Sect. 6.

2 Preliminaries

Let $\mathbb{N}$ denote the set of natural numbers, including 0. For $m, n \in \mathbb{N}$, let $[m..n] := [m, n] \cap \mathbb{N}$, and let $[m] := [1..m]$. We consider undirected, finite graphs. Given such a graph G, let $V(G)$ denote its set of nodes and $E(G)$ its set of edges. We denote the minimum and maximum node degree of G by $d_{\min}(G)$ and $d_{\max}(G)$, respectively, dropping G if it is clear from context.

We use big-O notation only in combination with a graph G. The asymptotics of the notation are then with respect to $|V(G)|$ (which we usually call n). Additionally, the notation $\widetilde{O}$ allows for factors polylogarithmic in $|V(G)|$. In the same context, a *constant* is a value $\Theta(1)$, that is, a value bounded independently of $|V(G)|$.

An event A occurs *asymptotically almost surely (a.a.s.)* if and only if $\Pr[A] = 1 - o(1)$.

2.1 Percolation Processes

We introduce the perturbed percolation process, which is a superposition of two classical bootstrap percolation processes, using different edges and thresholds.

Bootstrap Percolation. Let G be a graph with n nodes, $r \in \mathbb{N}_{>0}$, and $I \subseteq V(G)$. The *bootstrap percolation process P on G with threshold r and initial active set I* is a deterministic discrete-time process on $V(G)$ in which each node is either *active* or *inactive*. In each round, each node adjacent to at least r active nodes becomes active. Let $(A_t)_{t\in\mathbb{N}}$ denote the sequence of sets of active nodes over time. Note that $A_0 = I$ and that, for all $t \in \mathbb{N}$ with $t \geq n-1$, $A_{t+1} = A_t$. We say that P *percolates completely* if and only if $|A_{n-1}| = n$.

Let $t_1, t_2 \in \mathbb{N}$, and let $T = \min\{t \in \mathbb{N} \mid A_t = A_{n-1}\}$. We say that P has a *polynomial activation rate for* $[t_1..t_2]$ if and only if there is a constant $c > 0$ such that for all $t \in [t_1..t_2]$ it holds that $|A_t| = \widetilde{O}(t^c + 1)$. Further, P has a *rapid activation rate for* $[t_1..t_2]$ if and only if $t_2 = t_1 + O(\log n)$, there is a constant $c \in (0,1)$ such that $|A_{t_1}| \leq cn$, and $|A_{t_2}| = n$. We say P has a *polynomial* (resp. *rapid*) *activation rate* if and only if it has a polynomial (resp. rapid) activation rate for $[0..T]$. Last, we say that P has a *polynomial-to-rapid activation rate* if it has both a *polynomial activation rate* and *rapid activation rate*. Note that this is equivalent to the existence of $t_1, t_2 \in \mathbb{N}$ and a constant $c > 0$ such that $t_1 \in \Omega(n^c)$ and that P has a polynomial activation rate for $[0..t_1]$ and a rapid activation rate for $[t_2..T]$.

Perturbed Percolation. Let $G = (V, E)$ be a graph decomposable into a *local graph* $G_\ell = (V, E_\ell)$ and a *global graph* $G_g = (V, E_g)$ (each possibly random), i.e., $E = E_\ell \cup E_g$. Further, let $r \in \mathbb{N}_{>0}$ and $I \subseteq V$. The *perturbed percolation process* P *on* G *with threshold* r *and initial active set* I is the union of the bootstrap percolation process on G_ℓ with threshold 1 and the one on G_g with threshold r, both with initial active set I. That is, in each round, each node with an active neighbor in G_ℓ or at least r active neighbors in G_g becomes active. The notion of polynomial/rapid activation rate from bootstrap percolation naturally extends to P.

We introduce *randomization* into the connections via a random permutation of the nodes. To this end, we assume w.l.o.g. that there exists a bijective labeling $\ell\colon V(G) \to [1..n]$. Let σ be a permutation over $[1..n]$, chosen uniformly at random, independently of any other potential random choices, and let G'_g be identical to G_g. Then $E_g = \{\{\sigma(\ell(u)), \sigma(\ell(v))\} \in V(G)^2 \mid \{u, v\} \in E(G'_g)\}$. Technically, G_g is random (due to σ), and G'_g represents a (possibly deterministic) isomorphic representation of G_g. However, throughout the paper, we refer to both graphs as the global graph. When talking about the graph itself, we refer to G'_g, which can be deterministic. In contrast, if we refer to its edges, we refer to the set $E(G_g)$, which is random. Without randomization, there always exist perturbed percolation processes with (solely) rapid activation rates, due to possible dependencies between G_ℓ and G_g. In particular, there are graphs G_ℓ and G_g in the graph classes below such that the perturbed percolation process ends within $O(\log n)$ rounds. Randomization eliminates such cases. In case that G_ℓ and G_g are independent, randomization does not change anything. In particular, it is not required for our results concerning random graphs.

Throughout the paper, we assume the following order of events: 1. Fix G_ℓ and G_g in some order. 2. Randomize G_g as described above. 3. Fix an initial active set of nodes. Note that this implies that the initial active set can be chosen adversarially w.r.t. the realizations of the resulting graph of the perturbed percolation process.

2.2 Graph Classes

As local graphs, we consider graphs with polynomially expanding neighborhoods. As global graphs, we consider expanders, especially random regular graphs and Erdős–Rényi graphs.

Polynomial-Neighborhood Graphs. For a *connected* graph $G = (V, E)$, let $d_G : V^2 \to \mathbb{N}$ denote the distance between all pairs of nodes in G. That is, for all $u, v \in V$, the value $d_G(u, v)$ is the length of a shortest path from u to v. Further, for all $u \in V$ and all $h \in [0..|V| - 1]$, let $B_h(u) = \{v \in V \mid d_G(u, v) \leq h\}$ denote the ball of distance at most h around u.

Let $c > 0$ be a constant. We say that G is a *polynomial-neighborhood graph (PNG) of growth* c if and only if for all $u \in V$ and all $h \in [0..|V| - 1]$ it holds that $|B_h(u)| = \widetilde{O}(h^c + 1)$.

Examples of PNGs include grid graphs (with and without looping boundaries), cycles, and, a.a.s., random geometric graphs with expected node degree polylogarithmic in n.

Expanders. We call a graph an *expander* if and only if its spectral expansion λ is bounded away from 1 from above and below (see Sect. 4.2 for more details). We note that expanders can be deterministic or random. It is well-known that both Erdős–Rényi Graphs [19] and random d-regular graphs are expanders [25] (see Theorems 4 and 5).

Random Regular Graphs. Let $n \in \mathbb{N}_{>0}$, let $d \in [3..n-1]$, and let $\mathcal{G}_{n,d}$ denote the class of all (deterministic) d-regular graphs with n nodes. Each uniform sample G from $\mathcal{G}_{n,d}$ is a *random d-regular graph with n nodes*, denoted as $G_{n,d}$.

Erdős–Rényi Graphs. Let $n \in \mathbb{N}_{>0}$ and $p \in [0, 1]$. A graph G is an *Erdős–Rényi graph with n nodes and edge probability p*, denoted as $G_{n,p}$, if and only if $|V(G)| = n$ and each $e \in V^2 \setminus \{(v, v) \mid v \in V\}$ is in $E(G)$ with probability p, independent of all other choices.

3 Extreme Thresholds

We consider perturbed percolation on PNGs with n nodes as local graphs for the extreme cases of $r \geq n$ and $r = 1$, where r is the threshold of the global graph.

Case $r \geq n$. This case is equivalent to *bootstrap* percolation on PNGs with a threshold of 1. We show that regardless of the (bootstrap) threshold, the rate of the process on PNGs is polynomial if the initially active set is constant. We note that the perturbed percolation process percolates completely if and only if the local graph is connected.

Theorem 1. *Let $c > 0$ be a constant, and let G be a PNG of growth c. Further, let $I \subseteq V(G)$ such that $|I| = \Theta(1)$, and let $r' \in [n-1]$. Then the* bootstrap *percolation process on G with threshold r' and initial active set I has a polynomial activation rate.*

Proof. Let $t \in \mathbb{N}$, and recall that A_t is the set of active nodes at the end of round t. From each $u \in I$, the bootstrap percolation process reaches at most $B_t(u)$, that is, it holds that $|A_t| \leq \sum_{u \in I} |B_t(u)|$. Since G is a PNG of growth c and since $|I| = \Theta(1)$, it follows that $|A_t| = \widetilde{O}(|I| \cdot (t^c + 1)) = \widetilde{O}(t^c + 1)$, which concludes the proof.

Case $r = 1$. It follows from the literature that the rate is rapid from the start (Corollary 1) if the global graph is an Erdős–Rényi graph, as the diameter of the graph is logarithmic.

Theorem 2 [34, **Theorem 4**]. *Let $n \in \mathbb{N}_{>0}$, $\varepsilon > 0$ be a constant, and let G be a graph with n nodes that is decomposable into a connected local graph and into a $G_{n,\varepsilon/n}$ as a global graph. Then a.a.s., G has a diameter of $O(\log n)$.*

For d-regular expanders, it is well-known that the diameter is $O(\log n)$ [29, page 455].

The following statement immediately follows (as it only requires that the diameter is $O(\log n)$), noting that the diameter of a $G_{n,p}$ does not increase when p increases.

Corollary 1. *Let $G = (V, E)$ be a graph with $n \in \mathbb{N}_{>0}$ nodes that is decomposable into a connected local graph and into a global graph G_g. Further, let $c \in (0, 1)$ be a constant, and let $I \subseteq V(G)$ such that $I \neq \emptyset$ and $|I| \leq cn$.*

1. *Let G_g be $G_{n,p}$ with $p \in [\Omega(1/n), 1]$. Then a.a.s., the perturbed percolation process on G with threshold 1 and initial active set I has a rapid activation rate.*
2. *For $d \in [3..n-1]$, let G_g be a d-regular expander with n nodes. Then a.a.s., the perturbed percolation process on G with threshold 1 and initial active set I has a rapid activation rate.*

4 Polynomial-to-Rapid Activation Rate

We prove the emergence of a polynomial-to-rapid activation rate for suitable values of r between the extreme cases considered above. Our main result is the following.

Corollary 2. *Let G be a graph with $n \in \mathbb{N}_{\geq 3}$ nodes that is decomposable into a PNG as local graph and into a graph with spectral expansion $\lambda \in \mathbb{R}_{>0}$ and $d_{\max} = O(d_{\min})$ as global graph. Let $d = 2|E(G)|/n$, and let $r \in [2..(1-\lambda)d_{\min}^2/(4d)]$. Then a.a.s., there exists a $V' \subseteq V(G)$ with $|V'| = n - n^{3/4}$ such that for all $v \in V'$, the perturbed percolation process on G with threshold r and initial active set $\{v\}$ has a polynomial-to-rapid rate.*

We prove this result by applying a general framework for proving that a perturbed percolation process P has a transition from polynomial to rapid rate on a graph $G = (V, E)$ with $|V| = n$. To this end, let G_ℓ denote the local graph that P acts on, and let G_g denote the global graph. Further, let P_ℓ and P_g denote the bootstrap percolation processes on G_ℓ and G_g, respectively. Last, let $(A_t)_{t\in[0..n-1]}$ denote the set of active nodes of P_ℓ after each round, and for all $v \in V$ and $U \subseteq V$, let $\Gamma_g(v, U) = \{u \in U \mid \{u, v\} \in E(G_g)\}$.

Framework. The framework comprises the following three independent properties:

1. For all $v \in V(G)$, the process P_ℓ with initial active set $\{v\}$ has a polynomial activation rate and percolates completely.
2. There are constants $c_1, c_2 \in (0, 1)$ and a set $|V'| \geq n - n^{1-c_1}$ such that for all $v \in V'$, having initial active set $\{v\}$ implies that for all $u \in V$, $|\Gamma_g(u, A_{n^{c_2}})| < r$.
3. There exists a constant $c_3 > 1$ such that for all $I \subseteq V$ with $|I| \geq n/c_3$, the process P_g with initial active set I has a rapid activation rate.

Properties 1 and 3 consider exclusively P_ℓ and P_g, respectively, whereas Property 2 connects P_ℓ with the global graph. Our framework yields the following general theorem.

Theorem 3 (Polynomial-to-rapid rate). *Let $n \in \mathbb{N}_{\geq 3}$, let $r \in [2..n-1]$, and let G be a graph with n nodes, decomposable into a local graph and into a global graph. Assume that P is a perturbed percolation process on G with threshold r and some initial active set such that Properties 1 to 3 are all satisfied. Then P has a polynomial-to-rapid rate.*

Proof. By Property 2, there exists a $c_2 > 0$ such that during the initial n^{c_2} rounds of P, all activations are exclusively due to the local graph. By Property 1, it follows that P has a polynomial activation rate for $[0..n^{c_2}]$.

Now consider the first round t^* such that the number of active nodes is at least n/c_3, where c_3 is from Property 3. Note that such a t^* exists, as the number of active nodes strictly increases each round until complete percolation, since the process on the local graph percolates completely. Further note that, due to Property 3, the number of active nodes in round $t^* - 1$ is less than n/c_3. By Property 3, for any set of active nodes in round t^*, the process P percolates completely in $O(\log n)$ rounds. Thus, the process P has a rapid activation rate, starting from round $t^* - 1$, which concludes the proof.

In the following, we prove the properties of our framework separately. As Theorem 1 already proves Property 1, we are left to consider Properties 2 and 3.

4.1 Polynomial Rate

We show that Property 2 is satisfied for PNGs as local graph and for global graphs with a bounded maximum degree, which includes expanders and, a.a.s., Erdős–Rényi graphs.

Lemma 1. *Let $n \in \mathbb{N}_{\geq 3}$, $r \in [2..n-1]$, and $c_1, c_2 \in \mathbb{R}_{>0}$ with $c_2 < 1/3$ be constants. Further, let G be a graph with n nodes, decomposable into a PNG with growth c_1 as local graph and into a global graph G_g with $d_{\max}(G_g) \leq n^{c_2}$. Then with probability at least $1 - n^{-1/12}$, there exists a $V' \subseteq V(G)$ with $|V'| = n - n^{3/4}$ such that for all $v \in V'$, the perturbed percolation process on G with threshold r and initial active set $\{v\}$ has a polynomial activation rate for $[0..n^{(1/3-c_2)/c_1}]$.*

Proof. By monotonicity, it suffices to consider the case $r = 2$. Pick any node $v \in V$ as the initially active node. Let B_v be all nodes that get activated in the local graph G_ℓ after $O(n^{(1/3-c_2)/c_1})$ rounds. Hence, $|B_v| = O(n^{1/3-c_2})$. Note that within the graph G_g, due to the random labeling of the nodes, we can regard the subset B_v in G_g as a random set of size $|B_v|$. In particular, the events of any two nodes x, y being in B_v are negatively correlated. Now let $Z_v \subseteq V$ be the set of nodes in $V(G_g)$ that have at least 2 neighbors in B_v. Then,

$$\mathrm{E}[|Z_v|] \leq n \cdot \binom{d_{\max}(G_g)}{2} \cdot \left(\frac{|B_v|}{n}\right)^2 \leq \frac{n^{2c_2} \cdot n^{2/3-2c_2}}{n} = n^{-1/3}.$$

Hence by Markov's inequality, the probability of any activation occurring via global edges is $\Pr[|Z_v| \geq 1] \leq n^{-1/3}$.

Now define $Y := \{v \in V \mid |Z_v| \geq 1\}$. Then $\mathrm{E}[|Y|] \leq n^{2/3}$, and by another application of Markov's inequality, $\Pr[|Y| \geq n^{3/4}] \leq n^{-1/12}$.

4.2 Rapid Rate on the Global Graph

We show that expander graphs satisfy Property 3 (Theorem 6). For the special case of Erdős-Rényi graphs, we prove an even stronger bound, showing complete percolation in $O\big(\log(n)/\log\log n\big)$ rounds (Theorem 8). We note that due to our assumption that the random graphs are revealed *before* the initial active set is chosen, our theorems show that a.a.s. the global graphs have immediately a rapid activation rate for *arbitrary* sufficiently large initial active sets. This includes cases where the initial set is chosen adversarially w.r.t. the global graph. In contrast, classic results typically fix the global graph *after* or independent of the initial set [11,31], thus not allowing for adversarially chosen initial sets.

Expanders. For any graph G, for all $v \in V(G)$, let $\deg(v)$ be the degree of v, let $d = 2|E(G)|/n$ denote the average degree, and, for all $S \subseteq V(G)$, let $\mathrm{vol}(S) := \sum_{u \in S} \deg(u)$. We define the *normalized Laplacian matrix of G* by

$$\boldsymbol{L}_{u,v} = \begin{cases} 1 & \text{if } u = v, \\ -\frac{1}{\sqrt{\deg(u) \cdot \deg(v)}} & \text{if } \{v, w\} \in E(G), \\ 0 & \text{otherwise.} \end{cases}$$

We denote by $0 = \lambda_1 \leq \lambda_2 \leq \cdots \leq \lambda_n \leq 2$ the n eigenvalues of $\boldsymbol{L}$. Further, $\lambda := \max_{i \geq 2} |1 - \lambda_i|$ denotes the *spectral expansion*. A graph is called an *expander* if $\lambda \leq 1-c$ for some constant $c > 0$ (in other words, all eigenvalues are sufficiently far away from 0 and 2).

The following result shows that Erdős-Rényi graphs are expanders.

Theorem 4 [19, **Theorem 1.2**]. *Let $G = G_{n,p}$ be an Erdős-Rényi graph with expected degree $p(n-1) \geq c_1 \cdot \ln(n)$ for a sufficiently large constant $c_1 > 0$. Then a.a.s., the spectral expansion of $\boldsymbol{L}$ satisfies $\lambda(G) = O((p(n-1))^{-1/2})$.*

A similar result was shown by Friedman [25] for random regular graphs (for simplicity, we only state a slightly weaker version of his main result, which suffices for our purposes).

Theorem 5 [25, **Theorem A**]. *Let G be a $G(n, 2d)$ random $2d$-regular graph. Then for all $d = O(1)$, a.a.s., the spectral expansion of $\boldsymbol{L}$ satisfies $\lambda(G) = O(d^{-1/2})$.*

Our main result of this section is the rapid activation rate of expanders.

Theorem 6. *Let $n \in \mathbb{N}_{\geq 3}$, and let $G = (V, E)$ with $|V| = n$, with spectral expansion $\lambda > 0$, and with $d_{\max} = O(d_{\min})$. Further, let $d = 2|E(G)|/n$, let $r \in [2..(1-\lambda)d_{\min}^2/(4d)]$, and let $I \subseteq V$ with $|I| \geq 4\frac{r-1}{(1-\lambda)\cdot d_{\min}^2/d} \cdot n$. Then the bootstrap percolation process on G with threshold r and initial active set I percolates completely after $O(\frac{\log n}{1-\lambda})$ rounds.*

In case of Erdős-Rényi graphs with $p = \Omega(\log n/n)$ or random $2d$-regular graphs, $1-\lambda$ is bounded below by a positive constant, and thus the process percolates rapidly. We remark that the result and proof of Theorem 6 share some ideas with the work by [20], who investigate the size of *smallest* contagious sets in various classes of expander graphs. However, one key difference is that Theorem 6 provides a guarantee so that *all* sets of a certain size percolate, and it additionally establishes a bound on the number of steps until complete percolation.

We use the following version of the expander mixing-lemma to show Theorem 6.

Lemma 2 (Non-regular-expander mixing-lemma). *For all $S \subseteq V$ of a graph with spectral expansion λ, denoting with $e(S, V \setminus S)$ the number of edges between S and $V \setminus S$, we have*

$$\left| e(S, V \setminus S) - \frac{\mathrm{vol}(S) \cdot \mathrm{vol}(V \setminus S)}{\mathrm{vol}(G)} \right| \leq \lambda \cdot \frac{\mathrm{vol}(S) \cdot \mathrm{vol}(V \setminus S)}{\mathrm{vol}(G)}.$$

Proof (Proof of Theorem 6). We establish the result in two stages, depending on whether $|S|$ is greater or smaller than $n/2$. In the first stage, we show that whenever the set of active nodes S with $|S| = \varepsilon \cdot n$ satisfies $4\frac{r-1}{(1-\lambda)\cdot d_{\min}^2/d} \leq \varepsilon \leq 1/2$, then the number of active nodes increases by a factor of $1 + \Omega(1-\lambda)$. Applying Lemma 2 with S yields

$$e(S, V \setminus S) \geq (1-\lambda) \cdot \frac{\mathrm{vol}(S) \cdot \mathrm{vol}(V \setminus S)}{\mathrm{vol}(G)} \geq (1-\lambda) \cdot \frac{d_{\min}^2 \varepsilon n(1-\varepsilon)}{d}.$$

Now define $N := \{v \in V \setminus S \mid \deg_S(v) \geq r\} \subseteq V \setminus S$, which are the nodes that get activated by S in the next round. By decomposing $e(S, V \setminus S) = e(S, N) + e(S, (V \setminus S) \setminus N)$,

$$e(S, V \setminus S) \leq |N| \cdot d_{\max} + (|V \setminus S| - |N|) \cdot (r - 1),$$

and rearranging gives

$$|N| \geq \frac{e(S, V \setminus S) - (n - |S|) \cdot (r - 1)}{d_{\max}} \geq \frac{(1 - \lambda) \cdot \frac{d_{\min}^2}{d} \varepsilon (1 - \varepsilon) n - n \cdot (r - 1)}{d_{\max}}. \tag{1}$$

Hence, if $1/2 \geq \varepsilon \geq 4 \frac{r-1}{(1-\lambda) \cdot d_{\min}^2 / d}$, we conclude that

$$|N| \geq \frac{\left(\frac{1-\lambda}{2} \frac{d_{\min}^2}{d} \varepsilon - \frac{1-\lambda}{4} \frac{d_{\min}^2}{d} \varepsilon\right) \cdot n}{d_{\max}} \geq \frac{\frac{1-\lambda}{4} \frac{d_{\min}^2}{d} \varepsilon}{d_{\max}} \cdot n = \frac{1 - \lambda}{4} \cdot \frac{d_{\min}^2}{d_{\max} \cdot d} \cdot |S|.$$

Recall that we assumed $d_{\max} = O(d_{\min})$. Thus in the next step, we can replace S by $S \cup I$ and obtain an at least exponential growth (with factor $\Theta(1 - \lambda)$) in the number of active nodes until $|S| > n/2$.

Consider now the second stage, where we assume $|S| > n/2$ (thus $\varepsilon > 1/2$). As before, we infer in the same way $e(S, V \setminus S) \geq (1 - \lambda) \cdot d_{\min}^2 \varepsilon n (1 - \varepsilon)/d$. Recalling that $N = \{v \in V \setminus S \mid \deg_S(v) \geq r\}$, we obtain the following refined version of (1), using that $\varepsilon \geq 1/2$,

$$\begin{aligned} |N| &\geq \frac{(1 - \lambda) \frac{d_{\min}^2}{d} \varepsilon (1 - \varepsilon) n - (|V \setminus S| - |N|) \cdot (r - 1)}{d_{\max}} \\ &\geq \frac{\frac{1-\lambda}{2} \frac{d_{\min}^2}{d} (1 - \varepsilon) n - (1 - \varepsilon) \cdot n \cdot (r - 1)}{d_{\max}}. \end{aligned}$$

Hence, if $r - 1 \leq \frac{(1-\lambda) d_{\min}^2}{4d}$, we conclude that

$$|N| \geq \frac{\frac{1-\lambda}{4} \frac{d_{\min}^2}{d} (1 - \varepsilon) n}{d_{\max}} = \frac{1 - \lambda}{4} \cdot \frac{d_{\min}^2}{d_{\max} \cdot d} \cdot |V \setminus S|.$$

Thus, if $|S| > n/2$, the set of inactive nodes decreases exponentially in each round.

Erdős-Rényi Graphs. We first prove an upper bound for the time until complete percolation for bootstrap processes on Erdős-Rényi graphs, showing Property 3, which is better than the one following from Theorem 6. Then, we show that there exists an initial active set such that the time needed for complete percolation matches this bound. We make use of the well-known Chernoff bounds.

Theorem 7 (Chernoff bounds [3, Theorems A.1.12 and A.1.13]). *Let $n \in \mathbb{N}_{>0}$, $p \in [0,1]$, and $X \sim \mathrm{Bin}(n,p)$. Then*

1. *for all $\beta > 1$, it holds that $\Pr[X \geq \beta np] \leq (e^{\beta-1}\beta^{-\beta})^{np}$, and*
2. *for all $a \in (0, np]$, it holds that $\Pr[X < np - a] < \exp\big(-a^2/(2np)\big)$.*

The following bound shows a rapid activation rate for sufficiently large initial active sets.

Theorem 8. *Let $n \in \mathbb{N}_{\geq 3}$, $p \geq 20\ln(n)/n$, $r \in [2..\ln n]$. Further, let $I \subseteq V(G_{n,p})$ with $|I| = r \cdot n/\ln n$. Then a.a.s., the bootstrap percolation process on $G_{n,p}$ with threshold r and initial active set I percolates completely in at most $(1+o(1))\ln(n)/\ln\ln n$ rounds.*

Proof. We prove several claims about $G = G_{n,p}$, which ultimately show Theorem 8.

Claim (8.1). The minimum degree of a node of G is a.a.s. at least $13\ln n$.

Proof. The degree of each node v is a binomial random variable with parameters $n-1$ and p. By assumption $(n-1)p \geq (1-o(1))20\ln n$ and, by Theorem 7, Item 2, the probability that it is smaller than $13\ln n$ is at most

$$e^{-(1+o(1))(49/40)\ln n} = \frac{1}{n^{49/40-o(1)}}.$$

The assertion of the claim thus follows from the union bound.

Claim (8.2). Asymptotically almost surely, for every two disjoint sets C and B in G, with $|C| = n/2$ and $|B| = rn/\ln n$, there is a node c in C that has at least r neighbors in B.

Proof. Fix two disjoint sets B and C as above. Clearly it suffices to prove the claim for $p = 20\ln(n)/n$. For every node $v \in C$, the expected number of neighbors of v in B is $p|B| = 20r$. By Theorem 7, Item 2, the probability it has less than r neighbors in B is at most (with room to spare) $e^{-19^2r^2/(40r)} < \frac{1}{100}$. These events for distinct nodes $v \in C$ are pairwise independent, hence the probability that there is no node $v \in C$ as above is at most $(1/100)^{n/2}$. As there are less than 4^n pairs of sets B, C as above, the result follows by the union bound, since $4^n/100^{n/2} = o(1)$.

Claim (8.3). Asymptotically almost surely, for any two disjoint sets of nodes B and C, where $|B| \geq n/2$, $n - |B| \geq 12\ln n$ and $|C| = (n-|B|)/2$, there is a node in C that has at least r neighbors in B.

Proof. As before, fix two disjoint sets B, C as above, and note that we may assume that $p = 20\ln(n)/n$. For every fixed $v \in C$ the expected number of neighbors of v in B is $p|B| \geq 10\ln n$. As $r \leq \ln n$, the probability that v has less than r neighbors in B is at most $e^{-81\ln(n)/20} < n^{-4}$, by Theorem 7, Item 2. Therefore the probability that this is the case for every $v \in C$ is smaller than $(1/n^4)^{|C|}$. The number of possible pairs of sets B and C as above is smaller than $n^{3|C|}$ (as the number of choices for the complement of B is $\binom{n}{2|C|} \leq n^{2|C|}$), and the claim follows by the union bound.

Claim (8.4). Asymptotically almost surely, for every $y \leq \frac{n}{100 \ln n}$, no set of y nodes of G spans more than $y \ln n$ edges.

Proof. Fix a set Y of y nodes. The expected number of edges in it is $\binom{y}{2}p \leq \frac{y^2 10 \ln n}{n}$. By Theorem 7, Item 1, with

$$\beta = \frac{y \ln n}{(y^2 10 \ln n)/n} = \frac{n}{10y} \ (> 10 \ln n)$$

the probability that Y spans at least $y \ln n$ edges is at most

$$(e^{\beta-1}/\beta^{\beta})^{(y^2 10 \ln n)/n} \leq \beta^{-0.9\beta(y^2 10 \ln n)/n} = (10y/n)^{0.9 y \ln n} < e^{-2y \ln n}.$$

The number of sets of size y is $\binom{n}{y} \leq e^{y \ln n}$. We conclude by noting that the probability that there is a set Y spanning $y \ln n$ edges for any $y \leq \frac{n}{100 \ln n}$ is at most

$$\sum_{y=1}^{n/(100 \ln n)} e^{y \ln n} \cdot e^{-2y \ln n} = o(1).$$

Claim (8.5). Asymptotically almost surely, for every set B of nodes of size $n-x$, where $12 \ln n \leq x \leq n/1000$, the number of nodes outside B that do not have at least $\ln n$ $(\geq r)$ neighbors in B is smaller than $10x/\ln n$.

Proof. Fix a set B as above and a subset C of $10x/\ln n$ nodes in its complement. We bound the probability that no node of C has at least $\ln n$ neighbors in B as follows. By Sect. 4.2, a.a.s. each node in the graph has degree at least $13 \ln n$. Assume this is the case. Then every node of C has at least $12 \ln n$ neighbors in the complement B' of B (as it has at most $\ln n$ neighbors in B). By Sect. 4.2, a.a.s., the number of edges spanned by the set C is at most $|C| \ln n$. Thus the number of edges between C and $B' \setminus C$ has to be at least $10|C| \ln n = 100x$. The expected number of edges is

$$|C|(|B'| - |C|)p \leq \frac{10x}{\ln n} x \frac{20 \ln n}{n} = \frac{200x^2}{n}.$$

Applying Theorem 7, Item 1, with

$$\beta = \frac{100x}{200x^2/n} = \frac{n}{2x} \geq 500 \ (> e^5)$$

we conclude that the probability of having that many edges is at most

$$(e^{\beta-1}\beta^{-\beta})^{200x^2/n} \leq \beta^{-0.8\beta 200x^2/n} = (2x/n)^{80x}.$$

The number of choices for the sets B' and C is smaller than $\binom{n}{x}^2 \leq (en/x)^{2x}$. Thus, by the union bound the probability that there are sets B, C violating the claim is at most

$$\sum_{x \geq 12 \ln n} (en/x)^{2x} (2x/n)^{80x}.$$

Since $x \leq n/1000$, $2x/n \leq 1/500$ and hence $(2x/n)^{80x} \leq (x/(250n))^{40x} < (x/(250n))^{2x}$ showing that the sum above is at most $\sum_{x \geq 12 \ln n} (e/250)^{2x} = o(1)$.

We now prove that the number of rounds until complete percolation is a.a.s. $(1+o(1))\ln(n)/\ln\ln n$. Assuming that all claims hold, starting with any set A of $rn/\ln n$ nodes, by Claim 8.2, in one round at least $n/2$ nodes become active. By Claim 8.3, in 9 additional rounds the number of inactive nodes drops to at most $n/2^{10} < n/1000$. By Claim 8.5, in each round from now on, the number of inactive nodes drops by a factor of at least $\ln n/10$, as long as this number is above $12\ln n$. Once below $12\ln n$, one final step activates all remaining nodes, as the minimum degree is at least $13\ln n$, by Claim 8.1. This completes the proof.

Note that the bound in Theorem 8 is optimal for $p = \Theta(\log(n)/n)$ in the sense that there is an initial active set A such that the process takes, for some $\varepsilon \in (0,1]$, at least $(1-\varepsilon)\ln(n)/\ln\ln n$ rounds. This is the case since a.a.s. $d_{\max}(G) = O(\log n)$ (similar to Sect. 4.2). Assuming this is the case, for every node v the number of nodes within distance at most t is at most $\left(O(\log n)\right)^t$. For $t = (1-\varepsilon)\ln(n)/\ln\ln n$, this number is smaller than $n/2$. Hence there is a set A of $n/2 > rn/\ln n$ nodes so that the distance between A and v exceeds t. Thus, when starting with A of active nodes, t rounds do not suffice to activate v.

The following remark implies this is the same number of rounds the perturbed percolation with $r = 1$ for $p = \Theta(\log(n)/n)$ takes when starting from an active set of constant size.

Remark 1 [18, *Theorem 4]*. Let $n \in \mathbb{N}_{>0}$ and $p = \Theta(\log(n)/n)$. Then a.a.s., $G_{n,p}$ has a diameter of $\Theta(\log(n)/\log\log n)$.

5 Experimental Results

In this section, we provide empirical results on the polynomial-to-rapid activation rate both on the graphs analyzed above, and on further global-graph models. Our findings are consistent with our theoretical results as well as the expected behavior of the perturbed percolation process on such graph models. The Python implementation uses the libraries NetworKit [36] and igraph [21], collections of tools for generating and analyzing graphs. In particular, they provide implementations for several random graph models. All experiments were run on a machine with 4 Intel i7-7500U cores and 8 GB RAM. However, note that we are not concerned with wall clock times, and all experiments were finished within minutes.

5.1 Erdős–Rényi Graphs

Corollary 2 shows that a.a.s. there is a polynomial-to-rapid activation rate for a PNG local graph combined with a $G_{n,p}$, for some parameter range. In Fig. 1, we consider such configurations satisfying these conditions, in particular, with a two-dimensional torus on $n = 10^6$ nodes as local graph. All runs are on the same random $G_{n,p}$ with $p = 20\ln(n)/n$, as it is generated once for consistent comparison. One can see the linear increase of the number of activations per round on the local graph. After 500 rounds, the number of new active nodes per round starts decreasing, as the set of active nodes wraps around the torus.

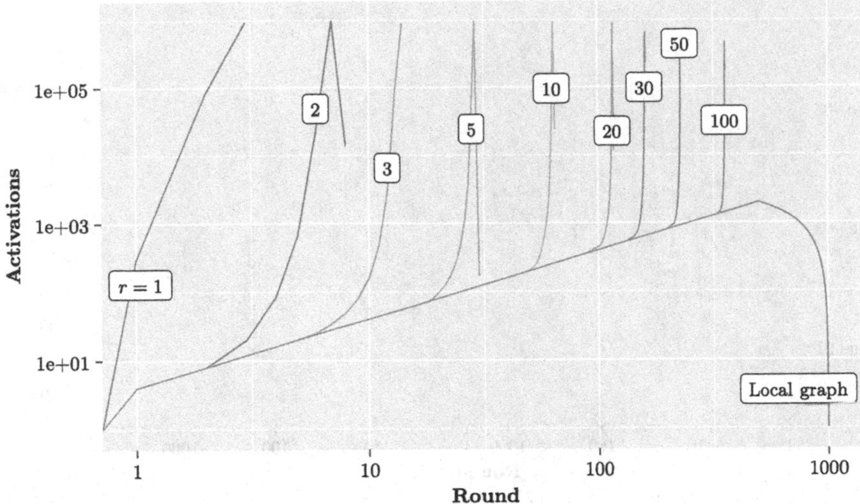

Fig. 1. The number of new active nodes in every round for different configurations. All runs are on $n = 10^6$ nodes with one initially active node. In one run ("Local graph"), we only consider the 2-dimensional torus on n nodes, while all other runs are on both the torus and a $G_{n,p}$ with $p = 20\ln(n)/n$ and a threshold r on the $G_{n,p}$. Note that both axes are logarithmic.

With the introduction of the $G_{n,p}$ with $r = 1$, the process completely percolates within three rounds, reflecting the rapid percolation. However, as r is increased, the effect of the global graph is withheld until some number of nodes are activated in the polynomial phase. Only then does the change to a rapid rate arise, and the process quickly percolates within few rounds, driven by the global edges. Even at $r = 100$, this effect is still observed.

5.2 Other Global-Graph Models

While our results only apply to the $G_{n,p}$ and expander graphs as global graph, we have strong reason to believe the same behavior can be observed for other global-graph models. We focus on two such models: (1) The Barabási-Albert (BA) model [2] uses a preferential-attachment approach, where nodes are iteratively added and connect to a fixed number of previous nodes proportional to their degree. (2) The hyperbolic random graph (HRG) model [33] randomly places nodes in a hyperbolic disk according to some probability distribution, and connects them if and only if they are close to each other. Both models exhibit small diameter and a power-law degree distribution, which should be beneficial for fast percolation. However, due to the underlying geometry, the HRG model has a large clustering coefficient, i.e., the neighbors of a node are likely to be neighbors of each other. We expect this feature to further accelerate the process, as this makes global edges more likely to hit the same nodes.

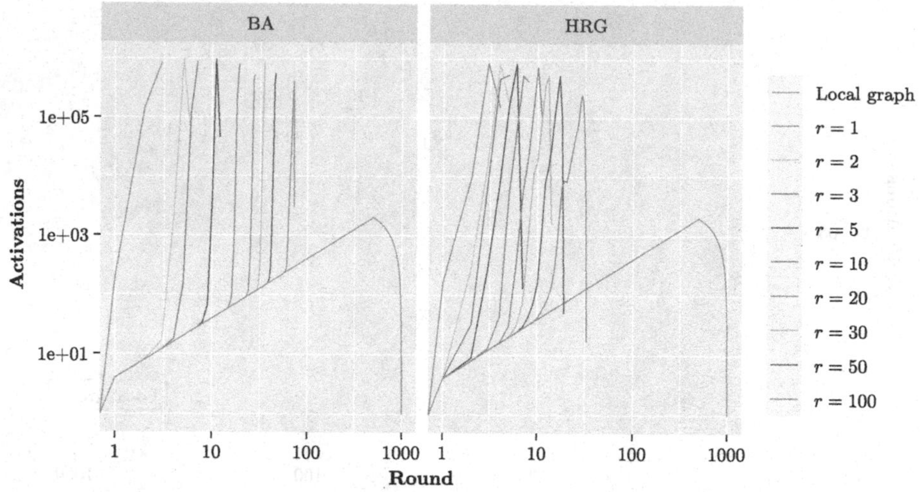

Fig. 2. The number of new active nodes in every round for different configurations. All runs are on $n = 10^6$ nodes with one initially active node. In one run ("Local graph"), we only consider the 2-dimensional torus on n nodes, while all other runs are on both the torus and a Barabási–Albert random graph (left), or hyperbolic random graph (right). Note that both axes are logarithmic.

The experiment setup is analogous to that described in Sect. 5.1, with the two-dimensional torus on n nodes as local graph. For the BA model, the number of attachments is chosen such that the expected average degree is $20 \ln n$. For the HRG model, we configure an expected power-law exponent of 3, and an expected average degree of $20 \ln n$. We consider the threshold model, i.e., a temperature of $T = 0$.

Our results of one run are depicted in Fig. 2. Again, the process for $r = 1$ reflects the rapid percolation, and for increasing r, the effect of the global edges is delayed until a threshold is reached. However, this threshold is reached earlier than in the $G_{n,p}$ version. For example, for $r = 100$, the $G_{n,p}$ has no activation by global edges until round 328, while with the BA model, this first happens in round 57. This can be explained by high-degree nodes in the BA model being more probable to reach the threshold r quickly.

For the HRG model, this effect is even stronger, with the first activation through global edges occurring in round 10 for $r = 100$. Even though both the BA and the HRG graph share the average degree and power-law exponent, the HRG graph has more high-degree nodes, in particular, those close to the center of the disk. Such nodes turn active very early, and then (through their high degree and high clustering) quickly activate the remaining nodes.

6 Outlook

With Lemma 1 and Theorems 6 and 8, we have shown bounds on the length of the initial (polynomial) and final (rapid) phase. It would be interesting to further analyze and tighten this gap. Our experiments (see Fig. 1) suggest that the transition is rather sharp once the first global activation occurs. Additionally, our experiments suggest that this behavior is very similar for other global-graph models, although we believe that the polynomial phase might be much shorter in the presence of a heavy-tailed degree distribution. Rigorously proving activation rates on such graph models would increase our understanding even further.

Acknowledgments. We thank David Peleg for suggesting this research direction and for multiple discussions, and Noga Alon for suggesting the proof of Theorem 8. This project has received funding from the European Union's Horizon 2020 research and innovation program under the Marie Skłodowska-Curie grant agreement No. 945298-ParisRegionFP.

References

1. Ajazi, F., Napolitano, G.M., Turova, T.: Phase transition in random distance graphs on the torus. J. Appl. Probabil. 1278–1294 (2017)
2. Albert, R., Barabási, A.-L.: Statistical mechanics of complex networks. Rev. Mod. Phys. **74**, 47–97 (2002)
3. Alon, N., Spencer, J.H.: The Probabilistic Method, 4th edn. Wiley, Hoboken (2016)
4. Abdullah, M.A., Fountoulakis, N.: A phase transition in the evolution of bootstrap percolation processes on preferential attachment graphs. Random Struct. Algorithms **52**(3), 379–418 (2018)
5. Amini, H., Fountoulakis, N.: Bootstrap percolation in power-law random graphs. J. Stat. Phys. **155**(1), 72–92 (2014)
6. Ball, F.: Stochastic and deterministic models for SIS epidemics among a population partitioned into households. Math. Biosci. **156**(1–2), 41–67 (1999)
7. Ball, F., Neal, P.: A general model for stochastic SIR epidemics with two levels of mixing. Math. Biosci. **180**(1–2), 73–102 (2002)
8. Ball, F., Neal, P.: Network epidemic models with two levels of mixing. Math. Biosci. **212**(1), 69–87 (2008)
9. Balogh, J., Bollobás, B.: Bootstrap percolation on the hypercube. Probab. Theory Relat. Fields **134**, 624–648 (2012)
10. Balogh, J., Bollobás, B., Duminil-Copin, H., Morris, R.: The sharp threshold for bootstrap percolation in all dimensions. Trans. Am. Math. Soc. **364**(5), 2667–2701 (2012)
11. Balogh, J., Pittel, B.G.: Bootstrap percolation on the random regular graph. Random Struct. Algorithms **30**(1–2), 257–286 (2007)
12. Bartal, A., Pliskin, N., Tsur, O.: Local/global contagion of viral/non-viral information: analysis of contagion spread in online social networks. PLoS ONE **15**(4), e0230811 (2020)
13. Bhansali, R., Schaposnik, L.P.: A trust model for spreading gossip in social networks: a multi-type bootstrap percolation model. Proc. Roy. Soc. A **476**(2235), 20190826 (2020)

14. Bradonjić, M., Saniee, I.: Bootstrap percolation on random geometric graphs. Probab. Eng. Inf. Sci. **28**(2), 169–181 (2014)
15. Candellero, E., Fountoulakis, N.: Bootstrap percolation and the geometry of complex networks. Stochast. Process. Appl. **126**(1), 234–264 (2016)
16. Centola, D.: The spread of behavior in an online social network experiment. Science **329**(5996), 1194–1197 (2010)
17. Chalupa, J., Leath, P.L., Reich, G.R.: Bootstrap percolation on a Bethe lattice. J. Phys. C Solid State Phys. **12**(1), L31–L35 (1979)
18. Chung, F., Lu, L.: The diameter of sparse random graphs. Adv. Appl. Math. **26**(4), 257–279 (2001)
19. Coja-Oghlan, A.: On the Laplacian eigenvalues of $G_{n,p}$. Comb. Probab. Comput. **16**(6), 923–946 (2007)
20. Coja-Oghlan, A., Feige, U., Krivelevich, M., Reichman, D.: Contagious sets in expanders. In: 26th Annual ACM-SIAM Symposium on Discrete Algorithms (SODA 2015), pp. 1953–1987. SIAM (2015)
21. Csardi, G., Nepusz, T.: The iGraph software package for complex network research. Int. J. Compl. Syst. **1695** (2006)
22. Doerr, B., Fouz, M., Friedrich, T.: Social networks spread rumors in sublogarithmic time. In: Proceedings of STOC, pp. 21–30 (2011)
23. Doerr, B., Fouz, M., Friedrich, T.: Why rumors spread so quickly in social networks. Commun. ACM **55**(6), 70–75 (2012)
24. Ebrahimi, R., Gao, J., Ghasemiesfeh, G., Schoenebeck, G.: Complex contagions in Kleinberg's small world model. In: 6th Conference on Innovations in Theoretical Computer Science (ITCS 2015), pp. 63–72 (2015)
25. Friedman, J.: On the second eigenvalue and random walks in random d-regular graphs. Combinatorica **11**, 331–362 (1991)
26. Gaffney, D.: #iranElection: Quantifying online activism. In: Proceedings of WebSci (2010)
27. Ghasemiesfeh, G., Ebrahimi, R., Gao, J.: Complex contagion and the weakness of long ties in social networks: revisited. In: 14th ACM Conference on Electronic Commerce (EC 2013), pp. 507–524 (2013)
28. González-Bailón, S., Borge-Holthoefer, J., Rivero, A., Moreno, Y.: The dynamics of protest recruitment through an online network. Sci. Rep. **1**(197), 1–7 (2011)
29. Hoory, S., Linial, N., Wigderson, A.: Expander graphs and their applications. Bull. Amer. Math. Soc. (N.S.) **43**(4), 439–561 (2006)
30. Jacquez, J.A., Simon, C.P., Koopman, J.: Structured mixing: heterogeneous mixing by the definition of activity groups. In: Castillo-Chavez, C. (ed.) Mathematical and Statistical Approaches to AIDS Epidemiology. LNB, vol. 83, pp. 301–315. Springer, Heidelberg (1989). https://doi.org/10.1007/978-3-642-93454-4_15
31. Janson, S., łuczak, T., Turova, T., Vallier, T.: Bootstrap percolation on the random graph $G_{n,p}$. Ann. Appl. Probabil. **22**(5), 1989–2047 (2012)
32. Koch, C., Lengler, J.: Bootstrap percolation on geometric inhomogeneous random graphs. In: 43rd International Colloquium on Automata, Languages, and Programming (ICALP 2016), pp. 147:1–147:15. Schloss Dagstuhl-Leibniz-Zentrum fuer Informatik (2016)
33. Krioukov, D., Papadopoulos, F., Kitsak, M., Vahdat, A., Boguñá, M.: Hyperbolic geometry of complex networks. Phys. Rev. E **82**, 036106 (2010)
34. Krivelevich, M., Reichman, D., Samotij, W.: Smoothed analysis on connected graphs. SIAM J. Discret. Math. **29**(3), 1654–1669 (2015)
35. Min, B., Miguel, M.S.: Competing contagion processes: complex contagion triggered by simple contagion. Sci. Rep. **8**(1), 1–8 (2018)

36. Staudt, C.L., Sazonovs, A., Meyerhenke, H.: NetworKit: a tool suite for large-scale complex network analysis (2015)
37. Turova, T.S., Vallier, T.: Bootstrap percolation on a graph with random and local connections. J. Stat. Phys. **160**(5), 1249–1276 (2015)

Phase Transition of the 3-Majority Dynamics with Uniform Communication Noise

Francesco d'Amore[1(✉)] and Isabella Ziccardi[2]

[1] Université Côte d'Azur, Inria, I3S, CNRS, Sophia Antipolis Cedex, France
francesco.d-amore@inria.fr
[2] Università degli Studi dell'Aquila, L'Aquila, Italy
isabella.ziccardi@graduate.univaq.it

Abstract. Communication noise is a common feature in several real-world scenarios where systems of agents need to communicate in order to pursue some collective task. In particular, many biologically inspired systems that try to achieve agreements on some opinion must implement resilient dynamics that are not strongly affected by noisy communications. In this work, we study the popular 3-MAJORITY dynamics, an opinion dynamics which has been proved to be an efficient protocol for the majority consensus problem, in which we introduce a simple feature of uniform communication noise, following (d'Amore et al. 2020). We prove that in the fully connected communication network of n agents and in the binary opinion case, the process induced by the 3-MAJORITY dynamics exhibits a phase transition. For a noise probability $p < 1/3$, the dynamics reaches in logarithmic time an almost-consensus metastable phase which lasts for a polynomial number of rounds with high probability. Furthermore, departing from previous analyses, we further characterize this phase by showing that there exists an attractive equilibrium value $s_{\text{eq}} \in [n]$ for the bias of the system, i.e. the difference between the majority community size and the minority one. Moreover, the agreement opinion turns out to be the initial majority one if the bias towards it is of magnitude $\Omega\left(\sqrt{n \log n}\right)$ in the initial configuration. If, instead, $p > 1/3$, no form of consensus is possible, and any information regarding the initial majority opinion is lost in logarithmic time with high probability. Despite more communications per-round are allowed, the 3-MAJORITY dynamics surprisingly turns out to be less resilient to noise than the UNDECIDED-STATE dynamics (d'Amore et al. 2020), whose noise threshold value is $p = 1/2$.

Keywords: Opinion dynamics · Consensus Problem · Communication Noise · Randomized Algorithms

M. Parter (Ed.): SIROCCO 2022, LNCS 13298, pp. 98–115, 2022.
https://doi.org/10.1007/978-3-031-09993-9_6

1 Introduction

The *consensus problem* is a fundamental problem in distributed computing [6] in which we have a system of agents supporting some opinions that interact between each other by exchanging messages, with the goal of reaching an agreement on some *valid* opinion (i.e. an opinion initially present in the system). In particular, many research papers focus on the *majority consensus problem* where the goal is to converge towards the initial majority opinion. The numerous theoretical studies in this area are justified by many different application scenarios, ranging from social networks [2,35], swarm robotics [5], cloud computing, communication networks [37], and distributed databases [19], to biological systems [23,24]. As for the latter, the goal of the majority consensus problem is to model some real-world scenarios where biological entities need to communicate and agree in order to pursue some collective task. Many biological entities in different real situations perform this type of process, e.g. molecules [12], bacteria [4], flock of birds [9], school of fish [38], or social insects [25], such as honeybees [36].

In such applicative scenarios, communication among agents is often affected by some form of noise. For this reason, one of the main goal in network information theory is to guarantee reliable communications in noisy networks [26]. In this context, error-correcting codes are very effective methods to reduce communication errors in computer systems [29,34], and this is why many theoretical studies of the (majority) consensus problem assume that communication between entities occurs without error, and instead consider some adversarial behavior (e.g., byzantine fault [8]). Despite their effectiveness in computer applications, error-correcting codes are quite useless if we want to model consensus in biological systems. Indeed, they involve sending complicated codes through communication links, and it is reasonable to assume that biological type entities communicate between each other in a simpler way. For this reason, in recent years many works have been focusing on the study of the (majority) consensus problem where the communication between entities is unreliable and subjected to uniform noise [16,17,23,24].

The first consensus dynamics that have been studied in the presence of noise communication are linear opinion dynamics, such as the VOTER dynamics and the AVERAGING dynamics. In particular, they were studied in the presence of uniform noise communication [31] or in the presence of some communities of *stubborn agents* (i.e. agents that never change opinion) [32,33,42]. In these settings, only *metastable* forms of consensus can be achieved, where a large subset of the agents agree on some opinion while other opinions remain supported by smaller subsets of agents, and this setting lasts for a relatively-long time. However, the VOTER model has a slow convergence time even in fully connected networks and a large initial bias towards some majority opinion [28], and the AVERAGING dynamics requires agents to perform non-trivial computation and, more importantly, to have large local memory. For these reasons, linear opinion dynamics struggle explaining the observed metastable consensus in multi-agent systems [11,15,22], and many research papers have begun to investigate new, more plausible, non-linear opinion dynamics.

To the best of our knowledge, the UNDECIDED-STATE dynamics is the first non-linear opinion dynamics analyzed in the presence of uniform communication noise [17]. It exhibits a phase-transition which depends on the noise parameter, and a metastable phase of almost-consensus is quickly reached and kept for long time when the noise isn't too high. It turns out to be a fast, very resilient dynamics, and this may explain why this type of process is adopted in some biological systems [36]. A description of the dynamics is given in Sect. 1.1.

In this work, we consider the popular 3-MAJORITY dynamics, which is based on majority update-rules, the latter being widely employed also in the biological research field [13,21]. In particular, we introduce in the system a uniform communication noise feature, following the definition of [17]. It has been proven that such dynamics, without communication noise, has a very similar behaviour to that of the UNDECIDED-STATE dynamics [6]. As we describe in the next section, the two dynamics behaves similarly (even if with crucial differences) even in presence of uniform noise, as both exhibit a phase transition. However, although 3-MAJORITY dynamics makes use of more per-round communications, it turns out to be less resilient to noise than UNDECIDED-STATE dynamics.

1.1 Our Results and Their Consequences

In this work, we study the 3-MAJORITY dynamics over a network of n agents, which induces a process that works as follows: at the beginning, each agent holds an opinion from a set Σ; at each subsequent discrete round, each agent pulls the opinions of three neighbor agents chosen independently uniformly at random and updates its opinion to the majority one, if there is any; otherwise, the agent adopts a random opinion among the sampled ones. This dynamics is a fast, robust protocol for the majority consensus problem in different network topologies (raging from complete graphs to sparser graphs) [6]. For a discussion about the origin and previous results of the 3-MAJORITY dynamics we defer the reader to Sect. 1.2.

We consider the dynamics in the binary opinion case over the fully connected network. We introduce in the process an *uniform communication noise* feature, following the definition in [17] and for which we give an equivalent formulation: for each communication with a sampled neighbor, there is probability $p \in (0, 1)$ that it is noisy, i.e. the received opinion is sampled u.a.r. between the possible opinions. Instead, with probability $1 - p$ the communication is unaffected by noise. As shown in [17], this noise model (over the complete network) is equivalent to a model without any communication noise and where two communities of stubborn agents (that is, they never change opinion) of equal size $\frac{pn}{2(1-p)}$ are present in the network, where each of the two community holds a different opinion. Even though the complete graph is a strong assumption for such communication networks, we remark that, at every round, an agent pulls an opinion from three neighbors: therefore, the round-per-round communication pattern results is a dynamic graph with $\mathcal{O}(n)$ edges. Furthermore, such a model can be used to capture the behavior of bio-inspired multi-agent systems in which mobile agents meet randomly at a relatively high rate. For more details about models for bio-inspired swarms of agents, we refer to [39].

In the aforementioned setting, we prove that the process induced by the 3-MAJORITY dynamics exhibits a phase-transition. Our results are summarized in the following theorem.

Theorem 1. *Let $\{s_t\}_{t\geq 0}$ be the bias of the process*[1] *induced by the* 3-MAJORITY *dynamics with uniform noise probability p. We prove the followings.*

- *If $p < 1/3$, let $s_0 = \Omega(\sqrt{n \log n})$ be the bias at the beginning of the process, $s_{eq} = \frac{n}{1-p}\sqrt{\frac{1-3p}{1-p}}$, and let $\varepsilon > 0$ be any sufficiently small constant. Then, there exists a time $\tau_1 = \mathcal{O}(\log n)$ such that, w.h.p.,*[2] *the process at time τ_1 reaches a metastable almost-consensus phase characterized by the equilibrium point s_{eq}, i.e.*
$$s_{\tau_1} \in [(1-\varepsilon)s_{eq}, (1+\varepsilon)s_{eq}].$$
Moreover, the bias is confined in such interval for $n^{\Theta(1)}$ rounds w.h.p.
- *If $p < 1/3$, let $s_0 = \mathcal{O}(\sqrt{n \log n})$ be the bias at the beginning of the process. Then, there exists a time $\tau_2 = \mathcal{O}(\log n)$ such that, w.h.p., the system becomes unbalanced towards an opinion, i.e.*
$$|s_{\tau_2}| = \Omega(\sqrt{n \log n}).$$
- *If $p > 1/3$, let $s_0 = \Omega(\sqrt{n \log n})$ be the bias at the beginning of the process. Then, there exists a time $\tau_3 = \mathcal{O}(\log n)$ such that, w.h.p. , at time τ_3 the majority opinion is lost, i.e. $s_{\tau_3} = \mathcal{O}(\sqrt{n})$. In addition, with constant probability, at time $\tau_3 + 1$ the majority opinion changes. Moreover, for $n^{\Theta(1)}$ additional rounds the absolute value of the bias is $\mathcal{O}(\sqrt{n \log n})$ w.h.p.*

Our result shows that 3-MAJORITY dynamics is less resilient to noise than the UNDECIDED-STATE dynamics, despite in the 3-MAJORITY dynamics more communication per-round are allowed. Indeed, the phase transition for the UNDECIDED-STATE dynamics turns out to be at the threshold $p = 1/2$ [17],[3] in the same setting as ours: since the threshold is higher than 1/3, the dynamics is able to solve the consensus problem even in the presence of more noise than the 3-MAJORITY dynamics.

We briefly recall the UNDECIDED-STATE dynamics: at each round, each agent pulls a single neighbor opinion x u.a.r. If the agent former opinion y is different from x, the agent becomes *undecided.* If the agent is undecided, then it simply adopts any opinion it sees. This two-phases update-rule turns out to be more resilient to noise and, hence, a swarm of agents would benefit from it. In [17],

[1] The bias s_t is the difference between the majority opinion community size and the minority opinion one at time t.

[2] An event holds *with high probability* (w.h.p. in short) with respect to n if the probability it occurs is at least $1 - n^{-\Theta(1)}$.

[3] In the cited work, an equivalent definition of noise model is given, and their formulation yields the threshold $p = 1/6$.

the authors prove that the dynamics exhibits a similar phase transition for the noise probability $p = 1/2$. Below the threshold, the dynamics w.h.p. rapidly breaks the symmetry and converges in logarithmic time to a metastable phase of almost-consensus that lasts for polynomial time, in which the majority opinion exceeds the minority one by a bias of order of $\Theta(n)$. Above the threshold, no form of consensus is possible, since the bias keeps bounded by $\mathcal{O}\left(\sqrt{n \log n}\right)$ for a polynomial number of rounds, w.h.p.

Nevertheless, we remark that our work shows technical novelties compared to [17]. A first difference lies in the fact that we find a precise equilibrium value s_{eq} that is attractive for the bias. Secondly, we characterize in detail what happens in the metastable almost-consensus phase: for every arbitrary small value $\varepsilon > 0$, we prove that the bias is confined in the interval $[(1-\varepsilon)s_{\mathrm{eq}}, (1+\varepsilon)s_{\mathrm{eq}}]$ for polynomial time w.h.p. Instead, in [17] no precise equilibrium value is found, and in the metastable-almost consensus phase the bias lies in an interval of width $\Theta(n)$, without arbitrarily approaching an equilibrium state; nevertheless, we remark that we think the UNDECIDED-STATE process should behave in such a way.

On the other hand, when the noise probability is above the threshold 1/3, we prove that no form of consensus is possible w.h.p. as in [17], but we also show that the majority opinion switches every $\mathcal{O}(\log n)$ rounds with constant probability. In order to prove this, some drift analysis results with super-martingale arguments are used [30].

As future directions, sparser topologies are worth to be investigated. We believe that, as long as the communication graph shows strong connection properties, similar phase transitions will be exhibited. Furthermore, it would be interesting to see whether the 3-MAJORITY dynamics with an arbitrary number of possible opinions, with the same noise model, has the exact same phase transition at the noise threshold value $p = 1/3$: in general, this corresponds to the fact that, for each node and at each round, exactly one communication among the three ones is noisy in expectation.

1.2 Related Works

Origin of the 3-MAJORITY *Dynamics.* The study of the 3-MAJORITY dynamics arises on the ground of the results obtained for the MEDIAN dynamics in [20]. The MEDIAN dynamics considers a totally ordered opinion set, in which each agent pulls two neighbor opinions i, j u.a.r. and then updates its opinion k to the median between i, j, and k. The dynamics turns out to be a fault-taulerant, efficient dynamics for the majority consensus problem. However, as pointed out in [6], the MEDIAN dynamics may not guarantee with high probability convergence to a valid opinion in case of the presence of an adversary, which is needed for the consensus problem. Moreover, the opinion set must have an ordering, property that might not be met by applicative scenarios such as biological systems [6]. These facts naturally lead researchers to look for efficient dynamics that satisfy the above requirements.

To the best of our knowledge, [1] is the first work analyzing the h-MAJORITY dynamics. In detail, in the h-MAJORITY dynamics we have n nodes and, at every round, every node pulls the opinion from h random neighbors and sets his new opinion to the majority one (ties are broken arbitrarily). More extensive characterizations of the 3-MAJORITY dynamics over the complete graph are given in [7,8,10,27]. We defer the reader to the full version for further related works [18].

Other Popular Non-linear Opinion Dynamics. Other important and efficient opinion dynamics for the majority consensus problem are the 2-CHOICES and the UNDECIDED-STATE dynamics. For an overview on the state of the art about opinion dynamics we defer the reader to [6]. We just want to quickly give the definitions of the 2-CHOICES dynamics (the UNDECIDED-STATE was already defined in the previous subsection). In the 2-CHOICES, each agent samples two neighbors u.a.r. and updates its opinion to the majority opinion among its former opinion and the two sampled neighbor opinions if there is any. Otherwise, it keeps its opinion. We just want to remark that the expected per-round behaviors of the 2-CHOICES dynamics and that of the 3-MAJORITY are the same, while the actual behaviors differ substantially in high probability [10]. This is why mean-field arguments are sometimes not sufficient to analyze such processes. For example, we have ran simple experiments that suggest that our uniform noise model on the 2-CHOICES dynamics yields a threshold noise value $p = 1/2$, just like the UNDECIDED-STATE dynamics.

As the 2-CHOICES and the 3-MAJORITY dynamics, the UNDECIDED-STATE dynamics turns out to be an efficient majority consensus protocol, with the difference that it requires only one communication per round for each agent. Further description is given in the previous section. It is worth mentioning the more recent work [3], which analyzes a variant of the UNDECIDED-STATE dynamics in the many-color case starting from any initial configuration.

Consensus Dynamics in the Presence of Noise or Stubborn Agents. The authors of [41] initiate the study of the consensus problem in the presence of communication noise. They consider the Vicsek model [40], in which they introduce a noise feature and a notion of robust consensus. Subsequently, dynamics for the consensus problem with noisy communications have received considerable attention. In particular, as mentioned in the introduction, this direction is motivated, among many reasons, by the desire to find models for the consensus problem in natural phenomena [23].

The communication noise studied in this type of problem can be divided in two types: uniform (or unbiased) and non-uniform (or biased). The uniform case wants to capture errors in communications between agents in real-world scenarios. The non-uniform communication noise instead describes the case in which agents have a preferred opinion. The authors of [23] are the first to explicitly focus on the uniform noise model. In detail, they study the broadcast and the majority consensus problem when the opinion set is binary. In their model of noise, every bit in every exchanged message is flipped independently with some probability smaller than 1/2. As a result, the authors give natural protocols that solve the aforementioned problems efficiently. The work [24] generalizes the above study to opinion sets of any cardinality.

As for the non-uniform communication noise case, in [16] it is considered the h-MAJORITY dynamics with a binary opinion set {ALPHA, BETA}, with a probability p that any received message is flipped towards a *fixed* preferred opinion, say BETA, while with probability $1-p$ the former message keeps intact. They suppose there is an initial majority agreeing on ALPHA, and they analyze the *time of disruption*, that is the time the initial majority is subverted. They prove there exists a threshold value $p^\star$ (which depends on h), such that 1) if $p < p^\star$, the time of disruption is at least polynomial, w.h.p., and 2) if $p > p^\star$, the time of disruption is constant, w.h.p. Their result holds for any sufficiently dense graph. We remark that our work differs from [16] in that there is no preferred opinion, and the noise affecting communications may result in any possible opinion.

The noise feature affecting opinion dynamics has been shown to be equivalent to a model without noise, in which communities of stubborn agents (i.e., they never change opinion) are added to the network [17]. For a discussion on related works considering such a model, we defer the reader to the full version of this work [18].

1.3 Structure of the Paper

The next section contains the preliminaries for the analysis and the result statements. Section 3 is devoted to the statements of the main theorems. In Sect. 4 we prove the theorems. Finally, for missing proofs and for the probabilistic results we use, we defer the reader to the full version of this work [18].

2 Preliminaries

The 3-Majority Dynamics. Let $G = (V, E)$ be a finite graph of n nodes (the agents), where each node is labelled uniquely with labels in $[n] := \{1, \ldots, n\}$. Furthermore, each node supports an opinion from a set of opinions Σ. The 3-MAJORITY dynamics defines a stochastic process $\{M_t\}_{t\in\mathbb{N}}$ which is described by the opinion of the nodes at each time step, i.e. $M_t = (i_1(t), \ldots, i_n(t)) \in \Sigma^n$ for every $t \geq 0$, where $i_j(t)$ is the opinion of node j at time t. The transition probabilities are characterized iteratively by the majority update rule as follows: given any time $t \geq 0$, let $M_t \in \Sigma^n$ be the state of the process at time t. Then, at time $t+1$, each node $u \in V$ samples three neighbors in G independently uniformly at random (with repetition) and updates its opinion to the majority one among the sampled neighbor opinions, if there is any. Otherwise, it adopts a random opinion among the sampled ones. For the sake of clarity, we remark that when u samples a neighbor node twice, the corresponding opinion counts twice.

Since M_t depends only on M_{t-1}, it follows that the process is a Markov chain. In the following, we will call the state of the process also by *configuration of the graph*.

The Communication Noise. We introduce an uniform communication noise feature in the dynamics, which is equivalent to that in [17]. Let $0 < p < 1$ be a constant. When a node pulls a neighbor opinion, there is probability p that the received opinion is sampled u.a.r. in Σ; instead, with probability $1 - p$, the former opinion keeps intact and is received.

3-MAJORITY *Dynamics in the Binary Opinion Case.* The communication network we focus on is the complete graph $G = K_n$ with self loops in the binary opinion case, i.e. $\Sigma = \{\text{ALPHA}, \text{BETA}\}$. For the symmetry of the network, the state of the process is fully characterized by the number of nodes supporting a given opinion, which implies that the nodes do not require unique IDs. Hence, we can write $M_t = (a_t, b_t)$, where a_t is the number of the nodes supporting opinion ALPHA at time t, and b_t is the analogous for opinion BETA. Moreover, since at each time t, $a_t + b_t = n$, it suffices to know $\{b_t\}_{t \geq 0}$ to fully describe the process.

We define the bias of the process at time t by

$$s_t = b_t - a_t = 2b_t - n, \tag{1}$$

which takes value in $\{-n, \dots, n\}$, and we notice that the process can also be characterized by the values of the bias alone, i.e. $\{s_t\}_{t \geq 0}$. We will use the latter sequence to refer to the process. We remark that $s_t > 0$ if the majority opinion at time t is BETA and $s_t < 0$ if it is ALPHA. We say that configurations having bias $s_t \in \{n, -n\}$ are monochromatic, meaning that every node supports the same opinion, while a configuration with $s_t = 0$ is symmetric. In the introduction, we took the bias to be $|s_t|$ but, for the sake of the analysis, we consider its *signed* version here. We finally remark that the random variable b_t (and, analogously, a_t) is the sum of i.i.d. Bernoulli r.v.s, which allows us to make use of the popular Chernoff bounds. In detail, if $X_i^{(t)}$ is the r.v. yielding 1 if node i adopts opinion BETA at round $t + 1$, and 0 otherwise, then $b_t = \sum_{i \in [n]} X_i^{(t)}$. Therefore, for (1),

$$s_t = 2 \sum_{i \in [n]} X_i^{(t)} - n = \sum_{i \in [n]} (2X_i^{(t)} - 1), \tag{2}$$

where $(X_i^{(t)} - 1)$ are i.i.d. taking values in $\{-1, 1\}$. For this reason, we can apply the Hoeffding bound to the bias.

Some Notation. For any function $f(n)$, we make use of the standard Landau notation $\mathcal{O}(f(n))$, $\Omega(f(n))$,$\Theta(f(n))$. Furthermore, for a constant $c > 0$, we write $\mathcal{O}_c(f(n))$,$\Omega_c(f(n))$, and $\Theta_c(f(n))$ if the hidden constant in the notation depends on c.

3 Results

We here show our three main theorems. The first one shows how the dynamics solves the majority consensus problem when $p < 1/3$, even if in a "weak" form (since only an almost-consensus is reached). Section 4.1 is devoted to the proof of this theorem.

Theorem 2 (Victory of the majority). *Let $\{s_t\}_{t\geq 0}$ be the process induced by the* 3-MAJORITY *dynamics with uniform noise probability $p < 1/3$. Let $\varepsilon > 0$ be any arbitrarily small constant (such that $\varepsilon < 1/3$ and $\varepsilon^2 \leq (1-3p)/2$) and let $\gamma > 0$ be any constant. Let $s_{eq} = \frac{n}{(1-p)}\sqrt{\frac{1-3p}{1-p}}$. Then, for any starting configuration s_0 such that $s_0 \geq \gamma\sqrt{n \log n}$ and for any sufficiently large n, the following holds w.h.p. :*

(i) there exists a time $\tau_1 = \mathcal{O}_{\gamma,\varepsilon,p}(\log n)$ such that $(1-\varepsilon)s_{eq} \leq s_{\tau_1} \leq (1+\varepsilon)s_{eq}$;
(ii) there exists a value $c = \Theta_{\gamma,\varepsilon,p}(1)$ such that, for all $k \leq n^c$, $(1-\varepsilon)s_{eq} \leq s_{\tau_1+k} \leq (1+\varepsilon)s_{eq}$.

Our second theorem shows how the dynamics is capable of quickly breaking the initial symmetry. By applying also Theorem 2, it shows that the consensus problem is solved. The proof of the theorem is shown in Sect. 4.2.

Theorem 3 (Symmetry breaking). *Let $\{s_t\}_{t\geq 0}$ be the process induced by the* 3-MAJORITY *dynamics with uniform noise probability $p < 1/3$, and let $\gamma > 0$ be any positive constant. Then, for any starting configuration s_0 such that $|s_0| \leq \gamma\sqrt{n \log n}$ and for any sufficiently large n, w.h.p. there exists a time $\tau_2 = \mathcal{O}_{\gamma,p}(\log n)$ such that $|s_{\tau_2}| \geq \gamma\sqrt{n \log n}$.*

Our last theorem shows that no form of consensus is possible when $p > 1/3$, and it is proved in Sect. 4.3.

Theorem 4 (Victory of noise). *Let $\{s_t\}_{t\geq 0}$ be the process induced by the* 3-MAJORITY *dynamics with uniform noise probability $p > 1/3$. Let $\varepsilon > 0$ be any arbitrarily small constant (such that $\varepsilon < \min\{1/4, (1-p), (3p-1)/2\}$) and let $\gamma > 0$ be any positive constant. Then, for any starting configuration s_0 such that $|s_0| \geq \gamma\sqrt{n \log n}$ and for any sufficiently large n, the following holds w.h.p.:*

(i) there exists a time $\tau_3 = \mathcal{O}_{\varepsilon,p}(\log n)$ such that $s_{\tau_3} = \mathcal{O}_{\varepsilon}(\sqrt{n})$ and, moreover, the majority opinion switches at the next round with probability $\Theta_{\varepsilon}(1)$;
(ii) there exists a value $c = \Theta_{\gamma,\varepsilon}(1)$ such that, for all $k \leq n^c$, it holds that $|s_{\tau_3+k}| \leq \gamma\sqrt{n \log n}$.

4 Analysis

In this section we analyze the process. We first give some preliminary results. Afterwards, in Sect. 4.1 we prove Theorem 2, in Sect. 4.2 we prove Theorem 3, while Sect. 4.3 is devoted to the proof Theorem 4.

We now give the expectation of the bias at time t, conditional on its value at time $t-1$.

Lemma 1. *Let $\{s_t\}_{t\geq 0}$ be the process induced by the* 3-MAJORITY *dynamics with uniform noise probability $p \in (0,1)$. The conditional expectation of the bias is*

$$\mathbb{E}\left[s_t \mid s_{t-1} = s\right] = \frac{s(1-p)}{2}\left(3 - \frac{s^2}{n^2}(1-p)^2\right). \tag{3}$$

The proof is omitted and can be found in the full version [18]. By the lemma above, we deduce that there are up to three equilibrium configurations in expectation. The first one corresponds to $s = 0$, and the other (possible) equilibrium correspond to the condition

$$\frac{1-p}{2}\left(3 - \frac{s^2}{n^2}(1-p)^2\right) = 1$$

The latter condition results in

$$s = \pm\frac{n}{(1-p)} \cdot \sqrt{\frac{3(1-p)-2}{(1-p)}} = \pm\frac{n}{(1-p)} \cdot \sqrt{\frac{1-3p}{1-p}},$$

which is well defined if only if $p \leq 1/3$. We will denote the absolute value of the latter two values by s_{eq}.

4.1 Victory of the Majority

The aim of this subsection is to prove Theorem 2: so, in each statement we assume that $\{s_t\}_{t\geq 0}$ is the process induced by the 3-MAJORITY dynamics with uniform noise probability $p < 1/3$.

We first show a lemma which states that, for any small constant $\varepsilon > 0$, whenever $s_{t-1} \notin [(1-\varepsilon)s_{\text{eq}}, (1+\varepsilon)s_{\text{eq}}]$, then s_t gets closer to the interval.

Lemma 2. *For any constant $\varepsilon > 0$ such that $\varepsilon^2 < (1-3p)/2$ and for any $\gamma > 0$, if $s \geq \gamma\sqrt{n \log n}$, the followings hold*

1. *if $s \leq (1-\varepsilon)s_{eq}$, then $\mathbb{P}\left[s_t \geq (1+3\varepsilon^2/4)s \mid s_{t-1} = s\right] \geq 1 - \frac{1}{n^{\gamma^2\varepsilon^4/32}}$;*
2. *if, $s \geq (1+\varepsilon)s_{eq}$, then $\mathbb{P}\left[s_t \leq (1-3\varepsilon^2/4)s \mid s_{t-1} = s\right] \geq 1 - \frac{1}{n^{\gamma^2\varepsilon^4/32}}$.*

Proof. We first notice that

$$(1-\varepsilon)s_{\text{eq}} \leq \frac{n}{1-p}\sqrt{\frac{1-3p-2\varepsilon^2}{1-p}}, \tag{4}$$

which holds since $\varepsilon^2 \leq (1-3p)/2$ and can be proved with simple calculations. For Lemma 1, if each $s \leq (1-\varepsilon)s_{\text{eq}}$, then

$$\begin{aligned}\mathbb{E}\left[s_t \mid s_{t-1} = s\right] &= \frac{s(1-p)}{2}\left(3 - \frac{s^2}{n^2}(1-p)^2\right) \\ &\geq s\left(\frac{3-3p}{2} - \frac{1-3p-2\varepsilon^2}{2}\right) = s(1+\varepsilon^2).\end{aligned}$$

where the inequality follows from (4). Since (2), for the Hoeffding bound, it holds that

$$\begin{aligned}&\mathbb{P}\left[s_t \leq s(1+\varepsilon^2) - s\varepsilon^2/4 \mid s_{t-1} = s\right] \leq e^{-s^2\varepsilon^4/(32n)} \\ &\leq e^{-\gamma^2\varepsilon^4 \log n/32} \leq \frac{1}{n^{\gamma^2\varepsilon^4/32}}.\end{aligned}$$

The second inequality in the lemma follows by a symmetric argument, observing that

$$(1+\varepsilon)s_{\text{eq}} \geq \frac{n}{1-p}\sqrt{\frac{1-3p+2\varepsilon^2}{1-p}},$$

for ε such that $\varepsilon^2 < (1-3p)/2$.

The following lemma serves to bound how far the bias can get from the interval $[(1+\varepsilon)s_{\text{eq}}, (1-\varepsilon)s_{\text{eq}}]$.

Lemma 3. *For any constants $\varepsilon > 0$ and $\gamma > 0$, if $s \geq \gamma\sqrt{n \log n}$, the followings hold*

1. *if $s \leq (1+\varepsilon)s_{eq}$, then $\mathbb{P}\left[s_t \geq (1-\varepsilon-\varepsilon^2)s \mid s_{t-1} = s\right] \geq 1 - \frac{1}{n^{\gamma^2\varepsilon^2/16}}$;*
2. *if $s \geq (1-\varepsilon)s_{eq}$ with $\varepsilon < 1$, then $\mathbb{P}\left[s_t \leq (1+\varepsilon)s \mid s_{t-1} = s\right] \geq 1 - \frac{1}{n^{\gamma^2\varepsilon^2 p^2}}$.*

The proof is similar to that of the previous lemma and can be found in the full version. We now provide another lemma to control the behavior of the bias. The proof consists again in the application of simple concentration bounds.

Lemma 4. *For any constant $k > 0$, the followings hold:*

1. *if $s \geq s_{eq}$, then $\mathbb{P}\left[s_t \geq 2s_{eq}/3 \mid s_{t-1} = s\right] \geq 1 - 1/n^k$.*
2. *if $0 \leq s \leq 2s_{eq}/3$, then $\mathbb{P}\left[s_t \leq s_{eq} \mid s_{t-1} = s\right] \geq 1 - 1/n^k$.*

We can piece together the above lemmas, which imply the following corollary, whose proof consists in many calculations and can be found in the full version.

Corollary 1. *For any constant $\varepsilon > 0$ such that $\varepsilon < 1/3$ and $\varepsilon^2 < (1-3p)/2$, the followings hold:*

1. *if $|s_{eq} - s| \leq (\varepsilon/4)s_{eq}$, then*

$$\mathbb{P}\left[|s_{eq} - s_t| \leq \varepsilon s_{eq} \mid s_{t-1} = s\right] \geq 1 - \frac{1}{n^{\gamma^2\varepsilon^2p^2/2^5}};$$

2. *if $(\varepsilon/4)s_{eq} \leq |s_{eq} - s| \leq s_{eq}/3$, then*

$$\mathbb{P}\left[|s_{eq} - s_t| \leq |s_{eq} - s| \cdot \left(1 - \frac{3\varepsilon^2}{2^5}\right) \,\middle|\, s_{t-1} = s\right] \geq 1 - \frac{1}{n^{\gamma^2\varepsilon^4p^2/(2^{18}3^2)}}.$$

We are finally ready to prove the theorem.

Proof (Proof of Theorem 2). We divide the proof in different cases. First, suppose that $(\varepsilon/4)s_{\text{eq}} \leq |s_{\text{eq}} - s| \leq \varepsilon s_{\text{eq}}$. Let $T_1 = n^{\gamma^2\varepsilon^4p^2/(2^{19}3^2)}$. Then, from Corollary 1.(i) and (ii), for the chain rule, we have that

$$\mathbb{P}\left[\bigcap_{k=1}^{T}\{|s_{\text{eq}} - s_{t+k}| \leq \varepsilon s_{\text{eq}}\} \,\middle|\, s_t = s\right] \geq 1 - \frac{1}{n^{\gamma^2\varepsilon^4p^2/(2^{20}3^2)}}.$$

Second, suppose that $\varepsilon s_{\text{eq}} \leq |s_{\text{eq}} - s| \leq s_{\text{eq}}/3$. Then, from Corollary 1.(ii), for the chain rule, a time T_2 exists, with

$$T_2 = \mathcal{O}\left(-\frac{\log n}{\log\left(1 - \frac{3\varepsilon^2}{2^5}\right)}\right) = \mathcal{O}\left(\log n/\varepsilon^2\right)$$

such that

$$\mathbb{P}\left[|s_{\text{eq}} - s_{t+T_2}| \leq \varepsilon s_{\text{eq}} \mid s_t = s\right] \geq 1 - \frac{1}{n^{\gamma^2\varepsilon^4 p^2/(2^{20}3^2)}}.$$

Third, suppose that $s \leq 2s_{\text{eq}}/3$. From Lemma 2.(i) and Lemma 4.(ii), for the chain rule and the union bound, there is a time

$$T_3 = \mathcal{O}\left(\frac{\log n}{\log\left(1 + \frac{3\varepsilon^2}{4}\right)}\right) = \mathcal{O}\left(\log n/\varepsilon^2\right)$$

such that

$$\mathbb{P}\left[2s_{\text{eq}}/3 \leq s_{t+T_3} \leq s_{\text{eq}} \mid s_t = s\right] \geq 1 - \frac{1}{n^{\gamma^2\varepsilon^4/2^6}}.$$

Then, we are in one of the first two cases, and we conclude for the chain rule. Fourth, suppose that $s \geq (1 + \frac{1}{3})s_{\text{eq}}$. From Lemma 2.(ii) and Lemma 4.(i), for the chain rule, a time T_4 exists, with $T_4 = \mathcal{O}(\log n)$, such that

$$\mathbb{P}\left[|s_{\text{eq}} - s_{T_4}| \leq s_{\text{eq}}/3 \mid s_t = s\right] \geq 1 - \frac{1}{n^{\gamma^2 3^4/2^6}}.$$

The theorem follows with $\tau_1 = \mathcal{O}(T_2 + T_3 + T_4)$.

4.2 Symmetry Breaking

The aim of this section is to prove Theorem 3: so, in each statement we assume that $\{s_t\}_{t\geq 0}$ is the process induced by the 3-Majority dynamics with uniform noise probability $p < 1/3$. The symmetry breaking analysis essentially relies on the following lemma which has been proved in [14].

Lemma 5. *Let $\{X_t\}_{t\in\mathbb{N}}$ be a Markov Chain with finite-state space Ω and let $f : \Omega \mapsto [0, n]$ be a function that maps states to integer values. Let c_3 be any positive constant and let $m = c_3\sqrt{n}\log n$ be a target value. Assume the following properties hold:*

1. *for any positive constant h, a positive constant $c_1 < 1$ (which depends only on h) exists, such that for any $x \in \Omega : f(x) < m$,*

$$\mathbb{P}\left[f(X_t) < h\sqrt{n} \mid X_{t-1} = x\right] < c_1;$$

2. *there exist two positive constants δ and c_2 such that for any $x \in \Omega : h\sqrt{n} \leq f(x) < m$,*

$$\mathbb{P}\left[f(X_t) < (1+\delta)f(X_{t-1}) \mid X_{t-1} = x\right] < e^{-c_2 f(x)^2/n}.$$

Then the process reaches a state x with $f(x) \geq m$ within $\mathcal{O}_{c_2,\delta,c_3}(\log n)$ rounds with probability at least $1 - 2/n$.

Our goal is yo apply the above lemma to the 3-MAJORITY process, which defines a Markov chain. In particular, we claim the hypothesis of Lemma 5 are satisfied when the bias of the system is $o\left(\sqrt{n \log n}\right)$, with $f(\boldsymbol{x}) = s\left(\boldsymbol{x}\right)$, $m = \gamma\sqrt{n}\log n$ for any constant $\gamma > 0$. Then, Lemma 5 implies the process reaches a configuration with bias greater than $\Omega\left(\sqrt{n \log n}\right)$ within time $\mathcal{O}\left(\log n\right)$, w.h.p. We need to prove that the two hypotheses hold.

Lemma 6. *For any constant $c_3 > 0$, let s be a value such that $|s| < c_3\sqrt{n}\log n$. Then,*

1. *for any positive constant $h > 0$, there exists a positive constant $c_1 < 1$ (which depends only on h), such that*

$$\mathbb{P}\left[s_t < h\sqrt{n} \;\middle|\; s_{t-1} = s\right] < c_1;$$

2. *two positive constants δ, c_2 exist (depending only on p), such that if $|s| \geq h\sqrt{n}$, then*

$$\mathbb{P}\left[s_t < (1+\delta)s \mid s_{t-1} = s\right] < e^{-\frac{c_2 s^2}{n}}.$$

The proof of this latter result is very interesting and makes use of the Berry-Essen inequality. We defer the reader [18]. The symmetry breaking is then a simple consequence of the above Lemma.

Proof. (Proof of Theorem 3). Apply Lemmas 6 and 5 with $h = c_3 = \gamma$.

4.3 Victory of Noise

In this subsection, we prove Theorem 4: so, in each statement, we assume that $\{s_t\}_{t\geq 0}$ is the process induced by the 3-MAJORITY dynamics with uniform noise probability $p > 1/3$.

We make use of tools from drift analysis to the absolute value of the bias of the process, showing that it reaches magnitude $\mathcal{O}\left(\sqrt{n}\right)$ quickly. Then, since the standard deviation of the bias is $\Theta\left(\sqrt{n}\right)$, we have constant probability that the majority opinion switches Lemma 9. Finally, with Lemma 10, we show that the bias keeps bounded in absolute value by $\mathcal{O}\left(\sqrt{n \log n}\right)$.

Lemma 7. *For any constant $\varepsilon > 0$ such that $\varepsilon < (1-p)$, if $s \geq 2\sqrt{n}/\left(\varepsilon^2\right)$, the following holds*

$$\mathbb{E}\left[|s_t| \mid s_{t-1} = s\right] \leq \mathbb{E}\left[s_t \mid s_{t-1} = s\right] \cdot \left(1 + \frac{\varepsilon}{2}\right).$$

The proof of this lemma can be found in [18] and makes use of estimation of the standard deviation of the bias to bound its expected absolute value.

With next lemma, we show that the absolute value of the process quickly becomes of magnitude $\mathcal{O}(\sqrt{n})$.

Lemma 8. *For any constant $\varepsilon > 0$ such that $\varepsilon < \min\{(1-p), (3p-1)/2\}$ we define $s_{min} = \sqrt{n}/\varepsilon^2$. Then, for any starting configuration s_0 such that $s_0 \geq s_{min}$, with probability at least $1 - 1/n$ there exists a time $\tau = \mathcal{O}_\varepsilon(\log n)$ such that $|s_\tau| \leq s_{min}$.*

Proof. Let $h(x) = \frac{\varepsilon \cdot x}{2}$ be a function. Let $X_t = |s_t|$ if $s_t \geq s_{\min}$, otherwise $X_t = 0$. We now estimate $\mathbb{E}[X_t - X_{t-1} \mid X_{t-1} \geq s_{\min}, \mathcal{F}_{t-1}]$, where $\mathcal{F}_t$ is the natural filtration of the process X_t. We have that

$$\begin{aligned}
&\mathbb{E}[X_t - X_{t-1} \mid X_{t-1} \geq s_{\min}, \mathcal{F}_{t-1}] \\
&= \mathbb{E}[X_t \mid X_{t-1} \geq s_{\min}, \mathcal{F}_{t-1}] - X_{t-1} \\
&\overset{(a)}{\leq} \mathbb{E}[|s_t| \mid s_{t-1} \geq s_{\min}, \mathcal{F}_{t-1}] - s_{t-1} \\
&\overset{(b)}{\leq} \mathbb{E}[s_t \mid s_{t-1} \geq s_{\min}, \mathcal{F}_{t-1}] \cdot \left(1 + \frac{\varepsilon}{2}\right) - s_{t-1} \\
&\overset{(c)}{\leq} s_{t-1}(1-\varepsilon)\left(1 + \frac{\varepsilon}{2}\right) - s_{t-1} \leq -\frac{\varepsilon \cdot s_{t-1}}{2},
\end{aligned}$$

where (a) holds because $X_t \leq |s_t|$, (b) holds for Lemma 7, and (c) holds for Lemma 1. Thus,

$$\mathbb{E}[X_{t-1} - X_t \mid X_{t-1} \geq s_{\min}, \mathcal{F}_{t-1}] \geq h(X_{t-1}).$$

Since $h'(x) = \varepsilon/2 > 0$, we can apply Corollary 3.(iii) in [30]. Let τ be the first time $X_t = 0$ or, equivalently, $|s_t| < s_{\min}$. Then

$$\begin{aligned}
\mathbb{P}[\tau > t \mid s_0] &< \exp\left[-\frac{\varepsilon}{2} \cdot \left(t - \frac{2}{\varepsilon} - \int_{s_{\min}}^{s_0} \frac{2}{\varepsilon \cdot y}\, \mathrm{d}y\right)\right] \\
&\leq \exp\left[-\frac{\varepsilon}{2} \cdot \left(t - \frac{2}{\varepsilon} - \int_{s_{\min}}^{n} \frac{2}{\varepsilon \cdot y}\, \mathrm{d}y\right)\right] \\
&= \exp\left[-\frac{\varepsilon}{2} \cdot \left(t - \frac{2}{\varepsilon} - \frac{2}{\varepsilon}(\log n - \log s_{\min})\right)\right] \\
&= \exp\left[-\frac{\varepsilon}{2} \cdot \left(t - \frac{2}{\varepsilon} - \frac{2}{\varepsilon}((\log n)/2 + 2\log \varepsilon)\right)\right] \\
&\leq \exp\left[-\frac{\varepsilon \cdot t}{2} + 1 + \frac{\log n}{2}\right].
\end{aligned}$$

If $t = 4(\log n)/\varepsilon$, then we get that $\mathbb{P}[\tau > t \mid s_0] < e^{-3(\log n)/2+1} < 1/n$.

Next lemma states that, whenever the absolute value of the bias is of order of $\mathcal{O}(\sqrt{n})$, then the majority opinion switches at the next round with constant probability. It is proved by applying the reverse Chernoff bound (an anti-concentration inequality).

Lemma 9. *For any constant $\varepsilon > 0$ such that $\varepsilon < 1/4$, and let s_{t-1} be a configuration such that $|s_{t-1}| = s \leq \sqrt{n}/\varepsilon$. Then, the majority opinion switches at the next round with constant probability.*

Next lemma shows that the signed bias decreases each round.

Lemma 10. *For any constant $\varepsilon > 0$ such that $\varepsilon \leq (3p-1)/2$, the followings hold*

1. *if $s \geq \frac{\gamma}{2}\sqrt{n\log n}$, then $\mathbb{P}[s_t \leq (1-3\varepsilon/4)s \mid s_{t-1} = s] \geq 1 - \frac{1}{n^{\gamma^2\varepsilon^2/2^7}}$;*
2. *if $s \geq 0$, then $\mathbb{P}\left[-\frac{\gamma}{2}\sqrt{n\log n} \leq s_t \leq s + \frac{\gamma}{2}\sqrt{n\log n} \mid s_{t-1} = s\right] \geq 1 - \frac{2}{n^{\gamma^2/8}}$.*

The proof is a simple application of Hoeffding bounds. We are ready to prove Theorem 4.

Proof (Proof of Theorem 4). Claim (i) follows directly from Lemmas 8 and 9. As for claim (ii), whenever the bias at some round $t = \tau + k$ becomes $|s_t| \geq (\gamma/2)\sqrt{n\log n}$, from Lemma 10.(ii) (and its symmetric statement), we have that $|s_t| \leq \gamma\sqrt{n\log n}$ with probability $1 - 2/n^{\frac{\gamma^2}{8}}$. Then, from Lemma 10.(i) it follows that the bias starts decreasing each round with probability $1 - 1/n^{\gamma^2\varepsilon^2/2^7}$ until reaching $(\gamma/2)\sqrt{\log n}$. This phase in which the absolute value of the bias keeps bounded by $|\gamma\sqrt{n\log n}|$ lasts for at least $n^{\gamma^2\varepsilon^2/2^8}$ with probability at least $1 - 1/(2n^{\gamma^2\varepsilon^2/2^8})$ for the chain rule.

Acknowledgement. We thank professor Andrea Clementi for the insightful discussions and advice.

References

1. Abdullah, M.A., Draief, M.: Global majority consensus by local majority polling on graphs of a given degree sequence. Discret. Appl. Math. **180**, 1–10 (2015). https://doi.org/10.1016/j.dam.2014.07.026
2. Acemoglu, D., Como, G., Fagnani, F., Ozdaglar, A.E.: Opinion fluctuations and disagreement in social networks. Math. Oper. Res. **38**(1), 1–27 (2013). https://doi.org/10.1287/moor.1120.0570
3. Bankhamer, G., et al.: Fast consensus via the unconstrained undecided state dynamics. In: Naor, J.S., Buchbinder, N. (eds.) Proceedings of the 2022 ACM-SIAM Symposium on Discrete Algorithms, SODA 2022. SIAM (2022). https://doi.org/10.1137/1.9781611977073.135
4. Bassler, B.L.: Small talk: cell-to-cell communication in bacteria. Cell **109**(4) (2002). https://doi.org/10.1016/S0092-8674(02)00749-3
5. Bayindir, L.: A review of swarm robotics tasks. Neurocomputing **172**, 292–321 (2016). https://doi.org/10.1016/j.neucom.2015.05.116
6. Becchetti, L., Clementi, A.E.F., Natale, E.: Consensus dynamics: an overview. SIGACT News **51**(1) (2020). https://doi.org/10.1145/3388392.3388403
7. Becchetti, L., Clementi, A., Natale, E., Pasquale, F., Silvestri, R., Trevisan, L.: Simple dynamics for plurality consensus. Distrib. Comput. **30**(4), 293–306 (2016). https://doi.org/10.1007/s00446-016-0289-4

8. Becchetti, L., Clementi, A.E.F., Natale, E., Pasquale, F., Trevisan, L.: Stabilizing consensus with many opinions. In: Krauthgamer, R. (ed.) Proceedings of the Twenty-Seventh Annual ACM-SIAM Symposium on Discrete Algorithms, SODA 2016, Arlington, VA, USA, 10–12 January 2016, pp. 620–635. SIAM (2016). https://doi.org/10.1137/1.9781611974331.ch46
9. Ben-Shahar, O., Dolev, S., Dolgin, A., Segal, M.: Direction election in flocking swarms. Ad Hoc Netw. **12**, 250–258 (2014). https://doi.org/10.1016/j.adhoc.2012.05.001
10. Berenbrink, P., Clementi, A.E.F., Elsässer, R., Kling, P., Mallmann-Trenn, F., Natale, E.: Ignore or comply?: on breaking symmetry in consensus. In: Schiller, E.M., Schwarzmann, A.A. (eds.) Proceedings of the ACM Symposium on Principles of Distributed Computing, PODC. ACM (2017). https://doi.org/10.1145/3087801.3087817
11. Boczkowski, L., Korman, A., Natale, E.: Minimizing message size in stochastic communication patterns: fast self-stabilizing protocols with 3 bits. Distrib. Comput. **32**(3), 173–191 (2018). https://doi.org/10.1007/s00446-018-0330-x
12. Carroll, M.C.: The complement system in regulation of adaptive immunity. Nat. Immunol. **5**, 981–986 (2004). https://doi.org/10.1038/ni1113
13. Chaouiya, C., Ourrad, O., Lima, R.: Majority rules with random tie-breaking in Boolean gene regulatory networks. PLOS ONE **8**(7), 1–14 (2013). https://doi.org/10.1371/journal.pone.0069626
14. Clementi, A.E.F., Ghaffari, M., Gualà, L., Natale, E., Pasquale, F., Scornavacca, G.: A tight analysis of the parallel undecided-state dynamics with two colors. In: Potapov, I., Spirakis, P.G., Worrell, J. (eds.) 43rd International Symposium on Mathematical Foundations of Computer Science, MFCS 2018, Liverpool, UK. LIPIcs, vol. 117, pp. 28:1–28:15. Schloss Dagstuhl - Leibniz-Zentrum für Informatik (2018). https://doi.org/10.4230/LIPIcs.MFCS.2018.28
15. Condon, A., Hajiaghayi, M., Kirkpatrick, D., Maňuch, J.: Approximate majority analyses using tri-molecular chemical reaction networks. Nat. Comput. **19**(1), 249–270 (2019). https://doi.org/10.1007/s11047-019-09756-4
16. Cruciani, E., Mimun, H.A., Quattropani, M., Rizzo, S.: Phase transitions of the k-majority dynamics in a biased communication model. In: ICDCN 2021: International Conference on Distributed Computing and Networking. ACM (2021). https://doi.org/10.1145/3427796.3427811
17. d'Amore, F., Clementi, A., Natale, E.: Phase transition of a non-linear opinion dynamics with noisy interactions. In: Richa, A.W., Scheideler, C. (eds.) SIROCCO 2020. LNCS, vol. 12156, pp. 255–272. Springer, Cham (2020). https://doi.org/10.1007/978-3-030-54921-3_15
18. D'Amore, F., Ziccardi, I.: Phase transition of the 3-majority dynamics with uniform communication noise. CoRR (2021). https://arxiv.org/abs/2112.03543
19. Dietzfelbinger, M., Goerdt, A., Mitzenmacher, M., Montanari, A., Pagh, R., Rink, M.: Tight thresholds for cuckoo hashing via XORSAT. In: Abramsky, S., Gavoille, C., Kirchner, C., Meyer auf der Heide, F., Spirakis, P.G. (eds.) ICALP 2010. LNCS, vol. 6198, pp. 213–225. Springer, Heidelberg (2010). https://doi.org/10.1007/978-3-642-14165-2_19
20. Doerr, B., Goldberg, L.A., Minder, L., Sauerwald, T., Scheideler, C.: Stabilizing consensus with the power of two choices. In: Rajaraman, R., auf der Heide, F.M. (eds.) SPAA 2011: Proceedings of the 23rd Annual ACM Symposium on Parallelism in Algorithms and Architectures. ACM (2011). https://doi.org/10.1145/1989493.1989516

21. Dong, J., Fernández-Baca, D., McMorris, F., Powers, R.C.: Majority-rule (+) consensus trees. Math. Biosci. **228**(1), 10–15 (2010). https://doi.org/10.1016/j.mbs.2010.08.002
22. Emanuele Natale: On the Computational Power of Simple Dynamics. Ph.D. Thesis, Sapienza University of Rome (2017)
23. Feinerman, O., Haeupler, B., Korman, A.: Breathe before speaking: efficient information dissemination despite noisy, limited and anonymous communication. Distrib. Comput. **30**(5), 339–355 (2015). https://doi.org/10.1007/s00446-015-0249-4
24. Fraigniaud, P., Natale, E.: Noisy rumor spreading and plurality consensus. Distrib. Comput. **32**(4), 257–276 (2018). https://doi.org/10.1007/s00446-018-0335-5
25. Franks, N., Pratt, S., Mallon, E., Britton, N., Sumpter, D.: Information flow, opinion polling and collective intelligence in house-hunting social insects. Philos. Trans. Roy. Soc. Lond. Ser. B Biol. Sci. **357**, 1567–83 (2002). https://doi.org/10.1098/rstb.2002.1066
26. Gamal, A.E., Kim, Y.: Cambridge University Press. Network Information Theory (2011). https://doi.org/10.1017/CBO9781139030687
27. Ghaffari, M., Lengler, J.: Nearly-tight analysis for 2-choice and 3-majority consensus dynamics. In: Newport, C., Keidar, I. (eds.) Proceedings of the 2018 ACM Symposium on Principles of Distributed Computing, PODC 2018. ACM (2018). https://dl.acm.org/citation.cfm?id=3212738
28. Hassin, Y., Peleg, D.: Distributed probabilistic polling and applications to proportionate agreement. Inf. Comput. **171**(2), 248–268 (2001)
29. Koetter, R., Kschischang, F.R.: Coding for errors and erasures in random network coding. IEEE Trans. Inf. Theory (2008). https://doi.org/10.1109/TIT.2008.926449
30. Lehre, P.K., Witt, C.: Concentrated hitting times of randomized search heuristics with variable drift. In: Ahn, H.-K., Shin, C.-S. (eds.) ISAAC 2014. LNCS, vol. 8889, pp. 686–697. Springer, Cham (2014). https://doi.org/10.1007/978-3-319-13075-0_54
31. Lin, W., Zhixin, L., Lei, G.: Robust consensus of multi-agent systems with noise. In: 2007 Chinese Control Conference (2007). https://doi.org/10.1109/CHICC.2006.4347503
32. Mobilia, M., Petersen, A., Redner, S.: On the role of zealotry in the voter model. J. Stat. Mech. Theory Exp. **2007**(08), P08029 (2007)
33. Mobilia, M.: Does a single zealot affect an infinite group of voters? Phys. Rev. Lett. **91**(2), 028701 (2003)
34. Moon, T.K.: Error Correction Coding: Mathematical Methods and Algorithms. Wiley, New York (2005)
35. Mossel, E., Neeman, J., Tamuz, O.: Majority dynamics and aggregation of information in social networks. Auton. Agents Multi-agent Syst. **28**(3), 408–429 (2013). https://doi.org/10.1007/s10458-013-9230-4
36. Reina, A., Marshall, J.A.R., Trianni, V., Bose, T.: Model of the best-of-n nest-site selection process in honeybees. Phys. Rev. E **95** (2017). https://doi.org/10.1103/PhysRevE.95.052411
37. Ruan, Y., Mostofi, Y.: Binary consensus with soft information processing in cooperative networks. In: Proceedings of the 47th IEEE Conference on Decision and Control, CDC 2008. IEEE (2008). https://doi.org/10.1109/CDC.2008.4738899
38. Sumpter, D.J., Krause, J., James, R., Couzin, I.D., Ward, A.J.: Consensus decision making by fish. Curr. Biol. **18**(22), 1773–1777 (2008). https://doi.org/10.1016/j.cub.2008.09.064

39. Valentini, G., Ferrante, E., Dorigo, M.: The best-of-n problem in robot swarms: formalization, state of the art, and novel perspectives. Front. Robot. AI **4**, 9 (2017). https://doi.org/10.3389/frobt.2017.00009
40. Vicsek, T., Czirók, A., Ben-Jacob, E., Cohen, I., Shochet, O.: Novel type of phase transition in a system of self-driven particles. Phys. Rev. Lett. **75**, 1226–1229 (1995). https://doi.org/10.1103/PhysRevLett.75.1226
41. Wang, L., Liu, Z.: Robust consensus of multi-agent systems with noise. Sci. China Ser. F Inf. Sci. (2009). https://doi.org/10.1007/s11432-009-0082-0
42. Yildiz, E., Ozdaglar, A., Acemoglu, D., Saberi, A., Scaglione, A.: Binary Opinion Dynamics with Stubborn Agents. ACM Trans. Econ. Comput. **1**(4) (2013)

A Meta-Theorem for Distributed Certification

Pierre Fraigniaud[1], Pedro Montealegre[2(✉)], Ivan Rapaport[3], and Ioan Todinca[4]

[1] IRIF, Université de Paris and CNRS, Paris, France
pierre.fraigniaud@irif.fr

[2] Facultad de Ingeniería y Ciencias, Universidad Adolfo Ibañez, Santiago, Chile
p.montealegre@uai.cl

[3] DIM-CMM (UMI 2807 CNRS), Universidad de Chile, Santiago, Chile
rapaport@dim.uchile.cl

[4] LIFO, Université d'Orléans and INSA Centre-Val de Loire, Orléans, France
ioan.todinca@univ-orleans.fr

Abstract. Distributed certification, whether it be *proof-labeling schemes*, *locally checkable proofs*, etc., deals with the issue of certifying the legality of a distributed system with respect to a given boolean predicate. A certificate is assigned to each process in the system by a non-trustable oracle, and the processes are in charge of verifying these certificates, so that two properties are satisfied: *completeness*, i.e., for every legal instance, there is a certificate assignment leading all processes to accept, and *soundness*, i.e., for every illegal instance, and for every certificate assignment, at least one process rejects. The verification of the certificates must be fast, and the certificates themselves must be small. A large quantity of results have been produced in this framework, each aiming at designing a distributed certification mechanism for specific boolean predicates. This paper presents a "meta-theorem", applying to many boolean predicates at once. Specifically, we prove that, for every boolean predicate on graphs definable in the monadic second-order (MSO) logic of graphs, there exists a distributed certification mechanism using certificates on $O(\log^2 n)$ bits in n-node graphs of bounded treewidth, with a verification protocol involving a single round of communication between neighbors.

Keywords: Proof-labeling scheme · Locally checkable proof · Fault-tolerance · Distributed decision

---

This work was partially done during the visit of the second and third authors to IRIF at Université de Paris, and LIFO at Université d'Orléans, partially supported by ANR project DUCAT and FONDECYT 1220142.

P. Fraigniaud—Additional support for ANR projects QuData and DUCAT.

P. Montealegre—This work was supported by Centro de Modelamiento Matemático (CMM), ACE210010 and FB210005, BASAL funds for centers of excellence from ANID-Chile, FONDECYT 11190482, and PAI 77170068.

M. Parter (Ed.): SIROCCO 2022, LNCS 13298, pp. 116–134, 2022.
https://doi.org/10.1007/978-3-031-09993-9_7

1 Introduction

Context. Distributed certification is a concept that serves many purposes in distributed computing. One is fault tolerance. Indeed, the ability to certify the legality of a system-state with respect to some boolean predicate in a distributed manner guarantees that at least one process can launch a recovery procedure in case the system enters into an illegal state. Another application of distributed certification is safety. Indeed, distributed certification is a mechanism that guarantees that distributed algorithms dedicated to systems satisfying some specific property (e.g., algorithms dedicated to planar networks) can safely be used because, in case the system does not satisfy this property, at least one process can raise an alarm, and stop the computation.

Different certification mechanisms have been studied (cf. the related work section), all sharing the same principle. Distributed certification protocols involve a centralized *prover*, and a distributed *verifier*. The prover has complete knowledge of the system. It is computationally unbounded but not trustable. Given a boolean predicate $\mathcal{P}$ on system states, the prover assigns *certificate* to the processes, whose aim is to convince the processes that the system satisfies $\mathcal{P}$. The verifier is a distributed algorithm that runs at every process in the system, and is bounded to return a verdict (*accept* or *reject*) at each process after a limited communication among the processes. For instance, in a network, every processing node is bounded to communicate only once with its neighbors in the network before emitting its verdict.

To be correct, a distributed certification protocol for a boolean predicate $\mathcal{P}$ on system states must satisfy two properties. (1) Completeness: If the system satisfies $\mathcal{P}$, then there must exist a certificate assignment by the prover to the processes such that the verifier accepts at all processes. (2) Soundness: If the system does not satisfy $\mathcal{P}$, then, for every certificate assignment by the prover to the processes, it must be the case that the verifier rejects in at least one process. Network bipartiteness yields a simple example of distributed certification, using 1-bit certificates. For every bipartite network, every processing node in the network can be given a certificate 0 or 1, so that every processing node has a certificate different from the certificates assigned to its neighbors. The processing nodes can check these certificates in a single round of communication, where every processing node merely checks that the certificate of each of its neighbors is different from its own certificate. Completeness is satisfied by construction. Soundness is also satisfied. Indeed, if the network is not bipartite, then it is not 2-colorable. As a consequence, for every certificate assignment with certificates in $\{0, 1\}$, there are at least two neighboring processing nodes that receive the same certificate. These two processes will reject.

The main criterion measuring the quality of distributed certification is the *size* of the certificates. Indeed, the verification of $\mathcal{P}$ is typically performed frequently, for regularly checking that the system does satisfy $\mathcal{P}$, with the aim of reacting quickly if the system stops satisfying $\mathcal{P}$. As a consequence, there are frequent exchanges of certificates between the processes. Using small certificates limits the communication overhead caused by these exchanges.

Objective. A large collection of results related to distributed certification have been derived over the last twenty years (see Related Work), each result concerning a specific predicate. This paper is inspired by what has been achieved in the context of sequential computing where, instead of focusing on the design of an efficient algorithm for one specific problem, and then for another one, and so on and so forth, efforts have been made for deriving "meta-theorems", that is, results applying directly to large classes of problems. One prominent example is Courcelle's theorem [12] stating that every graph property definable in the monadic second-order (MSO) logic of graphs can be decided in linear time on graphs of bounded treewidth[1]. That is, even NP-hard problems such as vertex-coloring, minimum dominating set, minimum vertex cover, etc., have linear-time algorithms in the vast class of graphs with bounded treewidth. Each algorithm depends on the problem, but Courcelle's theorem essentially says that *every* problem expressible in the MSO logic has a linear-time algorithm in the class of graphs with bounded treewidth.

The objective of this paper is to address the existence of similar meta-theorems in the context of distributed certification applied to distributed computing in networks. Concretely, the question we address here is the following: is there a (large) class of boolean predicates on graphs for which one can guarantee the existence of a distributed certification mechanism with small certificates, say poly-logarithmic in the number of vertices of the graphs, for graphs taken from a (large) class of graphs?

Our Results. We present an analog of the aforementioned Courcelle's theorem in the context of distributed certification. Specifically, for every integer $k \geq 1$ and every MSO property φ on graphs, we consider the following set:

$$\mathcal{P}_{k,\varphi} = \{\text{graph } G : (\mathsf{tw}(G) \leq k) \wedge (G \models \varphi)\},$$

where $\mathsf{tw}(G)$ is the treewidth of G. We provide a distributed certification mechanism for $\mathcal{P}_{k,\varphi}$ using certificates of poly-logarithmic size, as a function of the number n of vertices in the graphs. Specifically, given any network modeled as a connected simple graph $G = (V, E)$, with a process running at each vertex $v \in V$, our certification mechanism satisfies that $G \in \mathcal{P}_{k,\varphi}$ if and only if there is a certificate assignment to the vertices such that all vertices accept. The main result of the paper is the following.

Theorem 1 (Informal). *For every $k \geq 1$ and every MSO property φ on graphs, there exists a distributed certification protocol for $\mathcal{P}_{k,\varphi}$ using certificates on $O(\log^2 n)$ bits.*

In fact, our theorem can be extended to properties including certifying solutions to maximization or minimization problems whose admissible solutions are

[1] Treewidth can be viewed as a measure capturing "how close" a graph is from a tree; roughly, a graph of treewidth k can be decomposed by a sequence of cuts, each involving a separator of size $O(k)$.

defined by MSO properties. In the statement of Theorem 1, the big-O notation hides constants that depend only on k and φ. The theorem has many corollaries, as the universe of MSO properties is large. This includes predicates such as non 3-colorability, which is known to require certificates of quadratic size in arbitrary graphs [18], and diameter at most D, for a fixed constant D, which is known to require certificates of linear size in arbitrary graphs [11].

Corollary 1. *For every $c \geq 1$, there exists a distributed certification protocol for certifying* non c-colorability *in the family of graphs with bounded treewidth, using certificates on $O(\log^2 n)$ bits.*

For every $D \geq 1$, there exists a distributed certification protocol for certifying diameter at most D *in the family of graphs with bounded treewidth, using certificates on $O(\log^2 n)$ bits.*

Also, many natural graph families have bounded treewidth, as illustrated by the family of graphs excluding a planar graph as a minor, and thus we get the following corollary of Theorem 1.

Corollary 2. *For every planar graph H, and every MSO property φ on graphs, there exists a distributed certification protocol certifying φ in the family of H-minor-free graphs, using certificates on $O(\log^2 n)$ bits.*

Again, the big-O notation in the above statement hides constants that depend only on H and φ. Note that, as every 4-node graph is planar, Corollary 2 extends the recent results in [9], which applies to the families of graphs excluding a given 4-node graph H as a minor.

Interestingly, $\mathsf{tw}(G) \leq k$, and H-minor-freeness are themselves MSO properties for fixed k and H. It follows that Theorem 1 provides us with a distributed certification mechanism for treewidth and fixed-minor-freeness.

Corollary 3. *Let $k \geq 0$, and let H be a planar graph. There exist distributed certification protocols for certifying the class of graphs with treewidth at most k, and certifying the class of H-minor-free graphs, both using certificates on $O(\log^2 n)$ bits.*

Our Techniques. For establishing Theorem 1 we proceed in two steps. First, we provide a protocol for certifying 3-approximation of treewidth. Such a protocol satisfies the following: for any given $k \geq 1$, the protocol for k is such that, for every graph G,

$$\begin{cases} \mathsf{tw}(G) \leq k & \Rightarrow \text{there exists a certificate assignment s.t. all vertices accept;} \\ \mathsf{tw}(G) > 3k+2 & \Rightarrow \text{for every certificate assignment, at least one vertex rejects.} \end{cases}$$

Lemma 1 (Informal). *For every $k \geq 1$ there exists a distributed protocol certifying a 3-approximation of the treewidth using certificates on $O(k^2 \log^2 n)$ bits.*

The proof of this lemma relies on a particular choice of a tree decomposition, that we prove locally certifiable by "transferring" certificates between nodes that are far away from each other, which is typically the case of vertices in a same bag of the decomposition, without creating congestion.

Next, for any MSO property φ and integer k, we design a protocol which certifies $\mathcal{P}_{k,\varphi}$ on input graph G. The protocol exploits the tree decomposition in the proof of Lemma 1, for certifying a correct execution of a sequential dynamic programming algorithm for φ over this decomposition. Concretely, we design a distributed certification for a correct execution of a sequential dynamic programming algorithm a la Courcelle, using in fact the sequential MSO certification due to Borie, Parker and Tovey [8].

Lemma 2 (Informal). *For every $k \geq 1$ and every MSO property φ on graphs, assuming given the certification protocol for 3-approximation k of treewidth from Lemma 1, there exists a distributed certification protocol for $\mathcal{P}_{k,\varphi}$ using additional certificates on $O(\log^2 n)$ bits.*

Related Work. The ability to detect illegal configurations of a distributed system was originally motivated by the design of fault-tolerant algorithms, especially self-stabilizing algorithms [1,2,20]. The notion of distributed certification as used in this paper originated from the seminal paper [22] defining *proof-labeling schemes* (PLS). We actually use a slight variant of PLS called *locally checkable proofs* (LCP) [18], which enables exchanging not only the certificates between the processing nodes, but also local states, including their IDs. Another related notion is *non-deterministic local decision* (NLD) [16] in which the certificates must not depend on the IDs given to the processing nodes. Distributed certification has been extended to various directions, including randomized PLS [17], approximate PLS [11,14], local hierarchies [3,15], interactive proofs [21,24], and even, recently, zero-knowledge distributed certification [4]. All the aforementioned papers contain a vast collection of certification results for various graph problems. In these papers, each certification protocol is specific of the problem at hand. To our knowledge, the only "meta-theorem" in the context of distributed certification is the recent paper [10], which shows that every MSO formula can be locally certified on graphs with bounded *treedepth* using certificates on $O(\log n)$ bits. We show that the same result holds for the larger class of graphs with bounded *treewidth*, to the cost of slightly larger certificates, on $O(\log^2 n)$ bits. We are therefore partially answering the questions raised in [10], asking whether it is "possible to certify any MSO formula on bounded treewidth graphs", and "to certify that the graph itself has treewidth at most k", using small certificates.

In framework of sequential algorithms, there is a large literature on "meta-theorems" proving that large families of combinatorial properties (typically expressed using some form of logic formulae) can be efficiently decided on particular graph classes. In addition to Courcelle's (meta) theorem [12] on MSO properties on graphs with bounded treewidth, it is worth mentioning the recent results establishing that properties expressible in *first-order logic* can be verified in polynomial time on graphs of bounded *twinwidth* [7], as well as on *nowhere-dense* graphs [19]. Both graph classes include planar graphs, and thus include graphs with arbitrarily large treewidth. Our work is participating to the general objective of extending these results to the framework of distributed computing.

2 Preliminaries

Distributed Certification. We consider networks modeled as connected simple graphs. Every vertex is a processing element, and the vertices exchange messages along the edges of the graph. We systematically denote by n the number of vertices in the considered graph. The vertices of a network/graph $G = (V, E)$ are given distinct identifiers (IDs), and we denote by $\mathsf{ID}(v)$ the identifier of vertex $v \in V$. These identifiers are not necessarily between 1 and n, but we adopt the standard assumption stating that IDs can be stored on $O(\log n)$ bits.

We consider boolean predicates on labeled graphs, i.e., graphs for which every vertex v is given a label $\ell(v) \in \{0,1\}^*$. These labels may represent a way to mark vertices (e.g., those in a dominating set), a color (e.g., in graph coloring), or any value depending on the graph property at hand. Given a boolean predicate $\mathcal{P}$ on labeled graphs, a *locally checkable proof* [18] for $\mathcal{P}$ is a prover-verifier pair. The prover is a non-trustable oracle with unbounded computing power. Given any labeled graph (G, ℓ), the prover assigns a certificate $c(v) \in \{0,1\}^*$ to every vertex $v \in V$. The verifier is a 1-round distributed algorithm running at all vertices of the graph. Given a labeled graph (G, ℓ) with a certificate assigned at every vertex, the vertices exchange their identifiers, labels, and certificates, between neighbors, and compute an output, accept or reject. To be correct, the pair prover-verifier must satisfy two conditions:

Completeness: If $(G, \ell) \models \mathcal{P}$, then, for every ID-assignment to the vertices, there must exist a certificate assignment by the prover to the vertices such that the verifier accepts at all vertices.

Soundness: If $(G, \ell) \not\models \mathcal{P}$, then, for every ID-assignment to the vertices, and for every certificate assignment by the prover to the vertices, it must be the case that the verifier rejects in at least one vertex.

Tree Decompositions and Terminal Recursive Graphs. Let us recall the classical definition of treewidth and tree decompositions, due to Robertson and Seymour [25].

Definition 1. *A* tree decomposition *of a graph $G = (V, E)$ is a pair (T, B) where $T = (I, F)$ is a tree, and $B = \{B_i, i \in I\}$ is a collection of subsets of vertices of G, called* bags, *such that the following conditions hold:*

- *For every $v \in V$, there exists $i \in I$ such that $v \in B_i$;*
- *For every $e = \{u, v\} \in E$ there is $i \in I$ such that $\{u, v\} \subseteq B_i$;*
- *For every $v \in V$, the set $\{i \in I : v \in B_i\}$ forms a connected subgraph of T.*

The width *of a tree decomposition is the maximum size of a bag, minus one. The* tree-width *of a graph G, denoted by* $\mathsf{tw}(G)$*, is the smallest width of a tree decomposition of G.*

To facilitate the distinction between the original graph $G = (V, E)$ and the decomposition tree $T = (I, F)$, we will speak of the *nodes* $i \in I$ of T and of the *vertices* $v \in V$ of G.

We consider tree decompositions as rooted, i.e., we fix some node $r \in I$ as the root of $T = (I, F)$. For a node $i \in I \setminus \{r\}$, we denote by $p(i)$ its parent in T, and set $p(r) = \perp$. For $i \in I$, we denote by T_i the subtree of T rooted in i, and by V_i the subset of vertices of G in the bags of T_i, i.e., $V_i = \cup_{j \in V(T_i)} B_j$. Also, for $i \in I \setminus \{r\}$, we define $F_i = B_i \setminus B_{p(i)}$. For the root r, we set $B_{p(r)} = \perp$ and $F_r = B_r$. Given a rooted tree $T = (I, F)$, and two nodes of $i, j \in I$, we denote by $j \preceq i$ the property that j is a descendant of i in T.

Graphs of bounded treewidth can also be defined recursively, based on a graph grammar. Let w be a positive integer. A *w-terminal graph* is a graph (V, E) together with a *totally ordered* set $W \subseteq V$ of at most w distinguished vertices. Vertices of W are called the *terminals* of the graph, and we denote by $\tau(G)$ the number of its terminals. Since W is totally ordered, we can speak of the rth terminal, for $1 \leq r \leq w$. Since in our case vertices are given distinct identifiers, one can view W as ordered w.r.t. these identifiers.

The class of *w-terminal recursive graphs* is defined starting from *w-terminal base graphs* through a sequence of *composition operations.* A *w-terminal base graph* is a w-terminal graph of the form (V, W, E) with $W = V$. A *composition operation* f acts on one or two w-terminal graphs producing a new w-terminal graph as follows.

When f is of arity 2, graph $G = f(G_1, G_2)$ is obtained by firstly making disjoint copies of the two graphs G_1 and G_2, then "glueing" together some terminals of G_1 and G_2. The glueing performed by f is represented by a matrix $m(f)$ having $\tau(G) \leq w$ rows and two columns, with integer values between 0 and $\tau(G)$. At row r of the matrix, $m_{rc}(f)$ indicates which terminal of each $G_c, c \in \{1, 2\}$ is identified to terminal number r of graph G. If $m_{rc}(f) = 0$, then no terminal of G_c is identified to terminal r of G (in particular, if $m_{r1}(f) = m_{r2}(f) = 0$ it means that terminal r of G is a new vertex, but this situation will not occur in our constructions). Moreover, a terminal of G_c is identified to at most one terminal of G, i.e., each non-zero value in $1, \ldots, \tau(G_c)$ appears at most once in column c of $m(f)$.

When f is of arity 1, the corresponding matrix $m(f)$ has a unique column. Graph $G = f(G_1)$ is obtained as before, by identifying terminal m_{i1} of G_1 to terminal r of G. Note that in this case G and G_1 have exactly the same vertex and edge sets, and the terminals of G form a subset of the terminals of G_1.

We point out that the number of possible different matrices and hence of different operations is bounded by a function on w.

Proposition 1. (Theorem 40 in *[6]*). *Graph $H = (V, W, E)$ is $(w + 1)$-terminal recursive if and only if there exists a tree decomposition of $G = (V, E)$,* *of width at most w, having W as root bag. Hence the grammar of $(w+1)$-terminal recursive graphs constructs exactly the graphs of treewidth at most w.*

Let us sketch briefly here how a tree decomposition of $G = (V, E)$ of width w can be transformed into a $(w+1)$-expression of the same graph. To each node i of the tree decompositions, we associate three $(w+1)$-terminal graphs:

- $G_i^b = (B_i, B_i, E(G[B_i]))$, the $(w+1)$-terminal base graph corresponding to graph $G[B_i]$ induced by bag B_i;
- $G_i = (V_i, B_i, E(G[V_i]))$, corresponding to $G[V_i]$, with bag B_i as set of terminals;
- If i differs from the root, $G_i^+ = (V_i \cup B_{p(i)}, B_{p(i)}, E(G[V_i \cup B_{p(i)}]))$ corresponding to the graph induced by $V_i \cup B_{p(i)}$, with $B_{p(i)}$ as set of terminals.

Let us describe how to compute the $(w+1)$-expression of these graphs, by parsing bottom-up the tree decomposition.

When i is a leaf, $G_i = G_i^b$ is a $(w+1)$-terminal base graph. Assume now that i is not a leaf and let $Children(i)$ be the children of node i in the decomposition tree. For each $j \in Children(i)$, we already possess an expression of the $(w+1)$-terminal graph $G_j = (V_j, B_j, E(G[V_j]))$. Observe that G_j^+ is obtained from a glueing of G_j and the base graph G_i^b, where the terminals of G_j contained in $B_j \cap B_i$ are glued on the corresponding terminals of G_j^+, and the others become non-terminals. Eventually, if i has more than one child, then G_i is obtained by the consecutive glueing of all G_j^+, $j \in Children(i)$, where the glueing is performed on B_i by the same matrix $m(f)$ having $m_{r1}(f) = m_{r2}(f) = r$, for $1 \leq r \leq |B_i|$.

Regular Properties and MSO. We consider graph properties $\mathcal{P}(G)$ assigning to each graph G a boolean value. We have in mind properties expressible in Monadic Second Order Logic, like "G is not 3-colourable", "G does not contain a given minor", etc. Nevertheless, technically, we do not need the definition of MSO formulae, and the interested reader may refer to [13]; we only need the fact that MSO properties are *regular*, in the sense defined below. By Courcelle's theorem, such properties can be decided in linear (sequential) time on graphs of bounded treewidth, if the tree decomposition (or the corresponding expression as a terminal recursive graph) is part of the input.

Definition 2 (regular property). *A graph property $\mathcal{P}$ is called* regular *if, for any value w, we can associate a finite set $\mathcal{C}$ of* homomorphism classes *and a* homomorphism function h, *assigning to each w-terminal recursive graph G a class $h(G) \in \mathcal{C}$ such that:*

1. *If $h(G_1) = h(G_2)$ then $\mathcal{P}(G_1) = \mathcal{P}(G_2)$.*
2. *For each composition operation f of arity 2 there exists a function $\odot_f$: $\mathcal{C} \times \mathcal{C} \to \mathcal{C}$ such that, for any two w-terminal recursive graphs G_1 and G_2,*

$$h(f(G_1, G_2)) = \odot_f(h(G_1), h(G_2))$$

and for each composition operation f of arity 1 there is a function $\odot_f : \mathcal{C} \to \mathcal{C}$ such that, for any w-terminal recursive graph G,

$$h(f(G)) = \odot_f(h(G)).$$

We illustrate this definition on the property "G is not 3-colourable". We can choose, as homomorphism $h(G = (V, W, E))$, the set of all three-partitions (W_1, W_2, W_3) of the set W of terminals, such that graph G has, as three colouring, the one where each colour $i \in \{1, 2, 3\}$ intersects W exactly in the set W_i. Observe that graph G satisfies the property of not being 3-colourable if and only if its homomorphism class is the empty set. It is a matter of exercise to figure out how to compute the homomorphism class of a base w-terminal graph (by enumerating all its three-partitions into independent sets), and how to compute functions $\odot_f$ updating the class of the graph after a composition operation f.

The first condition of Definition 2 separates the classes into *accepting* ones (i.e., classes $c \in \mathcal{C}$ such that $h(G) = c$ implies that $\mathcal{P}(G)$ is true) and *rejecting* ones (i.e., classes $c \in \mathcal{C}$ such that $h(G) = c$ implies that $\mathcal{P}(G)$ is false). In full words, the second condition states that, if we perform a composition operation on two graphs (resp. one graph), the homomorphism class of the result can be obtained from the homomorphism classes of the graphs on which these operations are applied. Therefore, if a w-terminal recursive graph is given together with its expression in this grammar, and if moreover we know how to compute the homomorphism classes of the base graphs and the composition functions $\odot_f$ over all possible composition operations f, then the homomorphism class of the whole graph for a regular property $\mathcal{P}$ can be obtained by dynamic programming. We simply need to parse the expression from bottom to top and, at each node, we compute the class of the corresponding sub-expression thanks to the second condition of regularity. At the root, the property is true if and only if we are in an accepting class.

Proposition 2. *([8, 12]). Any property $\mathcal{P}$ expressible by a MSO formula is regular. Moreover, given the MSO formula φ and parameter w, one can explicitely compute the set of classes, the homomorphism function for all w-terminal base graphs as well as the composition functions $\odot_f$ of all possible composition operations f.*

Altogether, this provides an effective algorithm for checking property $\mathcal{P}(G)$ in $O(n)$ time, by a sequential algorithm, given the w-expression (or, equivalently, the tree decomposition of width $w-1$) of the input graph, by computing bottom-up the homomorphism classes.

The notions of MSO and regular properties extend to properties on graphs and vertex subsets, i.e., we can consider properties $\mathcal{P}(G, X)$ assigning to each graph G and vertex subset X of G a boolean value. This allows to capture properties as "X is an independent set of G", or "X is an dominating set of G". Moreover, the whole framework can capture the problem of computing a (or, in our case, certifying that) set X is of maximum weight among those satisfying $\mathcal{P}(G, X)$, for graphs with polynomial weights on their vertices.

Coherent Tree Decompositions. By a classic result of Bodlaender [5], an optimal tree decomposition of graph G can be transformed into a decomposition whose tree is of logarithmic depth, while the size of the bags is at most multiplied by 3. We strongly rely on such decomposition, plus a connectivity property that we call *coherence*. We say that a rooted tree decomposition of a graph $G = (V, E)$ is *coherent* if for every $i \in I$, the set F_i is non empty and the graph $G[V_i \setminus B_{p(i)}]$ is connected.

We show that such a decomposition exists and provide some of its properties used in our certification protocol. Due to space restrictions, the proofs of the results of this sub-section can be found in the full version.

Lemma 3. *Let $k \geq 1$, and let G be a connected n-vertex graph of treewidth at most k. Then, G admits a coherent tree decomposition of width at most $3k + 2$ and depth $\mathcal{O}(\log n)$.*

In our protocol we must be able to communicate, for any node i of the decomposition, some information about V_i to a vertex in the bag corresponding to the parent node $p(i)$, more precisely, to some vertex of $F_{p(i)}$. The following lemma shows the existence a vertex $\ell_i \in V_i \setminus B_{p(i)}$ adjacent in G to some vertex $w \in F_{p(i)}$.

Lemma 4. *Let $T = (I, F)$ be a coherent tree decomposition of $G = (V, E)$. Then, for every $i \in I$ different from the root there exists a pair of vertices $\ell_i \in V_i \setminus B_{p(i)}$ and $w \in F_{p(i)}$ such that $\{w, \ell_i\} \in E$.*

Vertex ℓ_i is called the exit vertex of i, *and w is called the* vertex of $F_{p(i)}$ *in charge of node i.*

In our certification protocols, for each node i of the decomposition tree, the vertices of F_i as well as the exit vertex ℓ_i will receive from the prover some information concerning graph $G_i = G[V_i]$. We will need to ensure that ℓ_i and all vertices of F_i received the same information. For this purpose we use trees contained in $G[V_i \setminus B_{p(i)}]$, spanning ℓ_i and F_i.

Lemma 5. *Consider a coherent tree decomposition $T = (I, F)$ of graph $G = (V, E)$, of depth $O(\log n)$. For each node i of the decomposition tree, there is a subtree $S(i)$ of $G[V_i \setminus B_{p(i)}]$ spanning F_i and the exit vertex ℓ_i.*

Moreover each vertex of G appears $O(\log n)$ times in the family of trees $\mathcal{T}(G) = \{S(i) \mid i \in I\}$.

3 A Protocol Certifying a 3-Approximation of the Treewidth

In this section we describe a protocol certifying a 3-approximation of treewidth. More precisely, we prove the following Lemma.

Lemma 6. *For each $k \geq 1$ there is a distributed certification protocol that uses messages of size $O(k^2 \log^2 n)$ and ensures, for any input graph G, that:*

$$\begin{cases} \mathsf{tw}(G) \leq k & \Rightarrow \textit{there exists a certificate assignment s.t. all nodes accept;} \\ \mathsf{tw}(G) > 3k+2 & \Rightarrow \textit{for every certificate assignment, at least one node rejects.} \end{cases}$$

Let us describe the messages that the prover sends to each vertex of G, if $\mathsf{tw}(G) \leq k$. These messages describe a coherent tree decomposition of width at most $3k+2$ and of logarithmic depth, which exists by Lemma 3.

We identify node i of the decomposition tree with the number corresponding to a binary representation of the set of vertices B_i contained in its bag, so $1 \leq i \leq n^{\mathcal{O}(k)}$. In full words, a node is simply identified by the content of its bag, which is possible since coherent tree decompositions have pairwise disjoint bags.

Our protocol distinguishes two types of certificates, namely *main messages* and *auxiliary messages*. Each vertex receives one main message and $\mathcal{O}(\log n)$ auxiliary messages. Let us describe each one of them.

Main Messages. These messages are used to encode a tree decomposition, following Definition 1. Each vertex v receives as a certificate the following messages, that we denote $m(v)$:

1. A number $d = d(v)$, representing the depth of the node i such that $v \in F_i$
2. A list of sets $\mathcal{B}(v) = B_d(v), B_{d-1}(v), \ldots, B_1(v)$, representing the path of bags from node $i = B_d(v)$ to the root node.
3. The list of sets $\mathcal{F}(v) = F_d(v), F_{d-1}(v), \ldots, F_1(v)$, representing the sets $F_j(v) = B_j(v) \setminus B_{j-1}(v)$, for each $j \in \{1, \ldots, d\}$.
4. A list of sets $\mathcal{E}(v) = E_d(v), \ldots, E_1(v)$, where, for each $j \in \{1, \ldots, d\}$, $E_j(v) \subseteq \binom{B_j(v)}{2}$ represents the edge set of $G[B_j(v)]$.

Observe that the size of a main message is $\mathcal{O}(k^2 \log^2 n)$.

Auxiliary Messages. These messages allow to check the consistency of the main messages between vertices of a same set F_i, for each node i of the decomposition.

From Lemma 5, we have that for each node i there is a subtree $S(i)$ connecting all pair of vertices of F_i and the exit vertex ℓ_i. The vertices w of $S(i)$ are called *auxiliary vertices for* i. For a vertex w, let us call $Aux(w)$ the set of nodes i such that w is an auxiliary vertex for i. From Lemma 5, we know that for each $w \in V$, $|Aux(w)| = \mathcal{O}(\log n)$.

Each node w receives the set $Aux(w)$ and for each $i \in Aux(w)$ the message $m_{aux}(w, i)$ containing the following information where

- $d_{aux}(w, i)$ is the depth of node i.
- $\ell_i(w)$ is a vertex identifier of the exit vertex of F_i (cf. Lemma 4).
- $\alpha_i(w)$ is a vertex identifier of the vertex in $F_{p(i)}$ in charge of B_i (cf. Lemma 4).
- $F_i(w)$ is a set of vertices, representing F_i.

- $TreeCert(w)$ is the certificate that receives w in the protocol used to verify that $S(i)$ is a tree rooted at ℓ_i and spanning $F_i(w)$. More precisely $cert(F_i, w) = (parent(w), dist(w), sub(w))$, where:
  - $parent(w)$ represent the parent of w in $S(i)$ ($parent(w) = \perp$ if $w = \ell_i(w)$),
  - $dist(w)$ represents the distance from w to ℓ_i in $S(i)$, and
  - $sub(w)$ represents is the subset of $F_i(v)$ that are descendants of w in $S(i)$.

Observe that for any given vertex w and node i, the messages $m_{aux}(w, i)$ is of size $O(k \log n)$. Thanks to Lemma 5, a vertex w appears $\mathcal{O}(\log n)$ times as auxiliary vertex of some node i. Therefore, a vertex w receives in total $O(k \log^2 n)$ bits for auxiliary messages.

Verification Round. Given two vertices u and v such that $d(u) \leq d(v)$, we say that the main message of u is a d-suffix of the main message of v if $B_j(u) = B_j(v)$ and $E_j(u) = E_j(v)$ for each $j \in \{1, \ldots, d\}$.

Let $d = d(v)$. In the verification round, vertex v verifies the following conditions.

Consistency of the Tree Decomposition.

1. The size of each $B \in \mathcal{B}(v)$ is at most $3k + 3$.
2. The set $F_d(v)$ contains v.
3. For each $j \in \{2, \ldots, d\}$, the set $F_j(v)$ equals $B_j(v) \setminus B_{j-1}(v)$.
4. For each $w \in V(G)$ and $j_1, j_2 \in \{1, \ldots, d\}$ with $j_1 < j_2$, if $w \in B_{j_1} \cap B_{j_2}$, then $w \in B_{i_j}$ for every $j \in \{j_1 + 1, \ldots, j_2 - 1\}$.
5. For each $j_1, j_2 \in \{1, \ldots, d\}$, each pair of vertices $u_1, u_2 \in B_{j_1}(v) \cap B_{j_2}(v)$ satisfies that $\{u_1, u_2\} \in E_{j_1}(v) \iff \{u_1, u_2\} \in E_{j_2}(v)$.
6. For each $u \in B_d(v)$, v checks that $\{u, v\} \in E \iff \{u, v\} \in E_d(v)$.
7. For each $u \in N(v)$ such that $d(u) \geq d(v)$, v checks that $m(v)$ is a $d(v)$-suffix of $m(u)$.
8. For each $u \in N(v)$ such that $d(u) \leq d(v)$, v checks that $u \in B_d(v)$.
9. v checks that it is an auxiliary vertex for $B_d(v)$ and that it has a neighbor that is also an auxiliary vertex for $B_d(v)$.
10. For each vertex $w \in N(v) \cup \{v\}$ such that w is an auxiliary tree vertex for $B_d(v)$, v checks that $d_{aux}(w, B_d(w)) = d$ and $F_i(w) = F_d(v)$.

Consistency of the Auxiliary Trees and the Exit Vertex. The following conditions are used to verify that the nodes marked as auxiliary vertices for node i form an auxiliary subtree $S(i)$ rooted at ℓ_i and spanning F_i. At the same time, we check that all de nodes in $S(i)$ have the same auxiliary information, corresponding to the depth d_i of bag i, the contents of F_i, the identity of exit vertex ℓ_i, and the identity of the node of $F_{p(i)}$ responsable of i, and the same d_i-suffix of the main messages.

For each $i \in Aux(v)$, vertex v checks te following conditions

11. For each vertex $w \in N(v)$ such that w is an auxiliary tree vertex for i, v checks that

$$(d_{aux}(w,i), \ell_i(w), \alpha_i(w), F_i(w)) = (d_{aux}(v,i), \ell_i(v), \alpha_i(v), F_i(v))$$

12. $d_{aux}(v,i) \leq d(v)$.
13. Uses $TreeCert(F_i(v), v)$ to verify that there is an auxiliary tree $S(i)$ rooted in $\ell_i(v)$ and spanning $F_i(v)$. More precisely, v checks the following conditions:
   (a) If $v \neq \ell_i(v)$ then v has a neighbor with the label $parent(w)$ which is also an auxiliary vertex for i;
   (b) If $v \neq \ell_i(v)$, then $dist(parent(v)) = dist(v) - 1$;
   (c) If $v = \ell_i$ then $dist(v) = 0$, $sub(v) = F_i(v)$, v is adjacent to $\alpha_i(v)$ and $d(\alpha_i(v)) = d_{aux}(v,i) - 1$.
   (d) Set $sub(v)$ is the union of all sets $sub(w)$ over the children w of v in $S(i)$ (i.e., for all w such that $parent(w) = v$), plus vertex v itself if $v \in F_i$.

Soundness and Completeness. We now analyze the correctness of the protocol. The completeness follows directly by Lemmas 3, 4 and 5. In the following, we prove the soundness.

Soundness: Let us assume that all vertices accept a given certificate in the verification round. We now show that necessarily $\mathsf{tw}(G) \leq 3k+2$. For each node $v \in V$, let us call $B(v)$ and $F(v)$ the set $F_{d(v)}(v)$ and $B_{d(v)}(v)$, respectively. We say that a vertex v is in depth d if $d(v) = d$. The proof of the soundness is a consequence of the following claims.

Claim 1: *For each $i \in Aux(v)$, there is a tree $S(i)$ rooted in $\ell_i(v)$ spanning $F_i(v)$. Moreover, all the vertices in $S(i)$ are in a depth greater or equal than $d_{aux}(v,i)$, and their main messages have the same $d_{aux}(v,i)$-suffix. First, observe that by the verification of condition* **13 (a)-(c)**, *we have that $S(i)$ is defined by the set of all auxiliary vertices for i and the edges $\{w, parent(w)\}$. Since $S(i)$ is connected, by conditions* **10** *and* **11**, *all auxiliary vertices for node i agree in the same $F_i = F_i(v)$ and in the depth of i given by $d_{aux} = d_{aux}(v,i)$. By condition* **13 (c)-(d)**, *all vertices in F_i exist and are auxiliary vertices for node i. Finally, by condition* **12** *all nodes are in a depth greater or equal than d_{aux} and by condition* **7**, *the main messages of all vertices in $S(i)$ have the same d_{aux}-suffix.*

Claim 2: *For every vertex v, all nodes in $F(v)$ receive the same main messages as v. Let u be a vertex in $F(v)$. If u and v are adjacent the claim is true by condition* **7**. *Suppose then that $u \notin N(v)$. Since v verifies condition* **9**, *there is a set of auxiliary vertices for node $i = B(v)$. By* **Claim 1**, *$m(v)$ is a $d(v)$-suffix of $m(w)$, for every auxiliary vertex w for node i. Since all vertices in $F(v)$ are auxiliary vertices for i, we deduce that u has the same main messages than v.*

Claim 3: *For every pair of vertices* $u, v \in V$ *either* $F(v) = F(u)$ *or* $F(v) \cap F(u) = \emptyset$. *This is a direct corollary of* **Claim 2**. *Indeed, let us suppose that there exist a pair* $u, v \in V$ *such that* $F(v) \neq F(u)$ *but* $F(v) \cap F(u) \neq \emptyset$. *Then, without loss of generality, there is a node* $w \in F(v) \cap F(u)$ *such that* $F(w) \neq F(v)$, *which contradicts* **Claim 2**.

Claim 4: *For every vertex* v *such that* $d(v) > 1$, *there exist a node* u *such that* $m(u)$ *is a* $(d(v) - 1)$ *-suffix of* $m(v)$. *Let* $d = d(v)$. **Claim 1** *implies that the exit vertex* ℓ_i *for* $i = B_d(v)$ *exists and is the root of* $S(i)$, *which is in a depth greater or equal than* $d_{aux} = d$. *Condition* **13 (c)** *implies that* ℓ_i *is adjacent to a node* α_i *of depth* $d-1$. *Then, by condition* **7**, $m(\alpha_i)$ *is a* $d-1$*-suffix of* $m(\ell_i)$. *Since* $m(v)$ *is a* d*-suffix of* $m(\ell_i)$, *we deduce that* $m(\alpha_i)$ *is a* $d-1$*-suffix of* $m(v)$.

Claim 5: *For every* $u, v \in V$, *the sets* $F(u) \neq F(v)$ *if and only if* $B(u) \neq B(v)$. *First, observe that if* $F(u) = F(v)$, *then by condition* **2** *and* **Claim 2**, $B(v) = B(u)$. *For the reciprocal, let us suppose by contradiction that there exist* $u, v \in V$ *such that* $F(u) \neq F(v)$ *and* $B(u) = B(v)$. *Let us call* $d_1 = d(u)$ *and* $d_2 = d(v)$. *Since* $F(u) \neq F(v)$, *necessarily* $B_{d_1-1}(u) \neq B_{d_2-1}(v)$. *Let us assume, without loss of generality, that there exists a vertex* $w \in F(v) \setminus F(u)$. *Since* w *belongs to* $F(v)$, *we have that* $F(w) = F(v)$ *by* **Claim 2**, *and* w *does not belong to* $B_{d_1-1}(v)$. *Since* $w \notin F(u)$ *we have that* w *belongs to* $B_{d_2-1}(u)$.

Let us call d_3 *the maximum in* $\{1, \ldots, d_1 - 1\}$ *such that* $B_{d_3}(u)$ *belongs to* $\mathcal{B}(v)$. *Observe that* d_3 *exists, because applying condition* **7** *on all the vertices in* G *we deduce that* $B_1(u) = B_1(v)$. *If* $B_{d_3}(u)$ *contains* w, *then* v *fails to verify condition* **4**. *If* $B_{d_3}(u)$ *does not contain vertex* w, *there exists a* $d_4 \in \{d_1, \ldots, d_3 - 1\}$ *such that* $w \in F_{d_4}(u) = B_{d_4}(u) \setminus B_{d_4-1}(u)$. *Then,* **Claim 4** *applied to the vertices in the sequence* $F_{d_1}(u), F_{d_1-1}(u), ..., F_{d_4}(u)$ *implies that there is a node* w' *such that* $F(w') = F_{d_4}(u)$. *Then, by* **Claim 2**, $F(w) = F_{d_4}(u)$. *We deduce that* $B(v) = B_{d_4}(u)$, *which is a contradiction with the choice of* d_3.

Let us define I *as the set of indexes* $i \in [n^{\mathcal{O}(k)}]$ *for which there is a* $v \in V(G)$ *such that* i *is the binary representation of* $B(v)$. *By* **Claim 2, 3** *and* **5**, *we have a partition* $\{F_i\}_{i \in I}$ *of* $V(G)$, *such that, for each* $i \in I$, *all nodes in* F_i *receive the same main messages. In particular, for every vertex* v *in* F_i, *we have that* i *is the binary representation of* $B(v)$. *For each* $v \in F_i$, *we define* $p(i)$ *as the binary representation of* $B_{d(v)-1}(v)$ $(p(i) = \perp$ *if* $v \in B_1(v))$. *From* **Claim 4** *we know that the binary representation of* $B_{d(v)-1}(v)$ *is also in* I. *In other words, the nodes in* $F_{d(v)-1}(v)$ *have certificates that are consistent with the certificate of* v. *In particular, all vertices of* G *agree on the contents of the root node, that we call* B_1. *We then define the pair* $(T, \{B_i\}_{i \in I})$, *where* T *is defined by the tree with vertex set* I *and edge set* $\{i, p(i)\}$, *for each* $i \in I$ *different than the root.*

Claim 6: *The pair* $(T, \{B_i\}_{i \in I})$ *forms a tree decomposition of* G *of width* $3k+2$. *According to Definition 1 we have to check that the following three properties are satisfied:*

- For every $v \in V$, there exists $i \in I$ such that $v \in B_i$;
- For every $e = \{u, v\} \in E$ there is $i \in I$ such that $\{u, v\} \subseteq B_i$;
- For every $v \in V$, the set $\{i \in I : v \in B_i\}$ forms a connected subgraph of T.

The first two properties are directly verified as every vertex is given one bag that contains it in the main message. The second property is verified by condition **8**. Finally, for the third condition, let us suppose that there exists a vertex $v \in V$ such that $I_v = \{i \in I : v \in B_i\}$ is not connected. Let C_1 and C_2 be two different components of I_v, and let i_1 and i_2 be, respectively, the nodes in C_1 and C_2 of minimum depth. Observe that $F_{i_1} \neq F_{i_2}$ and by condition **3**, v must be contained in $F_{i_1} \cap F_{i_2}$, which contradicts **Claim 2**. We deduce that for every $v \in V$, the set $\{i \in I : v \in B_i\}$ forms a connected subgraph of T. We conclude that $(T, \{B_i\}_{i \in I})$ forms a tree decomposition of G. Finally, the width of the decomposition is verified by condition **1**.

We finish this section showing one more property of our verification algorithm, that is not required for the certification of the 3-approximation of the treewidth, but will be useful in the next section.

Claim 7: *For every $v \in V$ and every $j \in \{1, \ldots, d(v)\}$, the set $E_j(v)$ corresponds to the edges of graph induced by $B_j(v)$. We prove this claim by induction on $d(v)$. Suppose first that $d(v) = 1$. Since $F_1 = B_1$, we have that $F(u) = F(v)$ for every other vertex u in B_1. By* **Claim 2** *we obtain that v and u agree on the same set E_1. Then, by condition* **5** *on all the vertices in B_1, we deduce that $E_1 = E[G_1]$. Now suppose that the claim is true for every vertex of depth smaller than $d > 1$ and suppose that $d(v) = d$. By the induction hypothesis, for every $j \in \{1, \ldots, d-1\}$ the set $E_j(v)$ corresponds to the set of edges of $G[B_j(v)]$. Then, it remains to prove that $E_d(v)$ corresponds to the set of edges of $G[B_d(v)]$. Let w_1, w_2 be an arbitrary pair of vertices in $B(v)$, and call d_1 and d_2 the depth of w_1 and w_2, respectively. Without loss of generality assume that $d_1 \leq d_2$. By* **Claim 4** *applied to all vertices in the path of nodes between $B_d(v)$ and $B_{d_2}(w_2)$, we have that $E_{d_2}(w_2) = E_{d_2}(v)$. By condition* **6**, *we have that w_1, w_2 are adjacent if and only if $\{w_1, w_2\}$ belongs to $E_{d_2}(w_2)$. Suppose that $d_2 = d$. By* **Claim 2**, *we know that all nodes in $F(v)$ have the same main messages, in particular, they agree in the set $E_d(v)$. Then $E_{d_2}(v) = E_d(v)$. If $d_2 < d$, we have by condition* **5**, *that $w_1, w_2 \in E_{d_2}(v)$ if and only if $\{w_1, w_2\}$ belongs to $E_d(v)$. In both cases we deduce that $\{w_1, w_2\} \in E$ if and only if $\{w_1, w_2\} \in E_d(v)$.*

4 Certifying Regular Properties

In this section, we prove our main result, Theorem 1.

Theorem 2. *For every $k \geq 1$ and any regular graph property $\mathcal{P}(G)$, there exists a distributed certification protocol certifying that $\mathsf{tw}(G) \leq k$ and $\mathcal{P}(G)$ is true, using certificates on $O(\log^2 n)$ bits in n-node networks.*

For simplicity, we integrate the condition $\mathsf{tw}(G) \leq k$ to property $\mathcal{P}$, by setting $\mathcal{P}(G) = (\mathsf{tw}(G) \leq k) \wedge \mathcal{P}(G)$. The new property is regular because property $\mathsf{tw}(G) \leq k$ is regular (see, e.g., [23] for a discussion), and a conjunction of regular properties is regular by [8]. Basically, we enrich the protocol of Sect. 3 as follows. Either the protocol rejects because $\mathsf{tw}(G) > k$, or it constructs and certifies a tree decomposition at most $3k + 2$. In the latter case, we also certify property $\mathcal{P}$ using the tree decomposition of width $3k+2$ and the homomorphism classes $\mathcal{C}$ of the property on $(3k + 3)$-terminal graphs.

Fix the tree decomposition of width $3k + 2$. As in the sketch of proof of Proposition 1, for each node i of the decomposition tree, G_i denotes the $(3k+3)$-terminal graph corresponding to $G[V_i]$, with set of terminals B_i. Also, for each $w \in F_i$, let $Children(w)$ denote the set of children j of i such that w is in charge of node j (see Lemma 4 applied to j). In particular, the sets $Children(w)$ for $w \in F_i$ form a partition of the children nodes of i in the decomposition tree. Denote by $G_i[w]$ the $(3k+3)$-terminal graph obtained from $G[B_i \cup \bigcup_{j \in Children(w)} V_j]$ by choosing B_i as set of terminals. Note that if $Children(w)$ is empty, then $G_i[w]$ is simply the $3k + 3$-terminal base graph G_i^b corresponding to $G[B_i]$.

The **prover** appends two new pieces of information to the previous main messages of each vertex $v \in F_i$: the homomorphism class of G_i as well as the homomorphism class of $G_i[v]$. Moreover the homomorphism class of G_i is also added to the auxiliary message $m_{aux}(w, i)$ for every vertex w of the auxiliary tree $S(i)$. Note that this only ads a constant size to the previous main messages, since property $\mathcal{P}$ has a constant number of homomorphism classes. Auxiliary messages are increased by $O(\log n)$ bits, since each vertex w is in $O(\log n)$ auxiliary trees $S(i)$ by Lemma 5. Nevertheless, the constants here depend on k and on property $\mathcal{P}$.

We now update the **verification round** to exploit these new messages and check the property. As before, we use the auxiliary tree $S(i)$ to ensure that ℓ_i, and all vertices $v \in F_i$, have received from the prover the same isomorphism class for G_i.

It remains to check the consistency of the homomorphism classes for property $\mathcal{P}$ in the respective subgraphs.

Consistency of the Homomorphism Class of $G_i[v]$. Firstly, each vertex $v \in F_i$ in charge of some nodes must certify the homomorphism class of $G_i[v]$, in the sense that it compares the message received from the prover with the homomorphism class that he constructs from the nodes $j \in Children(v)$. Vertex v receives, for each $j \in Children[v]$, a message from ℓ_j with the homomorphism class of $\mathcal{P}$ restricted to the $(3k + 3)$-terminal graph G_j. Using Definition 2, it constructs the homomorphism class on G_j^+. Recall that $G_j^+ = f(G_j, G_i^b)$, i.e., G_j^+ is obtained by glueing G_j and the base graphs G_i^b induced by B_i, the glueing being performed by identifying the terminals of $B_j \setminus B_i$ in G_j to the corresponding vertices of B_i. Vertex v knows both sets B_i (which is in its initial message) and B_j (received from ℓ_j), so it has full knowledge of matrix $m(f)$ of the composition operation f. (There is a hidden technicality here. Node ℓ_j sends its main message to v in the unique communication round, and this message contains all bags $\mathcal{B}(\ell_j)$, in particular bag B_j. Node v can retrieve this bag, since its order in the list $\mathcal{B}(\ell_j)$, starting from the end of the list, is exactly the depth $i(v)$ of node i, plus one.) Then the homomorphism class of $h(G_j^+)$ is obtained as $\odot_f(h(G_j), h(G_i^b))$. Again v knows graph $G[B_i]$ hence it can compute its homomorphism class $h(G_i^b)$. It also knows $h(G_j^+)$ from ℓ_j, altogether v is able to compute the homomorphism class $h(G_j^+)$. Eventually, since $G[v]$ is obtained by glueing on B_i all graphs $G_j^+, j \in Children(v)$, v computes the homomorphism class of $G_i[v]$. If this class is not the same as the one received from the prover, vertex v rejects.

Consistency of the Homomorphism Class of G_i. Every vertex $v \in F_i$ checks the consistency between the message received from the prover as class of $\mathcal{P}$ on G_i, and the one it constructs from the glueing of all classes of $G_i[w]$ (that vertex w has received from the prover), for all $w \in F_i$, on B_i. Indeed, G_i is equal to the glueing, on B_i, of all graphs $G_i[w]$ with $w \in F_i$. Again, in case of inconsistency, vertex v it rejects.

Yes-instance. Every vertex belonging to F_r (the root node of the decomposition tree) accepts if the class of property $\mathcal{P}$ on G_r is an accepting one, otherwise it rejects.

Due to space restrictions, the soundness and completeness of the protocol are detailed in the full version. In a nutshell, the completeness is quite straightforward by construction of the messages. For the soundness, assume that all vertices accept. We proceed by induction on nodes i on the decomposition tree, from the leaves to the root, and show that the messages received by each $v \in F_i$ from the prover as homomorphism classes for $G_i[v]$ and G_i are correct. Eventually, since vertices of the root node accept, we conclude the homomorphism class of $\mathcal{P}$ on the whole graph is an accepting one, so $\mathcal{P}(G)$ is true.

5 Conclusion

To sum up, we proved that for every $k \geq 1$ and every MSO property on graphs, there exists a distributed protocol certifying that the input graph is of treewidth at most k and satisfies the required property, using certificates on $O(\log^2 n)$ bits. The result extends to optimisation problems, where we certify that a given vertex subset is of optimal weight (e.g., of maximum or of minimum size) for some MSO property, and the treewidth of the input graph is at most k.

The first natural question is whether we can reduce the size of certificate to $O(\log n)$ instead of $O(\log^2 n)$. We believe that such an improvement requires considerably different techniques, even for certifying that the treewidth of the input graph is at most k.

Another further research direction concerns certification versions for other algorithmic "meta-theorems". For example, given a graph property expressible by a first-order boolean formula, is there a distributed protocol certifying that the input graph is planar and satisfies the property, using certificates of logarithmic size?

Acknowledgment. The authors are thankful to Eric Remila for fruitful discussions on certification schemes related to the one considered in this paper.

References

1. Afek, Y., Kutten, S., Yung, M.: The local detection paradigm and its application to self-stabilization. Theor. Comput. Sci. **186**(1–2), 199–229 (1997)
2. Awerbuch, B., Patt-Shamir, B., Varghese, G.: Self-stabilization by local checking and correction. In: 32nd Symposium on Foundations of Computer Science (FOCS), pp. 268–277 (1991)
3. Balliu, A., D'Angelo, G., Fraigniaud, P., Olivetti, D.: What can be verified locally? J. Comput. Syst. Sci. **97**, 106–120 (2018)
4. Bick, A., Kol, G., Oshman, R.: Distributed zero-knowledge proofs over networks. In: ACM-SIAM Symposium on Discrete Algorithms (SODA) (2022)
5. Bodlaender, H.L.: NC-algorithms for graphs with small treewidth. In: 14th International Workshop on Graph-Theoretic Concepts in Computer Science (WG), vol. 344 of LNCS, pp. 1–10. Springer, 1988. https://doi.org/10.1007/3-540-50728-0_32
6. Bodlaender, H.L.: A partial k-arboretum of graphs with bounded treewidth. Theoret. Comput. Sci. **209**(1–2), 1–45 (1998)
7. Bonnet, É., Kim, E.J., Thomassé, S., Watrigant, R.: Twin-width I: tractable FO model checking. In: 61st IEEE Annual Symposium on Foundations of Computer Science, FOCS 2020, 16–19 November 2020, Durham, NC, USA, pp. 601–612. IEEE (2020)
8. Borie, R.B., Parker, R.G., Tovey, C.A.: Automatic generation of linear-time algorithms from predicate calculus descriptions of problems on recursively constructed graph families. Algorithmica **7**(5&6), 555–581 (1992)
9. Bousquet, N., Feuilloley, L., Pierron, T.: Brief announcement: local certification of graph decompositions and applications to minor-free classes. In: 35th International Symposium on Distributed Computing (DISC), vol. 209 of LIPIcs, pp. 49:1–49:4. Schloss Dagstuhl - Leibniz-Zentrum für Informatik (2021)
10. Bousquet, N., Feuilloley, L., Pierron, T.: Local certification of MSO properties for bounded treedepth graphs. arXiv:2110.01936 (2021)
11. Censor-Hillel, K., Paz, A., Perry, M.: Approximate proof-labeling schemes. Theor. Comput. Sci. **811**, 112–124 (2020)
12. Courcelle, B.: The monadic second-order logic of graphs. I. Recognizable sets of finite graphs. Inf. Comput. **85**(1), 12–75 (1990)
13. Courcelle, B., Engelfriet, J.: Graph Structure and Monadic Second-Order Logic. Cambridge University Press, Cambridge (2012)
14. Emek, Y., Gil, Y.: Twenty-two new approximate proof labeling schemes. In: 34th International Symposium on Distributed Computing (DISC), vol. 179 of LIPIcs, pp. 20:1–20:14. Schloss Dagstuhl - Leibniz-Zentrum für Informatik (2020)
15. Feuilloley, L., Fraigniaud, P., Hirvonen, J.: A hierarchy of local decision. Theor. Comput. Sci. **856**, 51–67 (2021)
16. Fraigniaud, P., Korman, A., Peleg, D.: Towards a complexity theory for local distributed computing. J. ACM **60**(5), 35:1–35:26 (2013)
17. Fraigniaud, P., Patt-Shamir, B., Perry, M.: Randomized proof-labeling schemes. Distrib. Comput. **32**(3), 217–234 (2019)
18. Göös, M., Suomela, J.: Locally checkable proofs in distributed computing. Theory Comput. **12**(1), 1–33 (2016)

19. Grohe, M., Kreutzer, S., Siebertz, S.: Deciding first-order properties of nowhere dense graphs. J. ACM **64**(3), 17:1–17:32 (2017)
20. Itkis, G., Levin, L.A.: Fast and lean self-stabilizing asynchronous protocols. In: 35th Annual Symposium on Foundations of Computer Science (FOCS), pp. 226–239 (1994)
21. Kol, G., Oshman, R., Saxena, R.R.: Interactive distributed proofs. In: ACM Symposium on Principles of Distributed Computing (PODC), pp. 255–264 (2018)
22. Korman, A., Kutten, S., Peleg, D.: Proof labeling schemes. Distrib. Comput. **22**(4), 215–233 (2010)
23. Liedloff, M., Montealegre, P., Todinca, I.: Beyond classes of graphs with "few" minimal separators: FPT results through potential maximal cliques. Algorithmica **81**(3), 986–1005 (2019)
24. Naor, M., Parter, M., Yogev, E.: The power of distributed verifiers in interactive proofs. In: ACM-SIAM Symposium on Discrete Algorithms (SODA), pp. 1096–1115 (2020)
25. Robertson, N., Seymour, P.D.: Graph minors. III. Planar tree-width. J. Comb. Theory Ser. B **36**(1), 49–64 (1984)

The Red-Blue Pebble Game on Trees and DAGs with Large Input

Niels Gleinig(✉) and Torsten Hoefler

ETH Zurich, 8092 Zurich, Switzerland
{niels.gleinig,torsten.hoefler}@inf.ethz.ch

Abstract. Data movements between different levels of a memory hierarchy (I/Os) are a principal performance bottleneck. This is particularly noticeable in computations that have low complexity but large amounts of input data, often occurring in "big data". Using the red-blue pebble game, we investigate the I/O-complexity of directed acyclic graphs (DAGs) with a large proportion of input vertices. For trees, we show that the number of leaves is a 2-approximation for the optimal number of I/Os. Similar techniques as we use in the proof of the results for trees allow us to find lower and upper bounds of the optimal number of I/Os for general DAGs. The larger the proportion of input vertices, the stronger those bounds become. For families of DAGs with bounded degree and a large proportion of input vertices (meaning that there exists some constant $c > 0$ such that for every DAG G of this family, the proportion p of input vertices satisfies $p > c$) our bounds give constant factor approximations, improving the previous logarithmic approximation factors. For those DAGs, by avoiding certain I/O-inefficiencies, which we will define precisely, a pebbling strategy is guaranteed to satisfy those bounds and asymptotics. We extend the I/O-bounds for trees to a multiprocessor setting with fast individual memories and a slow shared memory.

1 Introduction

Data movement between slow and fast memories (called I/O-transitions, or simply **I/Os**) are widely considered a principal performance bottleneck in computing [25]. Due to the ever increasing gap between the speed at which data can be processed and the speed at which it can be communicated, this phenomenon is particularly noticeable in computations that perform simple operations on large amounts of input data. This computation profile of **low arithmetic intensity + large amounts of input data** is very common and often referred to as *computations with memory bound performance.* For example, when performing inference on trained neural networks of fully connected layers (or other machine learning models), we need to read millions of values (mostly connection weights), but will use most of these values only once. Other problems that belong to this category can be found in Linear Algebra (addition of tensors or matrix-vector multiplication, either sparse or dense), Graph Theory (BFS, DFS, topological sorting), Statistics (computing moving averages, quantiles, covariances, linear

M. Parter (Ed.): SIROCCO 2022, LNCS 13298, pp. 135–153, 2022.
https://doi.org/10.1007/978-3-031-09993-9_8

regression, ANOVA, ANCOVA, etc.), and Image Processing (applying filters or affine transformations). Also, in general, most "big data" computations tend to belong to this category, because given the large amount of input data, it would not feasible to perform computational work of high complexity. Further, parallel summary statistics are a popular technique in the big data paradigm, naturally leading (on some level of granularity) to data dependencies that form trees.

In this paper, we will show how to perform I/Os efficiently in these computations with large input. As our main contribution, we present a set of rules on how to perform I/Os, together with several theorems that establish I/O-bounds and show: following these rules we are guaranteed to not use more than a constant number of times more I/Os than optimal. Interestingly, these rules are "local": they only require information about the neighbors. Hence they evade expensive precomputing of an I/O-optimal schedule and can be implemented at runtime.

1.1 Related Work

Communication-Efficient Algorithms. Communication-efficient algorithms have been developed for many concrete computational problems and models, including for example matrix multiplication [17,18], FFT [1,17], sorting [1], directed shortest paths [23], topological sorting [2], matrix transposition [1], the N-Body problem [14], QR- and LU-factorization [13], prime tables [5], and Cholesky decomposition [4]. In the blocked-I/O-model [1], using *time-forward processing* one can compute functions that have a given computation DAG $G = (V, E)$ with $O(\text{Sort}(|E|))$ I/Os [10]. In the particular case that the DAG is a tree, one can compute an Euler-tour of that tree with $O(\text{Sort}(|E|))$ I/Os and once the tree is laid out according to that Euler-tour, the function can be computed with $O(\text{Scan}(|E|))$ I/Os [21]. There are methods for I/O-efficient scheduling of tasks with tree-dependencies [22]. I/O-efficient algorithms are often closely tailored to the given problem. The red-blue pebble game allows to analyze and optimize the I/Os of *general* computations. For example, it has been used to optimize the I/Os of classical matrix multiplication [17,18], which can be considered very opposite to the computations of this paper as it allows extensive data reuse.

Pebble Games. There is a natural way to associate DAGs (directed acyclic graphs) to computations: vertices represent data and the direct predecessors of a vertex v are the data that are needed to compute v. Vertices with no incoming edges represent the input-data of the computation and those with no outgoing edges represent the output-data. Such a DAG is called a CDAG (computation DAG).

Pebble games are a family of combinatorial games that are played on DAGs [8,17]. By letting the DAG be the CDAG of a given computation, one can use them to analyze resource requirements of that computation: the black pebble game is for example used to model space complexity of general computations, whereas the reversible pebble game is used to model space complexity of reversible computations and the red-blue pebble game is used to model the requirement of I/Os.

Most pebble games are hard to solve (finding a pebbling strategy that is optimal with respect to some metric) or even approximate on general DAGs. It has been shown [16] that even the black pebble game, arguably the most simple pebble game, is PSPACE-complete to solve. Further, it is known that for many pebble games it is even PSPACE-hard to find approximate solutions [3,8,9,11]. It has been shown that the red-blue pebble game is PSPACE-complete and the red-blue pebble game without deletion is NP-complete [20].

There exists a polynomial time algorithm that approximates the solution of the red-blue pebble game (finding the minimal number of I/Os) by a multiplicative factor of $\log^{3/2}(n)$, using a number of red pebbles which is increased by a multiplicative factor of $\log^{3/2}(n)$ [7]. Yet, there is no known algorithm that approximates the solution of the red-blue pebble game by a constant multiplicative factor in polynomial time.

These hardness results suggest to search algorithms for pebble games on restricted classes of DAGs. This is furthermore motivated by the fact that the CDAGs of actual computations are a very restricted subset of all DAGs. They usually have a special structure (for example layered or planar), symmetries (large automorphism groups, i.e., they have many vertices that "look equal"), and other properties (for example regular or bounded in-degree) that can make it easier to find solutions. Hence, [12] raised the question whether for the red-blue pebble game there exist FPT algorithms for restricted classes of DAGs (such as bounded width graphs).

The main contribution of this paper are various theorems that give answers (some approximate, some exact) to the 3 problems that we present in the following section. We present lower and upper bounds that differ from each other by multiplicative constants that depend only on the degrees of a DAG and the proportion of input vertices, but not on the size of the DAG. In particular, for families of DAGs with bounded degrees and a large proportion of input vertices (meaning that there exists some constant $c > 0$ such that for every DAG G of this family, the proportion p of input vertices satisfies $p > c$), these bounds provide constant approximations.

1.2 The Red-Blue Pebble Game

The red-blue pebble game was introduced by Hong and Kung [17] to model the I/Os of a computation that has a given computation DAG $G = (V, E)$ and a fast memory of limited size $S \in \mathbb{N}^+$. Blue pebbles represent data that is stored in slow memory and red pebbles represent data that is stored in fast memory. There are no restrictions on the number of blue pebbles that can reside on G at any given time, but we can never have more than S red pebbles on G, where S represents the size of the fast memory. In the beginning, there is a blue pebble on each of the input vertices $V_{in} \subseteq V$ (those are the vertices with no incoming edges) and the game is completed when we have a blue pebble on each of the output vertices $V_{out} \subseteq V$ (those are the vertices with no outgoing edges). In order to obtain that goal, we are allowed to apply a sequence of the following actions

- R1 (Input): Replace a blue pebble by a red pebble.
- R2 (Output): Replace a red pebble by a blue pebble.
- R3 (Compute): Place a red pebble on a vertex for which all parents are carrying a red pebble (when the vertex has already a blue pebble, we replace it).
- R4 (Delete): Delete a red pebble.

We can think of action R3 as saying "we can compute a certain value, when all values that we need for that computation are in fast memory". Since the size of our fast memory is limited we will have to apply actions R2 and R4 sometimes.

Figure 1 shows a complete red-blue pebble game on a simple DAG. Keeping in mind that the I/Os in this model represent data movements that consume much time and energy, it becomes clear that we want to minimize their number.

Definition 11. *A **pebbling strategy** (or **computation strategy**) is a sequence of the actions R1, R2, R3, and R4 on the vertices of a DAG, which completes the red-blue pebble game (like for example* $[R1(v_1), R1(v_2), R3(v_3), R4(v_1)\ldots, R2(v_n)]$*). Each application of R1 or R2 counts as one **I/O** and we let* $rb(G, S)$ *denote the minimal number of I/Os of any pebbling strategy of the DAG G with S red pebbles, setting* $rb(G, S) = \infty$ *if there is no pebbling strategy of G with S red pebbles. A pebbling strategy which uses this minimal number* $rb(G, S)$ *of I/Os is called **I/O-optimal**. We say that it is possible to **pebble (or compute) G with k I/Os** if* $rb(G, S) \leq k$.

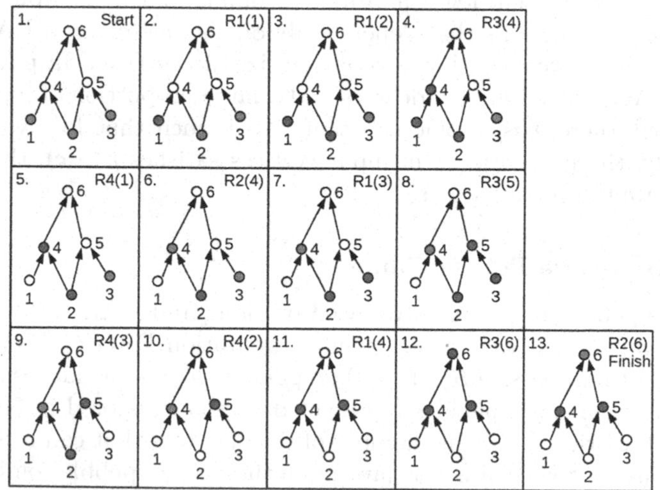

Fig. 1. Red-blue pebble game with $S = 3$ red pebbles. $Ri(j)$ denotes an application of action i on vertex j. This pebbling strategy is I/O-optimal and has a total of $rb(G, 3) = 6$ I/Os.

Problem 1. (Red-blue pebble game): given a DAG G and $S \in \mathbb{N}^+$, find the number $rb(G, S)$.

Solving problem 1 for general DAGs is PSPACE-complete [20], but this number can be bounded from above and below using different forms of optimal partitions and coverings of the DAG [6,15,17]. However, to the best of our knowledge, there is no known way to find those optimal partitions for general DAGs efficiently.

Problem 2: given a sequence of DAGs G_n, how fast does $rb(G_n, S)$ grow as $n \to \infty$?

If these DAGs G_n in problem 2 are the CDAGs of some algorithm for input size n, this asymptotic growth is generally known as the **I/O-complexity of the algorithm**. The minimum of the I/O-complexities of all algorithms that solve a given problem is known as the **I/O-complexity of the problem**. Here we are typically satisfied with answers that disregard constant multiplicative factors, i.e., pebbling strategies are called optimal if they use $\Theta(rb(G_n, S))$ I/Os. I/O-complexities have been studied for many algorithms and problems of practical importance [1,2,4,5,13,17,23,24], often in a more general model that considers additional parameters such as memory block size.

Now if the vertices of the CDAG of a computation represent values produced by constant-time operations and the motivation for studying Problems 1 or 2 (or analogous problems for other pebble games) is to improve the performance of that computation, it is important to note the following: the sequential runtime is given, up to constants, by the number of vertices $|V|$ and therefore any algorithm that solves Problems 1 or 2 in time above $\Omega(|V|)$ is of little practical value (unless one runs the same CDAG many times, letting the performance gains add up and justify a more expensive algorithm to solve problems 1 or 2). So for example algorithms that are based on spectral methods are likely to be inappropriate, since the time for computing eigenvalues or eigenvectors is at least of the order of $|V|^2$. Likewise, for the practical value of such an algorithm, it is also important that it does not use too many other resources. This imposes great restrictions on us and motivates us to ask if it is possible to know how to run an algorithm in a close to I/O-optimal way without having to do any pre-computations at all on the CDAG $G = (V, E)$. Could it be possible to find a set of simple rules, such that any computation strategy that follows these rules is close to I/O-optimal? More precisely, we are interested in the following question:

Problem 3: given a DAG G, $S \in \mathbb{N}^+$ and a constant $1 \leq \lambda \in \mathbb{R}$, is it possible to find a set of simple rules, such that any computation strategy for G which follows these rules does not use more than $\lambda \cdot rb(G, S)$ I/Os?

Note: our rules do not allow a vertex to have a blue and a red pebble at the same time. We chose this version because it makes the proofs more elegant. However, all of our results (both the bounds and the algorithms) still hold if we allow a vertex to hold both a blue and a red pebble (by using the word "place" instead of "replace" in rules R1 and R2 and removing the comment in the parentheses in R3). The upper bounds remain true, because by having this additional degree of freedom the pebble strategies can only improve. The lower bounds remain true, because even if we are allowed to have two pebbles on one vertex, we will still have to spend one I/O on each of the input vertices. Trees can be pebbled optimally without ever having to put two pebbles on one vertex.

The paper is organized as follows. In Sect. 2 we present and prove several lower and upper bounds for the number of I/Os for trees. In Sect. 3 we extend the bounds to more general classes of DAGs. In Sect. 4 we extend our bounds to a parallel setting. In Sect. 5 we present our experimental results.

1.3 Notations and Definitions

Let $G = (V, E)$ be a DAG. Throughout this paper we say that a vertex w is a child of the vertex v if there is an edge from v to w. A vertex v is a parent of w if and only if w is a child of v. We let $d_{out}(v)$ denote the number of children of v (we only consider simple DAGs), $d_{in}(v)$ the number of parents of v and $d(v) = d_{in}(v) + d_{out}(v)$ the degree of v. We let

$$\begin{aligned} \hat{d}_{out} &= \max_{v \in V} d_{out}(v), \\ \hat{d}_{in} &= \max_{v \in V} d_{in}(v) \\ \hat{d} &= \max_{v \in V} d(v) \end{aligned} \tag{1}$$

denote the maximal out-degree, maximal in-degree, and maximal degree respectively.

An in-tree is a directed tree in which all edges are directed in such a way that there is a unique vertex v (called the root or output vertex) which can be reached from any other vertex by a directed path (like in Fig. 2). Throughout this paper, all trees are in-trees. A vertex that has no in-coming edges is called a leaf and the set of all leaves is denoted $L(G)$.

2 Bounds for Trees

Notice that the condition of $S > \hat{d}_{in}$ which we require in our results is equivalent to letting S be large enough such that G can be computed.

Theorem 1. *Let $G = (V, E)$ be a tree with more than one vertex and $S > \hat{d}_{in}$. Then, the set of leaves $L(G)$ satisfies*

$$|L(G)| < rb(G, S) \leq 2|L(G)|. \tag{2}$$

Furthermore, when $|L(G)| > 1$ the second inequality is also strict.

Proof (Proof of Theorem 1*).* Since the computation depends on all of the leaves, we will have to put at least once a red pebble on each of the leaves. We furthermore will have to spend one I/O to store the output. This gives us the first inequality.

The second inequality we will prove by induction on the number of leaves $|L(G)|$. One can easily check that the inequalities are true for $|L(G)| = 1$ and $|L(G)| = 2$ (with strictness for $|L(G)| = 2$). Now let us assume that it also holds for trees with $n > 2$ leaves and let G be a tree with $n + 1$ leaves.

Without loss of generality the tree has no vertices of in-degree 1, because on trees such vertices can be removed (merging the incoming edge with the outgoing edge) without changing the number of I/Os. We can also assume that the root v has degree at least 2, because otherwise we could remove the root and the incoming edge without changing the number of I/Os. So let w be a parent of v and G^w be the sub-DAG below w (that is, all ancestors of w and w itself) and $\overline{G^w}$ be the DAG that we obtain by deleting from G all vertices and edges that are in G^w and the edge that goes from w to v (see Fig. 2). Then $|L(G^w)| \leq n$ and $|L(\overline{G^w})| \leq n$. So by induction hypothesis these trees can be pebbled with at most $2 \cdot |L(G^w)| - 1$ and $2 \cdot |L(\overline{G^w})| - 1$ respectively. Now consider the following pebbling strategy:

1. Pebble G^w I/O-optimally (getting a blue pebble on w)
2. Pebble $\overline{G^w}$ I/O-optimally and when you have red pebbles on all of the parents of v (and are about to put a red pebble on v), put a red pebble on w first and then put a red pebble on v and finish the game.

This is clearly a complete pebbling strategy and since $|L(G^w)|+|L(\overline{G^w})| = |L(G)|$ the number of I/Os is at most

$$(2 \cdot |L(G^w)| - 1) + (2 \cdot |L(\overline{G^w})| - 1) + 1 = 2 \cdot |L(G)| - 1. \tag{3}$$

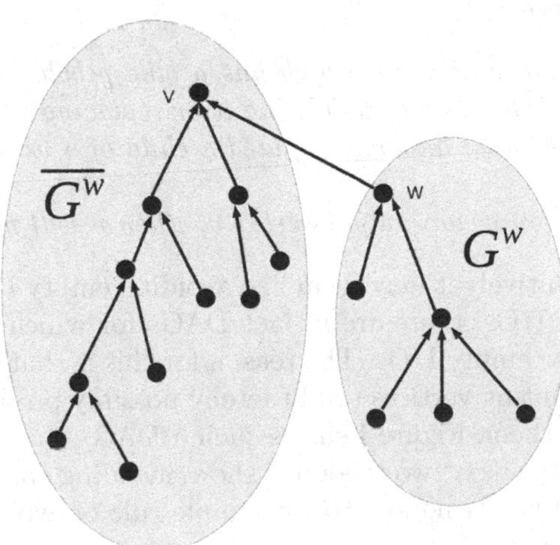

Fig. 2. This shows the subtrees G^w and $\overline{G^w}$ from the proof of Theorem 1.

Note: the multiplicative constants 1 and 2 from the inequalities in the previous theorem are tight:

- To see that the constant 1 is tight, notice that whenever $S > |L(G)|$, we have $rb(G, S) = |L(G)| + 1$. Also, for trees that consist of a line of depth n, with each of the first $n-1$ vertices on that line having one additional predecessor that is a leaf (like in Fig. 3), we have $rb(G, S) = |L(G)| + 1$ independently of S.
- If T_k^2, denotes a full binary tree of k levels, then

$$\lim_{k \to \infty} \frac{rb(T_k^2, 3)}{\mid L(T_k^2) \mid} = 2. \tag{4}$$

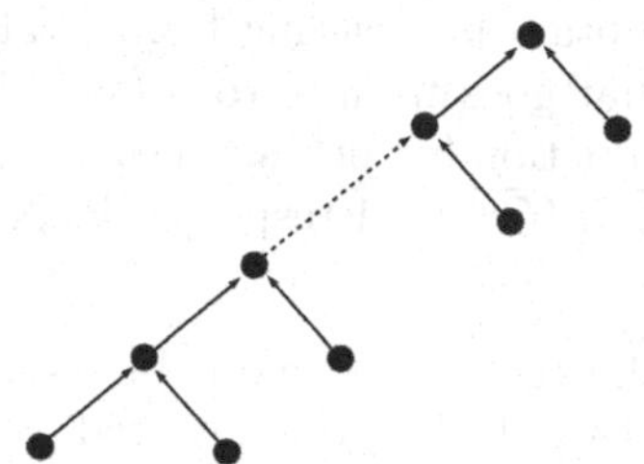

Fig. 3. Unbalanced trees require few I/Os. "Ladder trees" like this one, minimize the number of I/Os among all trees with a fixed number of leaves.

To present our next result we need the following definition.

Definition 21. *We say that a computation has* **empty I/Os** *if it does at least one of the following things:*

1. *Put a red pebble on a vertex v which has a blue pebble on it and eventually remove the red pebble (either by deleting it or replacing it with a blue one) or finish the game, without having computed a child of v which had not yet been computed before.*
2. *Put a blue pebble on a non-output vertex v, when v will not be needed again.*

Even though intuitively it may seem like avoiding empty I/Os always reduces the total number of I/Os, there are in fact DAGs for which every I/O-optimal pebbling strategy has empty I/Os. The reason for this is that one can sometimes save I/Os by recomputing vertices (and thereby possibly producing empty I/Os) rather than loading them. Figure 4 shows such a DAG.

Nevertheless, as the next two theorems show, avoiding empty I/Os does guarantee provably good performance. So the simple rule of avoiding empty I/Os is an answer to problem 3.

Theorem 2. *Let $G = (V, E)$ be a tree with more than one vertex and without vertices of in-degree* 1 *and $S > \hat{d}_{in}$. Then, any computation strategy of G which has no empty I/Os has a number of I/Os that is between $|L(G)|$ and $3 \cdot |L(G)|$ and hence optimal up to a multiplicative factor of 3.*

Proof (Proof of Theorem 2). That the number of I/Os is larger than $|L(G)|$ follows again from the fact that we will have to put at least once a red pebble on each of the leaves.

For the other inequality we recall that any tree which has no vertices of in-degree 1 satisfies

$$| I(G) | \leq | L(G) |, \tag{5}$$

where $I(G) = V \setminus L(G)$ denotes the set of inner vertices.

When we put a blue pebble on a vertex which has a red pebble on it, we will eventually put again a red pebble on it, because otherwise we would have a empty I/O of the second type. When we put a red pebble on a vertex v which has a blue pebble on it, we will compute *child*(v) before doing any other I/Os on v, because otherwise we would have a empty I/O of the first type. Once *child*(v) has been computed, we will not do any other I/Os on v, because v is no longer needed. It follows that we spend at most one I/O on any leaf and at most two I/Os on any inner vertex. Combining this with Eq. (5) we conclude that the total number of I/Os is less than

$$| L(G) | + 2 | I(G) | \leq 3 | L(G) | . \tag{6}$$

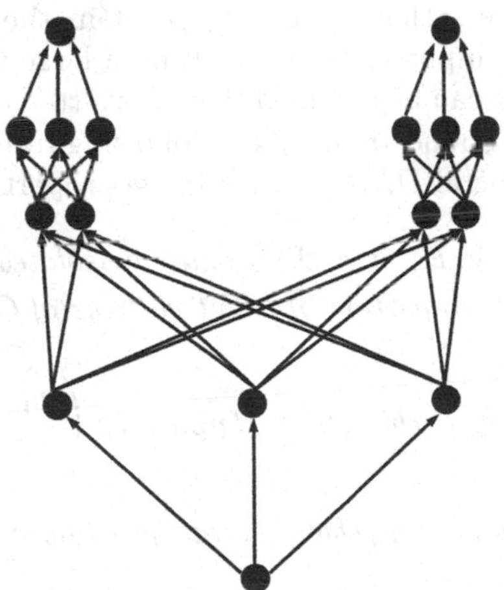

Fig. 4. When $S = 5$ this DAG can be pebbled with $rb(G, 5) = 5$ I/Os, but any pebbling strategy without empty I/Os has at least 7 I/Os (and when we allow a vertex to have two pebbles at the same time, the numbers become 4 and 6). The reason is that any pebbling strategy will need to have red pebbles on the three vertices of the second layer at two different moments. For that, the optimal strategy would reload the input and then recompute those three vertices, whereas a strategy without empty I/Os would have to reload the three vertices in order to avoid empty I/Os.

Note: when a pebbling strategy for trees without vertices of in-degree 1, besides from having no empty I/Os, also

1. "computes subtrees strictly one after another" (that is, whenever it started to perform computations on some subtree t of T it does not do any computations that are not needed for t, until it has finished with the computation of t),
2. whenever a vertex is computed, the pebbles from the parents are immediately deleted,
3. does not perform store operations unless there are S red pebbles on the DAG,

then it is a pebbling strategy like the one from the proof of Theorem 1 and hence it is optimal up to a multiplicative factor of 2.

This order ("computing the subtrees strictly one after another") is also known as *postorder*, and related to previous work [19,22]. Using postorders, one can also find exact solutions to the red-blue pebble game on trees. To do this, one needs to determine in which order the subtrees can be traversed most efficiently: depending on the order in which we traverse subtrees, it can be possible to hold the output of one subtree in fast memory while computing the next subtree, sparing 2 I/Os. Finding such an optimal order of the subtrees can be done with dynamic programming approaches. However, we will not discuss the details of that in this paper since our goal is to find general bounds for general DAGs.

3 Bounds for General DAGs

The key property of trees that we made use of in the proof of the previous results is that the set of input vertices makes up a large proportion of the set of all vertices (or at least it can be assumed that it is a tree with that property) and the out-degree is small (bounded by 1). The following result gives good estimates of $rb(G, S)$ for more general DAGs G with those properties.

Theorem 3. *Let $G = (V, E)$ be a DAG without isolated vertices and $S > \hat{d}_{in}$. Let $p = \frac{|Input(G)|}{|V|}$ be the proportion of input vertices of G. Then,*

$$|Input(G)| \leq rb(G, S) \leq |Input(G)| \left(\frac{2\hat{d}_{out}}{p} - 1 \right) \tag{7}$$

and the number of I/Os of any pebbling strategy without empty I/Os is between these bounds.

Proof (Proof of Theorem 3). Since any input vertex has to be loaded at least once we get $|Input(G)| \leq rb(G, S)$. For the other inequality notice that in any pebbling strategy without empty I/Os, with every [store, load] pair of I/Os on one vertex we compute at least one new child, because otherwise we would get a empty I/O. So we spend at most $2 \cdot \hat{d}_{out}$ I/Os on any non-input vertex. Likewise, one can see that we spend at most $2\cdot\hat{d}_{out}-1$ I/Os on any input vertex. Therefore the total number of I/Os is at most

$$\begin{aligned} &|Input(G)|\left(2\hat{d}_{out}-1\right)+|V\backslash Input(G)|2\hat{d}_{out} \\ &=2\hat{d}_{out}|V|-|Input(G)| \\ &=|Input(G)|\left(\frac{2\hat{d}_{out}}{p}-1\right). \end{aligned} \tag{8}$$

Theorem 4. *Let $G_n=(V_n,E_n)$ be a sequence of regular DAGs without isolated vertices and with constant in-degree $d_{in}<S$ (that is, any vertex of G_n which is not an input vertex has in-degree d_{in}) and constant out-degree d_{out} (that is, any vertex of G_n that is not an output vertex has out-degree d_{out}). If $d_{out}<d_{in}$ then $rb(G_n,S)$ grows asymptotically like $|V_n|$. More precisely,*

$$|V_n|\left(1-\frac{d_{out}}{d_{in}}\right)\leq rb(G_n,S)\leq|V_n|\left(\frac{2\cdot d_{out}}{1-\frac{d_{out}}{d_{in}}}-1\right) \tag{9}$$

and the number of I/Os of any pebbling strategy without empty I/Os is also between these bounds.

Proof (Proof of Theorem 4). Since the sum of the in-degrees over all vertices equals the sum of the out-degrees, we have

$$(|V_n|-|Input(G_n)|)\cdot d_{in}=(|V_n|-|Output(G_n)|)\cdot d_{out}. \tag{10}$$

From this we obtain

$$\begin{aligned} |Input(G_n)|&=|V_n|-(|V_n|-|Output(G_n)|)\cdot\frac{d_{out}}{d_{in}} \\ &\geq|V_n|\cdot\left(1-\frac{d_{out}}{d_{in}}\right), \end{aligned} \tag{11}$$

which means that the proportion of input vertices is $p\geq\left(1-\frac{d_{out}}{d_{in}}\right)$. The result follows now by substituting p by $\left(1-\frac{d_{out}}{d_{in}}\right)$ in the RHS of the inequalities of Theorem (3) and replacing the LHS of those inequalities by the RHS of (11).

A remarkable aspect of the previous results is that the bounds only require $S>\hat{d}_{in}$, but besides that they do not depend on S. So they identify families of DAGs for which the I/O-complexity does not depend on the size of the fast memory.

Theorem 5. *Let $G_n=(V_n,E_n)$ be a sequence of DAGs without isolated vertices and $S>\hat{d}^n_{in}$. Let G_n have average in-degree $\bar{d}^n_{in}$ (where the average is taken over all vertices that are not input vertices), average out-degree $\bar{d}^n_{out}$ (where the* *average is taken over all vertices that are not output vertices) and maximal out-degree* $\hat{d}^n_{out}$. *Then,*

$$|V_n| \left(1 - \frac{\bar{d}^n_{out}}{\bar{d}^n_{in}}\right) \leq rb(G_n, S) \leq |V_n| \left(\frac{2 \cdot \hat{d}^n_{out}}{1 - \frac{\bar{d}^n_{out}}{\bar{d}^n_{in}}} - 1\right) \tag{12}$$

and hence, if

$$\limsup_{n\to\infty} \bar{d}^n_{out}/\bar{d}^n_{in} < 1 \tag{13}$$

and $\hat{d}^n_{out} \leq c, \forall n$ *for some constant* $c \in \mathbb{N}$, *then* $rb(G_n, S)$ *grows asymptotically like* $|V_n|$. *Furthermore any pebbling strategy without empty I/Os satisfies these bounds.*

Proof (Proof of Theorem 5*).* Like in the previous proof the Eq. (10) still holds if we replace d_{in} by $\bar{d}^n_{in}$ and d_{out} by $\bar{d}^n_{out}$. The rest of the proof is analogous.

For algorithms whose CDAGs have vertices that represent values produced by constant-time operations, this previous theorem can be interpreted as saying: "if the out-degrees are smaller than the in-degrees (differing on average by a multiplicative constant which is bounded away from 1) and the maximal out-degree is uniformly bounded, then the I/O-complexity equals the time complexity (which is linear in the input size)."

4 I/O-Bounds in a Parallel Setting

In order to establish I/O-bounds in multiprocessor settings, we introduce a variation of the red-blue pebble game. We model a setting with P processors, each one having a fast memory of size S/P and access to a shared memory of infinite size. Throughout this section we assume S/P is an integer. There are P different types of red pebbles, corresponding to the fast memories of the P processors. We denote the colors of these pebbles $r_1, r_2, \ldots, r_P$. At any time step and for any processor $p \in \{1, 2, \ldots, P\}$ there cannot be more than S/P pebbles of color r_p on the DAG, and we can perform one of the following actions

- R1 (Input): replace a blue pebble by a pebble of color r_p.
- R2 (Output): replace a pebble of color r_p by a blue pebble.
- R3 (Compute): place a pebble of color r_p on a vertex for which all parents are carrying a pebble of color r_p (when the vertex has already a blue pebble, we replace it).
- R4 (Delete): delete a pebble of color r_p.

In this parallel setting we say that a pebbling strategy has empty I/Os if it does at least one the following:

1. Put a pebble of some color $r_p \in \{r_1, r_2, \ldots, r_P\}$ on a vertex v which has a blue pebble on it and eventually remove the red pebble (either by deleting it or replacing it with a blue one) or finish the game, without having computed a child of v which had not yet been computed before by any of the processors.
2. Put a blue pebble on a non-output vertex v, when v will not be needed again.

As in the sequential case, initially, there is a blue pebble on each of the input vertices, and the goal is to get blue pebbles on all output vertices. In contrast to the standard red-blue pebble game, we are now able to perform up to P actions at each time step, one for each processor. Again, we want to minimize the total number of applications of R1 and R2, but in this setting we also want to minimize another quantity: the total amount of time steps. Since it is possible to save I/Os by taking more time steps (for example, by doing all the work with one single of the P processors), we will enforce the pebbling strategy to use a minimal number of time steps. That is, we want to know: what is the minimal number of I/Os used by pebbling strategies that pebble G in a minimal number of time steps.

For a given DAG G, we now let $T_P(G)$ denote the minimal number of time steps used by any strategy that pebbles G with P processors. We define $rb_P(G, S)$ as the minimal number of I/Os used by any pebbling strategy that pebbles G in $T_P(G)$ time steps. The next theorem shows that enforcing a parallel schedule that uses a minimal number $T_P(G)$ of time steps (which is a restriction on the orders in which we can pebble the vertices), does not affect the validity of the I/O-bounds established by Theorem 2, and hence, does not increase the number of I/Os by more than a factor 3.

Theorem 6. *Let G be a tree with $\hat{d}_{in} < S/P$. Then,*

$$|L(G)| \leq rb_P(G, S) \leq 3 \cdot |L(G)|, \tag{14}$$

and the number of I/Os of any pebbling strategy without empty I/Os is between these bounds.

To prove this Theorem we need a Lemma that shows how we can remove empty I/Os from time optimal pebbling strategies without losing time-optimality.

Lemma 1. *Let G be a tree with $\hat{d}_{in} < S/P$ and consider a pebbling strategy that pebbles G in optimal time $T_P(G)$ (possibly having empty I/Os). Then, we can transform this pebbling strategy into another pebbling strategy that uses the same number of time steps (and is hence also time optimal), but does not have any empty I/Os.*

Proof. Given a pebbling strategy that pebbles G in $T_P(G)$ steps we can remove all empty I/Os of the second type and obtain another valid strategy that pebbles G in at most the same number of steps (and hence is still time optimal).

It remains to show that we can also remove empty I/Os of the first type. Notice that empty I/Os of the first type occur in two cases: 1) We load a value, but we do not compute the child afterwards, 2) We load a value and we *do* compute the child, but the child has already been computed earlier.

In the first case, we can again remove the I/O without losing the completeness or time-optimality of the pebbling strategy.

Now consider a situation of the second case, where we place a red pebble on some vertex v and we *do* compute its child $child(v)$, but $child(v)$ has already been computed earlier. In this case the steps of loading v and recomputing $child(v)$ (which together take 2 time steps), could be replaced by two other steps: storing the value of $child(v)$ and eventually reading $child(v)$ again (this replacement increases the number of I/Os by one, but this is irrelevant to this proof as it does not increase the number of time steps).

With these changes we can transform any time optimal pebbling strategy *with* empty I/Os into a pebbling strategy *without* empty I/Os.

Proof (Proof of Theorem 6). The lower bound is established with the same reasoning as in the sequential model: all leaves have to be read at least once.

Now consider any P processor pebbling strategy that pebbles G in $T_P(G)$ time steps and use Lemma 1 to transform it into another pebbling strategy that runs in $T_P(G)$ time steps and does not have empty I/Os. We can associate to this P processor pebbling strategy a sequential strategy without empty I/Os (which we will call the "sequentialized" pebbling strategy):

- In the first time step, perform the action that processor 1 performed in the first time step,
- In the second time step perform the action that processor 2 performed in the first time step,

...

- In the P-th time step, perform the action that processor P performed in the first time step,
- In the $(P+1)$-th time step, perform the action that processor 1 performed in the second time step,

...

Since for any $p \in \{1, 2, \ldots, P\}$ there are not more than S/P pebbles of color r_p on the tree at any moment, the sequentialized pebbling strategy uses not more than $P \cdot (S/P) = S$ red pebbles. Since it also does not perform empty I/Os (now we mean "empty I/Os" in the sense of the sequential model), it follows from Theorem 2 that it uses at most $3 \cdot |L(G)|$ I/Os. Since the parallel pebbling strategy uses the same number of I/Os as the sequentialized strategy, we obtain $rb_P(G, S) \leq 3 \cdot |L(G)|$.

5 Experiments

To illustrate the theoretical contributions of our paper, we ran several experiments. The purpose of the experiments is to show the effect that the pebbling techniques used in this paper (avoiding empty I/Os, computing subtrees one after another) have on concrete I/O-counts and to see how close to the bounds we get on different trees and random DAGs.

5.1 Regular Tree, Varying Memory Size

We considered a binary tree of depth 8 and pebbled it in two different orders of the vertices: **A**) Level after level, and **B**) subtrees one after another. In both cases we performed I/Os according to the following rules: 1.) We only put a red pebble on a vertex when we need it for the computation of the next vertex (in the given order), 2.) we only remove a red pebble when there is no more space but we need to place another red pebble for the next computation step, 3.) when we need to remove a red pebble we choose it uniformly at random among the red pebbles that we will not need in the next computation step, and 4.) after each computation step we delete all red pebbles that will not be needed again in any future computations. By following these rules we are guaranteed to not perform empty I/Os (rules 1, 2, and 3 imply that we do not perform an empty I/O of the first kind; rule 4 implies that we do not perform an empty I/O of the second type). Hence, according to Theorem 1, using order **A** we should have between $|L(G)| = 256$ and $2 \cdot |L(G)| = 512$ I/Os, and according to Theorem 3, with order **B** we should have between $|L(G)| = 256$ and $3 \cdot |L(G)| = 768$ I/Os. In Fig. 5a we show how the number of I/Os decreases as we increase the memory size S. For each memory size we pebbled the tree 10 times and show bars that represent the smallest and largest observed number of I/Os (the variation stemming from the random choice when removing red pebbles). We can see that for both computation orders, the respective bounds are satisfied for all memory sizes (as predicted by our theorems). Furthermore, when the memory size is close to zero, the number of I/Os is close to the respective upper bound. When we increase the memory size, the number of I/Os decreases and eventually meets the lower bound. With computation order **A** we already meet the lower bound with memory size $S = 9$, whereas with order **B** this minimum is only attained when $S = 129$.

5.2 Random DAGs with Large Inputs

To generate more general DAGs with large input, we used the random sampling method described in Algorithm 1 (the constants 4 and 8 in this sampling method are arbitrary constants; we just had to fix free parameters). Notice that for $l = 0$, Algorithm 1 generates a tree, but as we increase l, we obtain DAGs that differ increasingly from a tree. However, the total number of vertices and input vertices remains equal (because we only add edges (i, j) for which j is not an input). Figure 5b shows how the number of I/Os increases as we add more edges to the DAG. The shown numbers are the average over 10 sampled DAGs and the bars correspond to the smallest and largest observed values. The shown bounds are obtained from Theorem 3. We can see that the lower bound remains constant (which follows from the fact that the input size of these DAGs is constant), while the upper bound increases linearly with the number of edges that we add (as this impacts the degrees of the vertices). Using a memory size $S = 100$ in the shown range, we use a number of I/Os that matches the lower bound exactly. In the whole range, the lower and upper bound differ by a factor of less than 2.

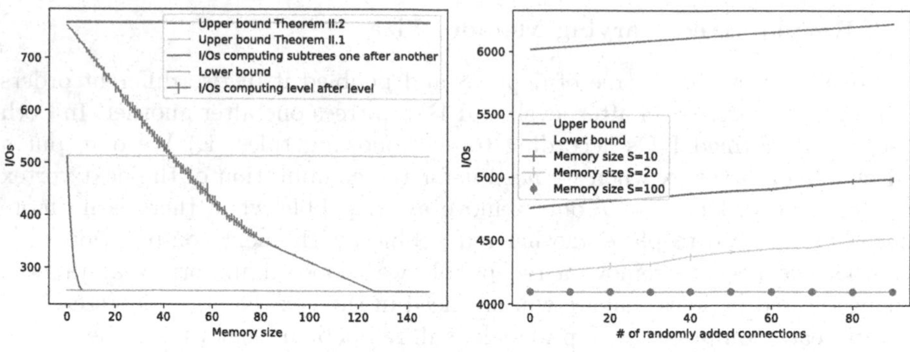

(a) Number of I/Os for varying memory sizes when pebbling a binary tree of depth 9.

(b) Starting with a regular tree of depth 4 and degree 8, we randomly add directed edges to it.

Fig. 5. (a) Number of I/Os for varying memory sizes when pebbling a binary tree of depth 9. (b) Starting with a regular tree of depth 4 and degree 8, we randomly add directed edges to it.

Algorithm 1. Input: a number $l \in \mathbb{N}$ /Output: a DAG with $8^5 - 1$ vertices and $8^5 - 2 + l$ edges.

Let G be a full 8-ary tree of depth 4 and identify V with the numbers $\{1, 2, \dots 8^5 - 1\}$, 1 being the root, $2, \dots, 9$ being the vertices of the first level, and so on;
for i from 1 to l **do**
  Choose uniformly at random vertices $i, j \in G$, such that j is not an input, $(i, j) \notin G$, and $j < i$;
  Add (i, j) to G;
end for

5.3 General Random DAGs with Varying Numbers of Vertices

Finally, we considered a second method for sampling random DAGs, defined in Algorithm 2 (also here the constants are arbitrary). We used this method with different values of l, to generate DAGs with varying numbers of vertices. In Fig. 6 we show how the number of I/Os grows as the number of vertices increases. We can clearly see that the I/O-complexity grows linearly with the number of vertices, as predicted by Theorem 5.

Algorithm 2. Input: a number $l \in \mathbb{N}$ /Output: a DAG with $3 \cdot l + 1$ vertices.

Let G be a graph with a single vertex
for i from 1 to l **do**
  Choose uniformly at random a vertex $v \in G$ with $d(v) < 9$;
  Add 3 new vertices to G;
  Add 3 edges, pointing from the 3 new vertices to v;
end for
Add 30 additional random edges (as in the previous subsection);

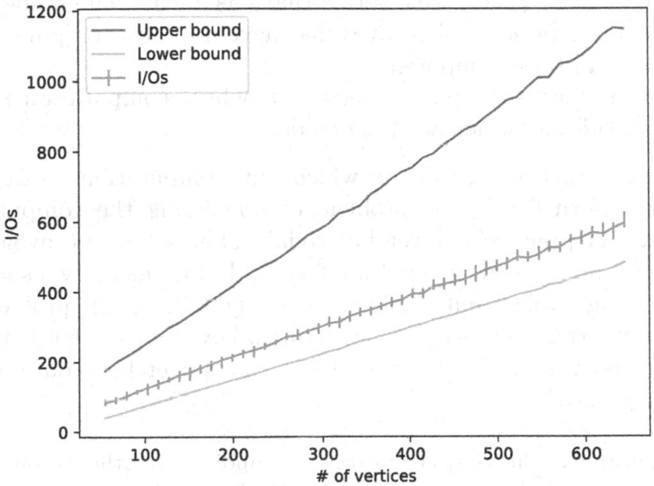

Fig. 6. Number of I/Os for varying sizes of randomly generated DAGs using memory size 32.

6 Conclusions

We introduce a set of I/O-rules that can be used as a fundamental tool in the design of I/O-efficient big data computations, both in a sequential and parallel setting. We prove several bounds for the optimal number of I/Os. These bounds are particularly strong when the DAGs have a large proportion of input vertices (meaning that there exists some constant $c > 0$ such that for every DAG G of this family, the proportion p of input vertices satisfies $p > c$). For these DAGs our bounds provide constant factor approximations, which improves the previous logarithmic approximation factors. Any pebbling strategies with exclusively useful (non-empty) I/Os have a number of I/Os within these bounds.

The rule of avoiding empty I/Os is a "local" rule: to decide whether or not it is possible to do any further non-empty I/Os on a given vertex v it suffices to have information about the children of v (have they all been computed or not). Yet we showed that any computation strategy that follows this I/O-rule is guaranteed to be I/O-optimal up to a multiplicative factor that is particularly small when the proportion of input vertices is large and the maximal out-degree is small. For trees, this multiplicative factor is 3 and it can be made 2 by adding some other I/O-rules (computing subtrees strictly one after another).

This raises some questions: could our results be improved if we add more sophisticated and "global" rules (rules that take global properties into account, such as depth, width, or average degree)? Could we get tighter bounds or bounds that are tight on larger classes of DAGs, if we set rules that regulate how we choose at each step the next vertex that we compute? The following rules would be natural candidates:

- "Always continue to compute a vertex that has a maximal number of parents with red pebbles (among all vertices that have not been computed and whose parents have all been computed)."
- "Always continue to compute a vertex for whose computation the smallest number of load-operations needs to be done."

For computations of CDAGs on which any computation strategy without empty I/Os performs well, the problem of scheduling the computation I/O-efficiently may in practice not yet be trivial. This is because avoiding empty I/Os is in fact non-trivial, given that fast and slow memory usually do not communicate single values but whole blocks of them. Here, the problem becomes a problem of external memory data structures: how should we lay out the data in external memory, such that for every block that we fetch, all data in this block is used at least once?

Acknowledgement. This project has received funding from the European Research Council (ERC) under the European Union's Horizon 2020 research and innovation programme grant agreement No 101002047 and from the European High-Performance Computing Joint Undertaking (JU) under grant agreement No.101034126.

References

1. Aggarwal, A., Vitter, J.S.: The input/output complexity of sorting and related problems. Commun. ACM **31**(9), 1116–1127 (1988)
2. Arge, L., Toma, L., Zeh, N.: I/o-efficient topological sorting of planar DAGs. In: Proceedings of the Fifteenth Annual ACM Symposium on Parallel Algorithms and Architectures, SPAA 2003, pp. 85–93, New York, NY, USA. ACM (2003)
3. Austrin, P., Pitassi, T., Wu, Y.: Inapproximability of treewidth, one-shot pebbling, and related layout problems. CoRR, abs/1109.4910 (2011)
4. Ballard, G., Demmel, J., Holtz, O., Schwartz, O.: Communication-optimal parallel and sequential cholesky decomposition. CoRR, abs/0902.2537 (2009)
5. Bender, M.A., et al.: The i/o complexity of computing prime tables, pp. 192–206 (2016)
6. Bilardi, G., Pietracaprina, A., D'Alberto, P.: On the space and access complexity of computation DAGs. In: Brandes, U., Wagner, D. (eds.) WG 2000. LNCS, vol. 1928, pp. 47–58. Springer, Heidelberg (2000). https://doi.org/10.1007/3-540-40064-8_6
7. Carpenter, T., Rastello, F., Sadayappan, P., Sidiropoulos, A.: Brief announcement: approximating the i/o complexity of one-shot red-blue pebbling. In: Proceedings of the 28th ACM Symposium on Parallelism in Algorithms and Architectures, SPAA 2016, pp. 161–163, New York, NY, USA. ACM (2016)

8. Chan, S.M.: Pebble Games and Complexity. PhD thesis, EECS Department, University of California, Berkeley, August 2013
9. Chan, S.M., Lauria, M., Nordström, J., Vinyals, M.: Hardness of approximation in PSPACE and separation results for pebble games, pp. 466–485 (2015)
10. Chiang, Y.-J., Goodrich, M., Grove, E., Tamassia, R., Vengroff, D., Vitter, J.: External-memory graph algorithms, May 1995
11. Demaine, E.D., Liu, Q.C.: Inapproximability of the standard pebble game and hard to pebble graphs, pp. 313–324 (2017)
12. Demaine, E.D., Liu, Q.C.: Red-blue pebble game: complexity of computing the trade-off between cache size and memory transfers. In: Proceedings of the 30th on Symposium on Parallelism in Algorithms and Architectures, SPAA 2018, pp. 195–204, New York, NY, USA. ACM (2018)
13. Demmel, J., Grigori, L., Hoemmen, M., Langou, J.: Communication-optimal parallel and sequential QR and LU factorizations. SIAM J. Sci. Comput. **34**(1), 206–239 (2012)
14. Driscoll, M., Georganas, E., Koanantakool, P., Solomonik, E., Yelick, K.: A communication-optimal n-body algorithm for direct interactions. In: 2013 IEEE 27th International Symposium on Parallel and Distributed Processing, pp. 1075–1084 (2013)
15. Elango, V., Rastello, F., Pouchet, L.-N., Ramanujam, J., Sadayappan, P.: On characterizing the data movement complexity of computational DAGs for parallel execution. CoRR, abs/1404.4767 (2014)
16. Gilbert, J.R., Lengauer, T., Tarjan, R.E.: The pebbling problem is complete in polynomial space, pp. 237–248 (1979)
17. Jia-Wei, H., Kung, H.T.: I/o complexity: the red-blue pebble game, pp. 326–333 (1981)
18. Kwasniewski, G., Kabić, M., Besta, M., VandeVondele, J., Solcà, R., Hoefler, T.: Red-blue pebbling revisited: near optimal parallel matrix-matrix multiplication. In: Proceedings of the International Conference for High Performance Computing, Networking, Storage and Analysis, SC 2019, New York, NY, USA (2019). Association for Computing Machinery
19. Liu, J.W.H.: On the storage requirement in the out-of-core multifrontal method for sparse factorization. ACM Trans. Math. Softw. **12**(3), 249–264 (1986)
20. Liu, Q.: Red-blue and standard pebble games : complexity and applications in the sequential and parallel models. Master's thesis, Department of Electrical Engineering and Computer Science, MIT, Massachusetts (2018)
21. Maheshwari, A., Zeh, N.: A survey of techniques for designing i/o-efficient algorithms, pp. 36–61, January 2002
22. Marchal, L., McCauley, S., Simon, B., Vivien, F.: Minimizing I/Os in out-of-core task tree scheduling. In: 2017 IEEE International Parallel and Distributed Processing Symposium Workshops (IPDPSW), pp. 884–893 (2017)
23. Meyer, U., Zeh, N.: I/O-efficient undirected shortest paths. In: Di Battista, G., Zwick, U. (eds.) ESA 2003. LNCS, vol. 2832, pp. 434–445. Springer, Heidelberg (2003). https://doi.org/10.1007/978-3-540-39658-1_40
24. Ranjan, D., Savage, J., Zubair, M.: Upper and lower I/O bounds for pebbling r-Pyramids. In: Iliopoulos, C.S., Smyth, W.F. (eds.) IWOCA 2010. LNCS, vol. 6460, pp. 107–120. Springer, Heidelberg (2011). https://doi.org/10.1007/978-3-642-19222-7_12
25. Unat, D., et al.: Trends in Data Locality Abstractions for HPC Systems. IEEE Trans. Parallel Distrib. Syst. (TPDS) **28**(10), 3007–3020 (2017)

Local Planar Domination Revisited

Ozan Heydt, Sebastian Siebertz(✉), and Alexandre Vigny

University of Bremen, Bremen, Germany
{heydt,siebertz,vigny}@uni-bremen.de

Abstract. We show how to compute an $(11 + \epsilon)$ - approximation of a minimum dominating set in a planar graph in a constant number of rounds in the LOCAL model of distributed computing. This improves on the previously best known approximation factor of 52, which was achieved by an elegant and simple algorithm of Lenzen et al. Our algorithm combines ideas from the algorithm of Lenzen et al. with recent work of Czygrinow et al. and Kublenz et al. to reduce to the case of bounded degree graphs. We can then apply an LP-based approximation in a constant number of rounds. We also study a distributed version of the classical greedy algorithm, which however falls short of achieving the best approximation ratio.

Keywords: Dominating set · LOCAL algorithms · Planar graphs

1 Introduction

A dominating set in an undirected and simple graph G is a set $D \subseteq V(G)$ such that every vertex $v \in V(G)$ either belongs to D or has a neighbor in D. The dominating set problem is a classical NP-complete problem [13] with many applications in theory and practice, see e.g. [8,19]. In this paper we study the distributed time complexity of finding dominating sets in planar graphs in the classical LOCAL model of distributed computing. In this model, a distributed system is modeled by an undirected (planar) graph G. Every vertex represents a computational entity and the vertices communicate through the edges of G. The vertices are equipped with unique identifiers and initially, every vertex is only aware of its own identity. A computation then proceeds in synchronous rounds. In every round, every vertex sends messages to its neighbors, receives messages from its neighbors and performs an arbitrary computation. The complexity of a LOCAL algorithm is the number of rounds until all vertices return their answer, in our case, whether they belong to a dominating set or not.

The problem of approximating dominating sets in the LOCAL model has received considerable attention in the literature. Since in general graphs it is not

This paper is part of a project that has received funding from the German Research Foundation (DFG) with grant agreement No. 444419611. We thank the anonymous referee for pointing out that the third phase of our algorithm can be improved by the use of LP-based techniques when the maximum degree is bounded.

M. Parter (Ed.): SIROCCO 2022, LNCS 13298, pp. 154–173, 2022.
https://doi.org/10.1007/978-3-031-09993-9_9

possible to compute a constant factor approximation in a constant number of rounds [16], much effort has been invested to improve the ratio between approximation factor and number of rounds on special graph classes. A very successful line of structural analysis of graph properties that can lead to improved algorithms was started by the influential paper of Lenzen et al. [17], who in particular proved that on planar graphs a 130-approximation of a minimum dominating set can be computed in a constant number of rounds. A careful analysis of Wawrzyniak [22] later showed that the algorithm computes in fact a 52-approximation. In terms of lower bounds, Hilke et al. [12] showed that there is no deterministic local algorithm (constant-time distributed graph algorithm) that finds a $(7-\epsilon)$-approximation of a minimum dominating set on planar graphs, for any positive constant ϵ. Better approximation ratios are known for some special cases, e.g. 32 if the planar graph is triangle-free [1, Theorem 2.1], 18 if the planar graph has girth five [2] and 5 if the graph is outerplanar (and this bound is tight) [5, Theorem 1].

In this work we tighten the gap between the best-known lower bound of 7 and the best-known upper bound of 52 on planar graphs by providing a new approximation algorithm computing an $(11+\epsilon)$-approximation (for every $\epsilon > 0$).

Our algorithm proceeds in three phases. The first phase is a preprocessing phase that was similarly employed in the algorithm of Lenzen et al. [17]. In a key lemma, Lenzen et al. proved that there are only few vertices whose open neighborhood cannot be dominated by at most six vertices. We improve this lemma and show that there are only slightly more vertices whose open neighborhood cannot be dominated by *three* other vertices. All these vertices are selected into an initial partial dominating set and as a consequence the open neighborhoods of all remaining vertices can be dominated by at most three vertices.

By defining the notion of *pseudo-covers*, Czygrinow et al. [6] provided a tool to carry out a fine grained analysis of the vertices that can potentially dominate the remaining neighborhoods. Using ideas of [14] and [20] we provide an even finer analysis for planar graphs on which we base the second phase of our distributed algorithm and compute a second partial dominating set.

For the third phase, call the number of non-dominated neighbors of a vertex v the *residual degree* of v. We prove that after the second phase we are left with a graph where every vertex has residual degree at most 30. In particular, every vertex from a minimum dominating set D can dominate at most 30 non-dominated vertices and we conclude that the set R of non-dominated vertices has size bounded by $31|D|$ (each vertex dominates its neighbors and itself). Hence, we could at this point pick all non-dominated vertices to add at most $31|D|$ vertices and conclude. Instead, we study two different ways to proceed.

Our first option is to apply an LP-approximation based on aresults of Bansal and Umboh [4], who showed that a very simple selection procedure leads to a constant factor approximation when the solution to the dominating set linear program (LP) is given. As shown by Kuhn et al. [15] we can approximate such a solution in a constant number of rounds when the maximum degree Δ of the graph is bounded. To apply these results we have to overcome two obstacles.

First, note that even though we have established that the maximum residual degree is bounded by 30, we may still have unbounded maximum degree Δ. We overcome this problem by keeping only a few representative potential dominators around the set R. By a simply density argument there can be only very few high degree vertices left that we simply select into the dominating set. The second obstacle, which is easily overcome, is that we do not need to dominate the whole remaining graph but only the set R. This requires a small adaptation of the LP-formulation of the problem and a proof that the algorithm of Bansal and Umboh still works for this slightly different problem. In total, in this third phase of the algorithm we add at most $(7+\epsilon)|D|$ vertices to the dominating set, leading to an $(11+\epsilon)$ - approximation in total (Theorem 1).

Our second option is to study a parallel distributed version of the classical greedy algorithm. We proceed in a greedy manner in 30 rounds as follows. In the first round, if a non-dominated vertex has a neighbor of residual degree 30, it elects one such neighbor into the dominating set (or if it has residual degree 30 itself, it may choose itself). The neighbors of the chosen elements are marked as dominated and the residual degrees are updated. Note that all non-dominated neighbors of a vertex of residual degree 30 in this round choose a dominator, hence, the residual degrees of all vertices of residual degree 30 are decreased to 0, hence, after this round there are no vertices of residual degree 30 left. In the second round, if a non-dominated vertex has a neighbor of residual degree 29, it elects one such vertex into the dominating set, and so on, until after 30 rounds in the final round every vertex chooses a dominator. Unlike in the general case, where nodes cannot learn the current maximum residual degree in a constant number of rounds, by establishing an upper bound on the maximum residual degree and proceeding in exactly this number of rounds, we ensure that we iteratively exactly choose the vertices of maximum residual degree. It remains to analyze the performance of this algorithm.

A simple density argument shows that there cannot be too many vertices of degree $i \geq 6$ in a planar graph. At a first glance it seems that the algorithm would perform worst when in every of the 30 rounds it would pick as many vertices as possible, as the constructed dominating set would grow as much as possible. However, this is not the case, as picking many high degree vertices at the same time makes the largest progress towards dominating the whole graph. It turns out that there is a delicate balance between the vertices that we pick in round i and the remaining non-dominated vertices that leads to the worst case. In total, this leads to a 20 - approximation (Theorem 2). While this approach falls short of achieving the best approximation ratio the algorithm is simple and interesting to study in its own right.

We then analyze our algorithm on more restricted graph classes, and prove that it computes the following approximations of factors: $(8+\epsilon)$ for triangle-free planar graphs, $(7+\epsilon)$ for bipartite planar graphs, $(8+\epsilon)$ for outerplanar graphs, and 7 for planar graphs of girth 5 (Theorems 3, 4, 6 and 7 respectively). This improves the currently best known approximation ratios of 32 and 18 for triangle-free planar graphs and planar graphs of girth 5, respectively.

2 Preliminaries

In this paper we study the distributed time complexity of finding dominating sets in planar graphs in the classical LOCAL model of distributed computing. We assume familiarity with this model and refer to the survey [21] for extensive background on distributed computing and the LOCAL model.

We use standard notation from graph theory and refer to the textbook [7] for extensive background. All graphs in this paper are undirected and simple. We write $V(G)$ for the vertex set of a graph G and $E(G)$ for its edge set. The *girth* of a graph G is the length of a shortest cycle in G. A graph is called *triangle-free* if it does not contain a triangle, that is, a cycle of length three as a subgraph. Equivalently, a triangle-free graph is a graph of girth at least four.

A graph is *bipartite* if its vertex set can be partitioned into two parts such that all its edges are incident with two vertices from different parts. More generally, the *chromatic number* $\chi(G)$ of a graph G is the minimum number k such that the vertices of G can be partitioned into k parts such that all edges are incident with two vertices from different parts. Hence, the bipartite graphs are exactly the graphs with chromatic number two. A set A is *independent* if all two distinct vertices $u, v \in A$ are non-adjacent. Every graph G contains an independent set of size at least $\lceil |V(G)|/\chi(G) \rceil$. Every bipartite graph is triangle-free.

A graph is *planar* if it can be embedded in the plane, that is, it can be drawn on the plane in such a way that its edges intersect only at their endpoints. By the famous theorem of Wagner, planar graphs can be characterized as those graphs that exclude the complete graph K_5 on five vertices and the complete bipartite $K_{3,3}$ with parts of size three as a minor. A graph H is a *minor* of a graph G, written $H \preceq G$, if there is a set $\{G_v : v \in V(H)\}$ of pairwise vertex disjoint and connected subgraphs $G_v \subseteq G$ such that if $\{u, v\} \in E(H)$, then there is an edge between a vertex of G_u and a vertex of G_v. We call $V(G_v)$ the *branch set* of v and say that it is *contracted* to the vertex v. We call H a *1-shallow minor*, written $H \preceq_1 G$, if $H \preceq G$ and there is a minor model $\{G_v : v \in V(H)\}$ witnessing this, such that all branch sets G_v have radius at most 1, that is, in each G_v there exists w adjacent to all other vertices of G_v. In other words, $H \preceq_1 G$ if H is obtained from G by deleting some vertices and edges and then contracting a set of pairwise disjoint stars. We refer to [18] for an in-depth study of the theory of sparsity based on shallow minors.

A graph is *outerplanar* if it has an embedding in the plane such that all vertices belong to the unbounded face of the embedding. Equivalently, a graph is outerplanar if it does not contain the complete graph K_4 on four vertices and the complete bipartite graph $K_{2,3}$ with parts of size 2 and 3, respectively, as a minor. If $J \preceq H$ and $H \preceq G$, then $J \preceq G$, hence a minor of a planar graph is again planar and a minor of an outerplanar graph is again outerplanar.

By Euler's formula, planar graphs are sparse: every planar n-vertex graph ($n \geq 3$) has at most $3n - 6$ edges (and a graph with at most two vertices has at most one edge). The ratio $|E(G)|/|V(G)|$ is called the *edge density* of G. In particular, every planar graph G has edge density strictly smaller than three.

Lemma 1. *Let G be a planar graph. Then the edge density of G is strictly smaller than 3 and $\chi(G) \leq 4$. Furthermore,*

1. *if G is bipartite, then the edge density of G is strictly smaller than 2 and $\chi(G) \leq 2$,*
2. *if G is triangle-free or outerplanar, then the edge density of G is strictly smaller than 2 and $\chi(G) \leq 3$.*

An orientation of a graph G is a directed graph $\overrightarrow{G}$ that for every edge $\{u, v\} \in E(G)$ has one of the arcs (u, v) or (v, u). The out-degree $d^+(v)$ of a vertex v in an orientation $\overrightarrow{G}$ of G is the number $|\{w : (v, w) \in E(\overrightarrow{G}\}|$. The following lemma is implicit in the work of Hakimi [11], see also [18, Proposition 3.3].

Lemma 2. *Let $d = \lceil \max_{H \subseteq G}\{|E(H)|/|V(H)|\}\rceil$. Then G has an orientation with maximum out-degree d.*

As every subgraph of a (triangle-free or outerplanar) planar graph is again a (triangle-free or outerplanar) planar graph we have the following corollary.

Corollary 3. *Let G be a planar graph. Then*

1. *G has an orientation with maximum out-degree 3.*
2. *Furthermore, if G is triangle-free or outerplanar, then G has an orientation with maximum out-degree 2.*

For a graph G and $v \in V(G)$ we write $N(v) = \{u : \{u, v\} \in E(G)\}$ for the *open neighborhood* of v and $N[v] = N(v) \cup \{v\}$ for the *closed neighborhood* of v. For a set $A \subseteq V(G)$ let $N[A] = \bigcup_{v \in A} N[v]$. A *dominating set* in a graph G is a set $D \subseteq V(G)$ such that $N[D] = V(G)$. We write $\gamma(G)$ for the size of a minimum dominating set of G. For $W \subseteq V(G)$ we say that a set $Z \subseteq V(G)$ *dominates* W if $W \subseteq N[Z]$.

In the following we mark important definitions and assumptions about our input graph in gray boxes and steps of the algorithm in red boxes.

> We fix a planar graph G and a minimum dominating set D of G with $\gamma := |D| = \gamma(G)$.

3 Phase 1: Preprocessing

As outlined in the introduction, our algorithm works in three phases. In phase i for $1 \leq i \leq 3$ we select a partial dominating set D_i and estimate its size in comparison to D. In the end we will return $D_1 \cup D_2 \cup D_3$. We will call vertices that have been selected into a set D_i *green*, vertices that are dominated by a green vertex but are not green themselves are called *yellow* and all vertices that still need to be dominated are called *red*. In the beginning, all vertices are marked red.

The first phase of our algorithm is similar to the first phase of the algorithm of Lenzen et al. [17]. It is a preprocessing step that leaves us with only vertices whose neighborhoods can be dominated by a few other vertices. Lenzen et al. proved that there exist less than 3γ many vertices v such that the open neighborhood $N(v)$ of v cannot be dominated by 6 vertices of $V(G) \setminus \{v\}$ [17, Lemma 6.3]. The lemma can be generalized to more general graphs, see [3]). We prove the following lemma, which is stronger in the sense that the number of vertices required to dominate the open neighborhoods is smaller than 6, at the cost of having slightly more vertices with that property.

We define D_1 as the set of all vertices whose neighborhood cannot be dominated by 3 other vertices.

$$D_1 := \{v \in V(G) : \text{ for all sets } A \subseteq V(G) \setminus \{v\} \text{ with } N(v) \subseteq N[A] \text{ we have } |A| > 3\}.$$

We prove a very general lemma that can be applied also for more general graph classes, even though we will apply it only for planar graphs. Hence, in the following lemma, G can be an arbitrary graph, while in the following lemmas G will again be the planar graph that we fixed in the beginning.

Lemma 4. *Let G be a graph, let D be a minimum dominating set of G of size γ and let ∇ be an integer strictly larger than the edge density of a densest bipartite 1-shallow minor of G. Let $\hat{D}$ be the set of vertices $v \in V(G)$ whose neighborhood cannot be dominated by $(2\nabla - 1)$ vertices of D other than v, that is,*

$$\hat{D} := \{v \in V(G) : \textit{ for all sets } A \subseteq D \setminus \{v\} \textit{ with } N(v) \subseteq N[A] \textit{ we have } |A| > (2\nabla - 1)\}.$$

Then $|\hat{D} \setminus D| < \chi(G) \cdot \gamma$.

Recall that minors of planar graphs are again planar, hence, the maximum edge density of a bipartite 1-shallow minor of a planar graph is smaller than 2 and hence we can choose $\nabla = 2$ for the case of planar graphs and we note the following corollary.

Corollary 5. *Let $\hat{D}$ be the set of vertices v whose neighborhood cannot be dominated by 3 vertices of D other than v, that is,*

$$\hat{D} := \{v \in V(G) : \textit{for all sets } A \subseteq D \setminus \{v\} \textit{ with } N(v) \subseteq N[A] \textit{ we have } |A| > 3\}.$$

Then $|\hat{D} \setminus D| < 4\gamma$.

Proof (of Lemma 4). Assume $D = \{b_1, \ldots, b_\gamma\}$. Assume that there are $\chi(G) \cdot \gamma$ vertices $a_1, \ldots, a_{\chi(G)\gamma} \notin D$ satisfying the above condition. As the chromatic number is monotone over subgraphs, the subgraph induced by the a_is is also $\chi(G)$-chromatic, so we find an independent subset of the a_is of size γ. We can hence assume that $a_1, \ldots, a_\gamma$ are not connected by an edge. We proceed towards a contradiction.

We construct a bipartite 1-shallow minor H of G with the following 2γ branch sets. For every $i \leq \gamma$ we have a branch set $A_i = \{a_i\}$ and a branch set $B_i = N[b_i] \setminus (\{a_1, \ldots, a_\gamma\} \cup \bigcup_{j<i} N[b_j] \cup \{b_{i+1}, \ldots, b_\gamma\})$. Note that the B_i are vertex disjoint and hence we define proper branch sets. Intuitively, for each vertex $v \in N(a_i)$ we mark the smallest b_j that dominates v as its dominator. We then contract the vertices that mark b_j as a dominator together with b_j into a single vertex. Note that because the a_i are independent, the vertices a_i themselves are not associated to a dominator as no a_j lies in $N(a_i)$ for $i \neq j$. Denote by $a_1', \ldots, a_\gamma', b_1', \ldots, b_\gamma'$ the associated vertices of H. Denote by A the set of the $a_i's$ and by B the set of the $b_j's$. We delete all edges between vertices of B. The vertices of A are independent by construction. Hence, H is a bipartite 1-shallow minor of G. By the assumption that $N(a_i)$ cannot be dominated by $2\nabla - 1$ elements of D, we associate at least 2∇ different dominators with the vertices of $N(a_i)$. Note that this would not necessarily be true if A was not an independent set, as all $a_j \in N(a_i)$ would not be associated a dominator.

Since $\{b_1, \ldots, b_\gamma\}$ is a dominating set of G and by assumption on $N(a_i)$, we have that in H, every a_i' has at least 2∇ neighbors in B. Hence, $|E(H)| \geq 2\nabla|V(A)| = 2\nabla\gamma$. As $|V(H)| = 2\gamma$ we conclude $|E(H)| \geq \nabla|V(H)|$. This however is a contradiction, as ∇ is strictly larger than the edge density of a densest bipartite 1-shallow minor of G.

Let us fix the set $\hat{D}$ for our graph G.

$$\hat{D} := \{v \in V(G) : \text{ for all sets } A \subseteq D \setminus \{v\} \text{ with } N(v) \subseteq N[A] \text{ we have } |A| > 3\}.$$

Note that $\hat{D}$ cannot be computed by a local algorithm as we do not know the set D. It will only serve as an auxiliary set in our analysis.

The first phase of the algorithm is to compute the set D_1, which can be done in 2 rounds of communication. Obviously, if the open neighborhood of a vertex v cannot be dominated by 3 vertices from $V(G) \setminus \{v\}$, then in particular it cannot be dominated by 3 vertices from $D \setminus \{v\}$. Hence $D_1 \subseteq \hat{D}$ and we can bound the size of D_1 by that of $\hat{D}$.

Lemma 6. *We have $D_1 \subseteq \hat{D}$, $|\hat{D} \setminus D| < 4\gamma$, and $|\hat{D}| < 5\gamma$.*

Proof. Lemma 6 follows the observation above together with Corollary 5.

From Lemma 6 we can conclude that $|D_1| < 5\gamma$. However, it is intuitively clear that every vertex that we pick from the minimum dominating set D is optimal progress towards dominating the whole graph. We will later show that this intuition is indeed true for our algorithm, that is, our algorithm performs worst when $D_1 \cap D = \varnothing$, which will later in fact allow us to estimate $|D_1| < 4\gamma$.

We mark the vertices of D_1 that we add to the dominating set in the first phase of the algorithm as green, the neighbors of D_1 as yellow and leave all other vertices red. Denote the set of red vertices by R, that is, $R = V(G) \setminus N[D_1]$. For $v \in V(G)$ let $N_R(v) := N(v) \cap R$ and $\delta_R(v) := |N_R(v)|$ be the *residual degree* of v, that is, the number of neighbors of v that still need to be dominated.

By definition of D_1, the neighborhood of every non-green vertex can be dominated by at most 3 other vertices. This holds true as well for the subset $N_R(v)$ of neighbors that still need to be dominated. Let us fix such a small dominating set for the red neighborhood of every non-green vertex.

For every $v \in V(G) \setminus D_1$, we fix $A_v \subseteq V(G) \setminus \{v\}$ such that:

$$N_R(v) \subseteq N[A_v] \text{ and } |A_v| \leq 3.$$

There are potentially many such sets A_v – we fix one such set arbitrarily. Let us stress that even though we could compute the sets A_v in a local algorithm (making decisions based on vertex ids), we only use these sets for our further argumentation and do not need to compute them.

4 Phase 2: Analyzing the Local Dominators

The second phase of our algorithm is inspired by results of Czygrinow et al. [6] and the greedy domination algorithm for biclique-free graphs of [20]. Czygrinow et al. [6] defined the notion of *pseudo-covers*, which provide a tool to carry out a fine grained analysis of vertices that can potentially belong to the sets A_v used to dominate the red neighborhood $N_R(v)$ of a vertex v. This tool can in fact be applied to much more general graphs than planar graphs, namely, to all graphs that exclude some complete bipartite graph $K_{t,\ell}$. A refined analysis for classes of bounded expansion was provided by Kublenz et al. [14]. We provide an even finer analysis for planar graphs on which we base a second phase of our distributed algorithm.

We first describe what our algorithm computes, and then provide bounds on the number of selected vertices. Intuitively, we select every pair of vertices with sufficiently many neighbors in common.

- For $v \in V(G)$ let $B_v := \{z \in V(G) \setminus \{v\} : |N_R(v) \cap N_R(z)| \geq 10\}$.
- Let W be the set of vertices $v \in V(G)$ such that $B_v \neq \emptyset$.
- Let $D_2 := \bigcup_{v \in W} (\{v\} \cup B_v)$.

Once D_1 has been computed in the previous phase, 2 more rounds of communication are enough to compute the sets B_v and D_2. Before we update the residual degrees, let us analyze the sets B_v and D_2. First note that the definition is symmetric: since $N_R(v) \cap N_R(z) = N_R(z) \cap N_R(v)$ we have for all $v, z \in V(G)$ if

$z \in B_v$, then $v \in B_z$. In particular, if $v \in D_1$ or $z \in D_1$, then $N_R(v) \cap N_R(z) = \varnothing$, which immediately implies the following lemma.

Lemma 7. *We have $W \cap D_1 = \varnothing$ and for every $v \in V(G)$ we have $B_v \cap D_1 = \varnothing$.*

Now we prove that for every $v \in W$, the set B_v cannot be too big, and has nice properties.

Lemma 8. *For all vertices $v \in W$ we have*

- *$B_v \subseteq A_v$ (hence $|B_v| \leq 3$), and*
- *if $v \notin \hat{D}$, then $B_v \subseteq D$.*

Proof. Assume $A_v = \{v_1, v_2, v_3\}$ (a set of possibly not distinct vertices) and assume there exists $z \in V(G) \setminus \{v, v_1, v_2, v_3\}$ with $|N_R(v) \cap N_R(z)| \geq 10$. As v_1, v_2, v_3 dominate $N_R(v)$, and hence also $N_R(v) \cap N_R(z)$, there must be some v_i, $1 \leq i \leq 3$, with $|N_R(v) \cap N_R(z) \cap N[v_i]| \geq \lceil 10/3 \rceil \geq 4$. Therefore, $|N_R(v) \cap N_R(z) \cap N(v_i)| \geq 3$, which shows that $K_{3,3}$ is a subgraph of G, contradicting the assumption that G is planar.

If furthermore $v \notin \hat{D}$, by definition of $\hat{D}$, we can find w_1, w_2, w_3 from D that dominate $N(v)$, and in particular $N_R(v)$. If $z \in V(G) \setminus \{v, w_1, w_2, w_3\}$ with $|N_R(v) \cap N_R(z)| \geq 10$ we can argue as above to obtain a contradiction.

Let us now analyze the size of D_2. For this we refine the set D_2 and define

1. $D_2^1 := \bigcup_{v \in W \cap D}(\{v\} \cup B_v)$,
2. $D_2^2 := \bigcup_{v \in W \cap (\hat{D} \setminus D)}(\{v\} \cup B_v)$, and
3. $D_2^3 := \bigcup_{v \in W \setminus (D \cup \hat{D})}(\{v\} \cup B_v)$.

Obviously $D_2 = D_2^1 \cup D_2^2 \cup D_2^3$. We now bound the size of the refined sets D_2^1, D_2^2 and D_2^3.

Lemma 9. $|D_2^1 \setminus D| \leq 3\gamma$.

Proof. We have

$$|D_2^1 \setminus D| = |\bigcup_{v \in W \cap D}(\{v\} \cup B_v) \setminus D| \leq |\bigcup_{v \in W \cap D} B_v| \leq \sum_{v \in W \cap D} |B_v|.$$

By Lemma 8 we have $|B_v| \leq 3$ for all $v \in W$ and as we sum over $v \in W \cap D$ we conclude that the last term has order at most 3γ.

Lemma 10. $D_2^2 \subseteq \hat{D}$ *and therefore* $|D_2^2 \setminus D| < 4\gamma$.

Proof. Let $v \in \hat{D} \setminus D$ and let $z \in B_v$. By symmetry, $v \in B_z$ and according to Lemma 8, if $z \notin \hat{D}$, then $v \in D$. Since this is not the case, we conclude that $z \in \hat{D}$. Hence $B_v \subseteq \hat{D}$ and, more generally, $D_2^2 \subseteq \hat{D}$. Finally, according to Lemma 6 we have $|\hat{D} \setminus D| < 4\gamma$.

Finally, the set D_2^3, which appears largest at first glance, was actually already counted, as shown in the next lemma.

Lemma 11. $D_2^3 \subseteq D_2^1$.

Proof. If $v \notin \hat{D}$, then $B_v \subseteq D$ by Lemma 8. Hence $v \in B_z$ for some $z \in D$, and $v \in D_2^1$.

Again, it is intuitively clear that the situation when the sets D_2^i are large does not lead to the worst case for the overall algorithm. For example, when D_2^1 is large we have added many vertices of the optimum dominating set D. For a formal analysis, we analyze the number of vertices of D that have been selected so far.

Let $\rho \in [0, 1]$ be such that $|(D_1 \cup D_2) \cap D| = \rho\gamma$.

Lemma 12. *We have* $|D_1 \cup D_2| < 4\gamma + 4\rho\gamma$.

Proof. By Lemma 11 we have $D_2^3 \subseteq D_2^1$, hence, $D_1 \cup D_2 = D_1 \cup D_2^1 \cup D_2^2$. By Lemma 6 we have $D_1 \subseteq \hat{D}$ and by Lemma 10 we also have $D_2^2 \subseteq \hat{D}$, hence $D_1 \cup D_2^2 \subseteq \hat{D}$. Again by Lemma 6, $|\hat{D} \setminus D| < 4\gamma$ and therefore $|(D_1 \cup D_2^2) \setminus D| < 4\gamma$.

We have $W \cap D \subseteq D_2^1 \cap D$, hence with Lemma 8 we conclude that

$$|D_2^1 \setminus D| \leq \Big| \bigcup_{v \in D \cap D_2^1} B_v \Big| \leq \sum_{v \in D \cap D_2^1} |B_v| \leq 3\rho\gamma,$$

hence $(D_1 \cup D_2) \setminus D < 4\gamma + 3\rho\gamma$. Finally, $D_1 \cup D_2 = (D_1 \cup D_2) \setminus D \cup ((D_1 \cup D_2) \cap D)$ and with the definition of ρ we conclude $|D_1 \cup D_2| < 4\gamma + 4\rho\gamma$.

The analysis of the next and final step of the algorithm will actually show that the worst case is obtained when $\rho = 0$.

We now update the residual degrees, that is, we update R as $V(G) \setminus N[D_1 \cup D_2]$ and for every vertex the number $\delta_R(v) = N(v) \cap R$ accordingly. We finally show that after the first two phases of the algorithm we are in the very nice situation where all residual degrees are small.

Lemma 13. *For all* $v \in V(G)$ *we have* $\delta_R(v) \leq 30$.

Proof. First, every vertex of $D_1 \cup D_2$ has residual degree 0. Assume that there is a vertex v of residual degree at least 31. As v is not in D_1, its 31 non-dominated neighbors are dominated by a set A_v of at most 3 vertices. Hence there is a vertex z (not in D_1 nor D_2) with $|N_R(v) \cap N_R[z]| \geq \lceil 31/3 \rceil = 11$, hence, $|N_R(v) \cap N_R(z)| \geq 10$. This contradicts that v is not in D_2.

5 Phase 3: LP-Based Approximation

5.1 LP-Based Approximation

In the light of Lemma 13, we could now simply choose D_3 as the set of elements not in $N[D_1 \cup D_2]$. We would get a constant factor approximation, but not a very good one. Instead, we now proceed with an LP-based approximation. The dominating set problem can be formulated as an integer linear program (ILP). Note that it remains to dominate the set R, which leads to the following ILP.

$$\begin{array}{ll} \text{Minimize} & \sum_{v \in V} x_v \\ \text{Subject to} & \sum_{u \in N[v]} x_u \geq 1 \ \forall v \in R \\ & x_v \in \{0,1\} \qquad \forall v \in V \end{array}$$

By relaxing the condition that $x_v \in \{0,1\}$ to $x_v \in [0,1] \subseteq \mathbb{R}$, we obtain the corresponding linear program (LP). By a result of Bansal and Umboh [4] one can obtain a constant factor approximation of a dominating set from a solution to the LP. The proof can easily be adapted to the problem of approximating a dominating set of the set R.

Lemma 14. *Let G be a graph that has an orientation with maximum out-degree d and let $R \subseteq V(G)$. Let $D_R \subseteq V(G)$ be a minimum dominating set of R. Let $(x_v)_{v \in V(G)}$ be a solution to the R-dominating set LP. Let $H := \{v \in V(G) : x_v \geq 1/(2d+1)\}$ and let $U := \{v \in R : v \notin N[H]\}$. Then $H \cup U$ dominates R and has size at most $(2d+1) \cdot |D_R|$.*

Observe that when given the solution $(x_v)_{v \in V(G)}$ to the R-dominating set LP the lemma gives rise to a simple LOCAL algorithm. First select all vertices v with $x_v \geq 1/(2d+1)$ into a dominating set and mark all their neighbors as dominated. Then select all non-dominated vertices of R into the dominating set. Clearly, $H \cup U$ is a dominating set of R. The rest of this section is devoted to the proof of the claimed approximation factor. The proof follows the presentation of Bansal and Umboh [4] with the improved bounds of Dvořák [9]. As every solution to the ILP is also a solution to the LP we have $\sum_{v \in V(G)} x_v \leq |D_R|$.

Consider an orientation of G such that the neighborhood of each vertex v is decomposed into $N^{in}(v)$ and $N^{out}(v)$, where $|N^{out}(v)| \leq d$.

Claim. For every vertex $v \in U$, we have $\left(\sum_{u \in N^{in}(v)} x_u\right) \geq d/(2d+1)$.

Proof. As v is not in H, $x_v < 1/(2d+1)$. As v is not in $N(H)$, for every vertex $u \in N^{out}(v)$ we have $x_u < 1/(2d+1)$. As $|N^{out}(v)| \leq d$, and by the first LP condition $\left(\sum_{u \in N^{in}(v)} x_u\right) \geq 1 - \frac{1}{2d+1} - \frac{d}{2d+1} \geq \frac{d}{2d+1}$.

We can now bound the size of U and H

Claim. $|H \cup U| \leq (2d+1) \sum_{v \in V} x_v$.

Proof. First, observe that $|H| \le (2d+1)\sum_{v\in H}\frac{1}{2d+1} \le (2d+1)\sum_{v\in H}(x_v)$. Then observe that $|U| \le \frac{2d+1}{d}\cdot\sum_{v\in U}\frac{d}{2d+1} \le \frac{2d+1}{d}\sum_{v\in U}\sum_{u\in N^{in}(v)} x_u \le \frac{2d+1}{d}\sum_{u\in N^{in}(U)}(d\cdot x_u) \le (2d+1)\sum_{u\in N^{in}(U)} x_u$.

By definition of U, we have that $N(U)$ and H are disjoint, this also holds for H and $N^{in}(U)$ hence $|H\cup U| \le (2d+1)\sum_{v\in V} x_v \le (2d+1)|D_R|$.

5.2 Solving LPs Locally

As shown by Kuhn et al. [15] we can locally approximate general covering LPs, in particular the above R-dominating set LP, when the maximum degree of the graph is bounded. More precisely, they show how to compute a $\Delta^{1/k}$-approximation in $\mathcal{O}(k^2)$ rounds. Assuming for a moment that Δ is bounded by an absolute constant we can choose k such that $\Delta^{1/k} = 1+\epsilon$, hence $k = (\log\Delta)/(\log(1+\epsilon))$, which is a constant depending only on Δ and ϵ in order to compute a $(1+\epsilon)$-approximation. Hence, it remains to establish the situation that not only the residual degrees are bounded but that Δ is bounded by an absolute constant and to choose ϵ appropriately.

Recall that we chose ρ such that $|(D_1\cup D_2)\cap D| = \rho\gamma$. When $\rho\gamma$ vertices of D were already chosen into the partial dominating set $D_1\cup D_2$ we have $|D_R| \le (1-\rho)\gamma$. With Corollary 3 we conclude the following corollary.

Corollary 15. *Let G be a graph that has an orientation with maximum out-degree d, let $R\subseteq V(G)$, let D_R be a minimum dominating set of R, and let $\epsilon > 0$. Then we can compute a set D' of size at most $(2d+1)(1+\epsilon)|D_R|$ that dominates R in $\mathcal{O}(\log\Delta/(\log(1+\epsilon))$ rounds in the LOCAL model.*

In particular, for our algorithm when

1. *G is planar, then $|D'| \le 7(1+\epsilon)|D_R| \le 7(1+\epsilon)(1-\rho)\gamma$ and when*
2. *G is planar and triangle-free or outerplanar, then $|D'| \le 5(1+\epsilon)|D_R| \le 5(1+\epsilon)(1-\rho)\gamma$.*

5.3 From Bounded Residual Degree to Bounded Degree

It remains to establish the situation that the maximum degree Δ of our graph is bounded. As argued, we have $|R| \le 31(1-\rho)\gamma$. As only the vertices of R need to be dominated it suffices to keep only the vertices that have a neighbor in R; other vertices are not useful as dominators. Also, when two vertices $u, v\in V(G)\setminus R$ have exactly the same neighbors in R, that is, $N_R(u) = N_R(v)$, it suffices to keep one of u and v. Note that we can locally decide whether $N_R(u) = N_R(v)$. For every set $N\subseteq R$ such that there is a vertex v with $N_R(v) = N$ we choose the one with the lowest identifier as a representative. We construct the graph G' consisting of R and all edges between vertices in R as well as the set of all representatives and a minimal set of edges such that $N_R(v)$ is equal in G and G' for all representatives v. Hence in G' we have $N_R(u)\ne N_R(v)$ for all $u\ne v\in V(G')\setminus R$. As argued above, every R-dominating set in G can be transformed into an R-dominating set of the same size in G' (by choosing the appropriate representative) and every R-dominating set in G' is an R-dominating set in G. We can hence continue to work with the graph G'. In order to avoid complicated notation we simply assume that $G = G'$.

Note that in general we could have $|V(G)| \in \Omega(2^{|R|})$, in a planar graph however, $|V(G)|$ is linear in $|R|$, which is crucial for our further argumentation.

Lemma 16. $|V(G)| \leq 12|R|$.

We follow the presentation of [10, Lemma 4.3]. The presented construction is not build by our algorithm, it simply shows that our graph G satisfies the above property.

Proof. Consider a sequence of graphs $G_0, G_1, \ldots, G_\ell$ such that G_i is a 1-shallow minor of G for all $0 \leq i \leq \ell$ as follows. Set $G_0 = G$, and for $0 \leq i \leq \ell - 1$ define G_{i+1} from G_i by choosing a vertex $v \in V(G_i) \setminus R$ such that $N_R(v)$ contains two non-adjacent vertices u, w in G_i and contract $\{u, v\}$ into the vertex u to obtain G_{i+1}. Note that contracting $\{u, v\}$ into u is equivalent to deleting vertex v and adding edges between each vertex in $N(v) \setminus u$ and u. Note that this contraction will only add edges to R and remove vertices from $V(G) \setminus R$. Hence, for $0 \leq i \leq \ell$, we maintain $R \subseteq V(G_i)$.

This process clearly terminates, and G_{i+1} has at least one more edge between vertices of R than G_i does. Note that G_i is a 1-shallow minor of G for $0 \leq i \leq \ell$, as the edges $e_1, \ldots, e_{i-1}$ that were contracted to vertices in R in order to construct dG_i had one endpoint each in R and $V(G) \setminus R$, the endpoint in $V(G) \setminus R$ being deleted after each contraction. Thus, $e_1, \ldots, e_{i-1}$ induce a set of stars in $V(G)$, and G_i is obtained from G by contracting these stars. We therefore conclude that G_i is a 1-shallow minor of G. In particular, this implies that G is planar and has at most $3|R| - 6$ edges between vertices of R.

Since there are at most $3|R| - 6$ edges between vertices of R, we have that $\ell \leq 3|R| - 6$. Since the iterative process stopped, we have that for every vertex $v \in V(G_\ell) \setminus R$, $N_R(v)$ is a clique in G_ℓ.

We conclude with a result of Wood [23, Corollary 2] showing that in an n-vertex planar graph there can be at most $8(n-2)$ many cliques. At most $3|R|-6$ vertices u with a possibly unique neighborhood $N_R(u)$ were contracted and at most $8(|R|-2)$ vertices v with different $N_R(v)$ that induce a clique in G_ℓ remain. We add the vertices of $|R|$ to conclude that $|V(G)| \leq 12|R|$.

By Lemma 13 we have $|R| \leq 31(1 - \rho)\gamma$, which immediately implies the following corollary.

Corollary 17. $|V(G)| \leq 372(1 - \rho)\gamma$.

5.4 Conclusion of the Algorithm

We now pick an arbitrary $\epsilon > 0$ and select all vertices with high degree Γ into our dominating set, where Γ is chosen such that there exist at most $(\epsilon/8)(1-\rho)\gamma$ vertices of degree at least Γ.

Let $\epsilon > 0$, $\epsilon' := \epsilon/8$, $\Gamma := 6\frac{372}{\epsilon'}$, and $D_3^1 := \{v \in V(G) \ : \ d(v) > \Gamma\}$.

Lemma 18. $|D_3^1| \leq \epsilon'(1-\rho)\gamma$.

Proof. We assume the opposite and count the number of edges and vertices of G. When we sum the degree of the vertices, we get twice the number of edges. Hence $2 \cdot |E(G)| \geq 6\frac{372}{\epsilon'} \cdot \epsilon'(1-\rho)\gamma$. Therefore, with Corollary 17, $|E(G)| \geq 3 \cdot 372(1-\rho)\gamma \geq 3|V(G)|$. This contradict the fact the graph is planar.

After picking D_3^1 into the dominating set, marking the neighbors of D_3^1 as dominated and updating the set R, we can delete the vertices of D_3^1. We are left with a graph of maximum degree Γ.

Let D_3^2 be the set computed by the LOCAL algorithm of Corollary 15 with parameter ϵ'.

Let $D_3 = D_3^1 \cup D_3^2$. We already noted that the definition of D_3 implies that $D_1 \cup D_2 \cup D_3$ is a dominating set of G. We now conclude the analysis of the size of this computed set.

Lemma 19. *We have that* $|D_3| \leq (7+\epsilon)(1-\rho)\gamma$

Proof. By Corollary 15 and Lemma 18 we have $|D_3^2| \leq 7(1+\epsilon')(1-\rho)\gamma$, and $|D_3^1| \leq \epsilon'(1-\rho)\gamma$. This with, $\epsilon' = \epsilon/8$ conclude the proof.

We can now conclude the proof of our main result.

Theorem 1. *There exists a distributed LOCAL algorithm that, for every* $\epsilon > 0$, *and every planar graph* G, *computes in a constant number of rounds a dominating set of size at most* $(11+\epsilon)\gamma(G)$.

Proof. First, D_1, D_2, and D_3 are computed locally, in a bounded number of rounds, and $D_1 \cup D_2 \cup D_3$ dominates G. Second, by Lemma 12 we have $|D_1 \cup D_2| < 4\gamma + 4\rho\gamma$. And, by Lemma 19 we have $|D_3| \leq (7+\epsilon)(1-\rho)\gamma$.

Therefore $|D_1 \cup D_2 \cup D_3| \leq \gamma(4 + 4\rho + 7 - 7\rho + \epsilon - \epsilon\rho) \leq \gamma(11 + \epsilon - 3\rho - \epsilon\rho)$. As $\rho \in [0, 1]$, this is maximized when $\rho = 0$. Hence $|D_1 \cup D_2 \cup D_3| \leq \gamma(11+\epsilon)$.

6 Alternative Phase 3: Greedy Domination in Planar Graphs of Maximum Residual Degree

We now study a second approach to compute a set D_3 to dominate the remaining set R. It does not lead to a better approximation factor (only in the case of girth 5 planar graphs), but is interesting to study in its own right. The approach is based on a simulation of the classical greedy algorithm, where in each round we select a vertex of maximum residual degree. Here, we let all non-dominated vertices that have a neighbor of maximum residual degree choose such a neighbor as its dominator (or if they have maximum residual degree themselves, they may choose themselves). In general this is not possible for a LOCAL algorithm, however, as we established a bound on the maximum degree we can proceed as follows. We let $i = 30$. Every red vertex that has at least one neighbor of residual degree 30 arbitrarily picks one of them and elects it to the dominating set. Then every vertex recomputes its residual degree and i is set to 29. We continue until i reaches 0 when all vertices are dominated. More formally, we define several sets as follows.

For $30 \geq i \geq 0$, for every $v \in R$ in parallel:

if there is some u with $\delta_R(u) = i$ and ($\{u, v\} \in E(G)$ or $u = v$), then
$dom_i(v) := \{u\}$ (pick one such u arbitrarily),
$dom_i(v) := \varnothing$ otherwise.

- $R_i := R$ *What currently remains to be dominated*
- $\Delta_i := \bigcup_{v \in R} dom_i(v)$ *What we pick in this step*
- $R := R \setminus N[\Delta_i]$ *Update red vertices*

Finally, $D_3 := \bigcup_{0 \leq i \leq 30} \Delta_i$.

Let us first prove that the algorithm in fact computes a dominating set.

Lemma 20. *When the algorithm has finished the iteration with parameter $i \geq 1$, then all vertices have residual degree at most $i - 1$.*

In particular, after finishing the iteration with parameter 1, there is no vertex with residual degree 1 left and in the final round all non-dominated vertices choose themselves into the dominating set. Hence, the algorithm computes a dominating set of G.

Proof. By induction, before the iteration with parameter i, all vertices have residual at most i. Assume v has residual degree i before the iteration with parameter i. In that iteration, all non-dominated neighbors of v choose a dominator (possibly v, then the statement is trivial), hence, are removed from R. It follows that the residual degree of v after the iteration is 0. Hence, after this iteration and before the iteration with parameter $i - 1$, we are left with vertices of residual degree at most $i - 1$.

We now analyze the sizes of the sets Δ_i and R_i. The first lemma follows from the fact that every vertex chooses at most one dominator.

Lemma 21. *For every $i \leq 30$, $\sum_{j \leq i} |\Delta_j| \leq |R_i|$.*

Proof. The vertices of R_i are those that remain to be dominated in the last i rounds of the algorithm. As every vertex that remains to be dominated chooses at most one dominator in one of the rounds $j \leq i$, the statement follows.

As the vertices of D that still dominate non-dominated vertices also have bounded residual degree, we can conclude that not too many vertices remain to be dominated.

Lemma 22. *For every $i \leq 30$, $|R_i| \leq (i+1)(1-\rho)\gamma$.*

Proof. First note that for every i, $D \setminus (D_1 \cup D_2 \cup \bigcup_{j>i} \Delta_j)$ is a dominating set for R_i; additionally each vertex in this set has residual degree at most i. And finally, this set is a subset of $D \setminus (D_1 \cup D_2)$. Hence by the definition of ρ, we get that $|D \setminus (D_1 \cup D_2 \cup \bigcup_{j>i} \Delta_j)| \leq (1-\rho)\gamma$. As every vertex dominates its residual neighbors and itself, we conclude $|R_i| \leq (i+1)(1-\rho)\gamma$.

The next lemma shows that we cannot pick too many vertices of high residual degree. This follows from the fact that planar graphs have bounded edge density.

Lemma 23. *For every $7 \leq i \leq 30$, $|\Delta_i| \leq \frac{3|R_i|}{i-6}$.*

Proof. Let $7 \leq i \leq 30$ be an integer. We bound the size of Δ_i by a counting argument, using that G (as well as each of its subgraphs) is planar, and can therefore not have too many edges.

Let $J := G[\Delta_i]$ be the subgraph of G induced by the vertices of Δ_i, which all have residual degree i. Let $K := G[\Delta_i \cup (N[\Delta_i] \cap R_i)]$ be the subgraph of G induced by the vertices of Δ_i together with the red neighbors that these vertices dominate.

As J is planar, $|E(J)| < 3|V(J)| = 3|\Delta_i|$. As every vertex of J has residual degree exactly i, we get $|E(K)| \geq i\Delta_i - |E(J)| > (i-3)|\Delta_i|$ (we have to subtract $|E(J)|$ to not count twice the edges of K that are between two vertices of J). We also have that $|V(K)| \leq |V(J)| + |R_i|$. We finally apply Euler's formula again to K and get that $|E(K)| < 3|V(K)|$ hence $(i-3)|\Delta_i| < 3|\Delta_i| + 3|R_i|$. Therefore $|\Delta_i| < \frac{3|R_i|}{i-6}$.

Finally, we can give a lower bound on how many elements are newly dominated by the chosen elements of high residual degree.

Lemma 24. *For every $1 \leq i \leq 29$, $|R_i| \leq |R_{i+1}| - \frac{(i-5)|\Delta_{i+1}|}{3}$.*

Proof. Similarly to the proof of Lemma 23 (by replacing i by $i+1$), we define $J := G[\Delta_{i+1}]$ and $K := G[\Delta_{i+1} \cup (N[\Delta_{i+1}] \cap R_{i+1})]$.

We then replace the bound $|V(K)| \leq |V(J)| + |R_{i+1}|$ by $|V(K)| \leq |V(J)| + |N[\Delta_{i+1}] \cap R_{i+1}|$.

We then get:

$$|E(K)| \leq 3|V(K)|,$$

$$(i+1)|\Delta_{i+1}| - 3|\Delta_{i+1}| \leq 3(|\Delta_{i+1}| + |N[\Delta_{i+1}] \cap R_{i+1}|), \text{ and}$$

$$|N[\Delta_{i+1}] \cap R_{i+1}| \geq \frac{(i+1-6)|\Delta_{i+1}|}{3}.$$

Now, as $R_i = R_{i+1} \setminus N[\Delta_{i+1}]$, we have $|R_i| \leq |R_{i+1}| - |N[\Delta_{i+1}] \cap R_{i+1}| \leq |R_{i+1}| - \frac{(i+1-6)|\Delta_{i+1}|}{3}$.

We now formulate (and present in the full version of the paper) a linear program to maximize $|D_3|$ under these constraints. As a result we conclude the following lemma.

Lemma 25. $|D_3| \leq 15.9(1-\epsilon)\gamma$.

We already noted that the definition of D_3 implies that $D_1 \cup D_2 \cup D_3$ is a dominating set of G. We now conclude the analysis of the size of this computed set. First, by Lemma 12 we have $|D_1 \cup D_2| < 4\gamma + 4\epsilon\gamma$. Then, by Lemma 25 we have $|D_3| \leq 15.9(1-\epsilon)\gamma$. Therefore $|D_1 \cup D_2 \cup D_3| < 19.9\gamma - 11.9\epsilon\gamma$. As $\epsilon \in [0,1]$, this is maximized when $\epsilon = 0$. Hence $|D_1 \cup D_2 \cup D_3| < 19.9\gamma$.

Theorem 2. *The LOCAL algorithm with greedy phase* 3 *computes for every planar graph G in a constant number of rounds a dominating set of size at most* $20\gamma(G)$.

7 Restricted Classes of Planar Graphs

In this section we further restrict the input graphs, requiring e.g. planarity together with a lower bound on the girth. Our algorithm works exactly as before, however, using different parameters. From the different edge densities and chromatic numbers of the restricted graphs we will then derive different constants and as a result a better approximation factor. Throughout this section we use the same notation as in the first part of the paper and state in the adapted lemmas with the same numbers as in the first part the adapted sizes of the respective sets. All proofs can be found in the full version of the paper.

As in the general case in the first phase we begin by computing the set D_1 and analyzing it in terms of the auxiliary set $\hat{D}$.

Adapted Corollary 5.

1. *If G is bipartite, then $|\hat{D} \setminus D| < 2\gamma$.*
2. *If G is triangle-free, outerplanar, or has girth 5, then $|\hat{D} \setminus D| < 3\gamma$.*

In case of triangle-free planar graphs (in particular in the case of bipartite planar graphs) we proceed with the second phase exactly as in the second phase of the general algorithm (Sect. 4), however, the parameter 10 is replaced by the parameter 7. In case of planar graphs of girth at least five or outerplanar graphs, we simply set $D_2 = \varnothing$.

If G is triangle-free:

- For $v \in V(G)$ let $B_v := \{z \in V(G) \setminus \{v\} : |N_R(v) \cap N_R(z)| \geq 7\}$.
- Let W be the set of vertices $v \in V(G)$ such that $B_v \neq \varnothing$.
- Let $D_2 := \bigcup_{v \in W} (\{v\} \cup B_v)$.

If G has girth at least 5 or G is outerplanar, let $D_2 = \varnothing$.

Adapted Lemma 12.

1. *If G is bipartite, then* $|D_1 \cup D_2| < 2\gamma + 4\rho\gamma$.
2. *If G is triangle-free, then* $|D_1 \cup D_2| < 3\gamma + 4\rho\gamma$.
3. *If G has girth at least 5 or is outerplanar, then* $|D_1 \cup D_2| < 3\gamma + \rho\gamma$.

Again, we update the residual degrees and proceed with the third phase.

Adapted Lemma 13.

1. *If G is triangle-free, then for all $v \in V(G)$ we have* $\delta_R(v) \leq 18$.
2. *If G has girth at least 5, then for all $v \in V(G)$ we have* $\delta_R(v) \leq 3$.
3. *If G is outerplanar, then for all $v \in V(G)$ we have* $\delta_R(v) \leq 9$.

7.1 LP-Based Approximation

When proceeding with the LP-based approximation as in Sect. 5 we conclude with Corollary 15 and Lemma 18 to obtain the following statement:

Adapted Lemma 19. *When G is triangle free or bipartite, then* $|D_3| \leq (5 + \epsilon)(1 - \rho)\gamma$.

We can then conclude each individual case.

Theorem 3. *There exists a distributed LOCAL algorithm that, for every $\epsilon > 0$ and every triangle free planar graph G, computes in a constant number of rounds a dominating set of size at most $(8 + \epsilon)\gamma(G)$.*

Theorem 4. *There exists a distributed LOCAL algorithm that, for every $\epsilon > 0$ and every bipartite planar graph G, computes in a constant number of rounds a dominating set of size at most $(7 + \epsilon)\gamma(G)$.*

As we will see, the greedy approach can improve the following theorem.

Theorem 5. *There exists a distributed LOCAL algorithm that, for every $\epsilon > 0$ and every planar graph G of girth at least 5, computes in a constant number of rounds a dominating set of size at most $(8 + \epsilon)\gamma(G)$.*

Also the following theorem is not optimal by the results of Bonamy et al. [5].

Theorem 6. *There exists a distributed LOCAL algorithm that, for every $\epsilon > 0$ and every outerplanar graph G, computes in a constant number of rounds a dominating set of size at most $(8 + \epsilon)\gamma(G)$.*

7.2 Greedy Approximation

When proceeding by computing a dominating set D_3 with the greedy algorithm presented in Sect. 6 in the respective number of rounds we get the following improvement of Theorem 5.

Theorem 7. *There exists a distributed LOCAL algorithm that, for every planar graph G of girth at least 5, computes in a constant number of rounds a dominating set of size at most $7\gamma(G)$.*

References

1. Alipour, S., Futuhi, E., Karimi, S.: On distributed algorithms for minimum dominating set problem, from theory to application. arXiv preprint arXiv:2012.04883 (2020)
2. Alipour, S., Jafari, A.: A local constant approximation factor algorithm for minimum dominating set of certain planar graphs. In: Proceedings of the 32nd ACM Symposium on Parallelism in Algorithms and Architectures, pp. 501–502 (2020)
3. Amiri, S.A., Schmid, S., Siebertz, S.: Distributed dominating set approximations beyond planar graphs. ACM Trans. Algorithms (TALG) **15**(3), 1–18 (2019)
4. Bansal, N., Umboh, S.W.: Tight approximation bounds for dominating set on graphs of bounded arboricity. Inf. Process. Lett. **122**, 21–24 (2017)
5. Bonamy, M., Cook, L., Groenland, C., Wesolek, A.: A tight local algorithm for the minimum dominating set problem in outerplanar graphs. In: DISC. LIPIcs, vol. 209, pp. 13:1–13:18. Schloss Dagstuhl - Leibniz-Zentrum für Informatik (2021)
6. Czygrinow, A., Hanckowiak, M., Wawrzyniak, W., Witkowski, M.: Distributed approximation algorithms for the minimum dominating set in k_h-minor-free graphs. In: 29th International Symposium on Algorithms and Computation (ISAAC 2018). Schloss Dagstuhl-Leibniz-Zentrum fuer Informatik (2018)
7. Diestel, R.: Graph Theory. Graduate Texts in Mathematics, 4th edn, vol. 173. Springer, Heidelberg (2012)
8. Du, D.Z., Wan, P.J.: Connected Dominating Set: Theory and Applications, vol. 77. Springer, New York (2012). https://doi.org/10.1007/978-1-4614-5242-3
9. Dvořák, Z.: On distance-dominating and-independent sets in sparse graphs. J. Graph Theory **91**(2), 162–173 (2019)
10. Gajarskỳ, J., et al.: Kernelization using structural parameters on sparse graph classes. J. Comput. Syst. Sci. **84**, 219–242 (2017)
11. Hakimi, S.: On the degree of the vertices of a directed graph. J. Franklin Inst. **279**(4), 290–308 (1965)
12. Hilke, M., Lenzen, C., Suomela, J.: Brief announcement: local approximability of minimum dominating set on planar graphs. In: Proceedings of the 2014 ACM Symposium on Principles of Distributed Computing, pp. 344–346 (2014)
13. Karp, R.M.: Reducibility among combinatorial problems. In: Miller, R.E., Thatcher, J.W., Bohlinger, J.D. (eds.) Complexity of Computer Computations. The IBM Research Symposia Series, pp. 85–103. Springer, Boston (1972). https://doi.org/10.1007/978-1-4684-2001-2_9
14. Kublenz, S., Siebertz, S., Vigny, A.: Constant round distributed domination on graph classes with bounded expansion. arXiv preprint arXiv:2012.02701 (2020)
15. Kuhn, F., Moscibroda, T., Wattenhofer, R.: The price of being near-sighted. In: SODA. vol. 6, pp. 1109557–1109666. Citeseer (2006)
16. Kuhn, F., Moscibroda, T., Wattenhofer, R.: Local computation: lower and upper bounds. J. ACM **63**(2), 17:1–17:44 (2016)
17. Lenzen, C., Pignolet, Y.A., Wattenhofer, R.: Distributed minimum dominating set approximations in restricted families of graphs. Distrib. Comput. **26**(2), 119–137 (2013)
18. Nešetřil, Jaroslav, Ossona de Mendez, Patrice: Sparsity. AC, vol. 28. Springer, Heidelberg (2012). https://doi.org/10.1007/978-3-642-27875-4
19. Sasireka, A., Kishore, A.N.: Applications of dominating set of a graph in computer networks. Int. J. Eng. Sci. Res. Technol. **3**(1), 170–173 (2014)

20. Siebertz, S.: Greedy domination on biclique-free graphs. Inf. Process. Lett. **145**, 64–67 (2019)
21. Suomela, J.: Survey of local algorithms. ACM Comput. Surv. (CSUR) **45**(2), 1–40 (2013)
22. Wawrzyniak, W.: A strengthened analysis of a local algorithm for the minimum dominating set problem in planar graphs. Inf. Process. Lett. **114**(3), 94–98 (2014)
23. Wood, D.R.: On the maximum number of cliques in a graph. Graphs Comb. **23**(3), 337–352 (2007)

Election in Fully Anonymous Shared Memory Systems: Tight Space Bounds and Algorithms

Damien Imbs[1], Michel Raynal[2(✉)], and Gadi Taubenfeld[3]

[1] LIS, Aix-Marseille University & CNRS & Univ. Toulon, Marseille, France
[2] Univ Rennes IRISA, Inria, CNRS, Paris, France
raynal@irisa.fr
[3] Reichman University, Herzliya, Israel

Abstract. This article addresses election in fully anonymous systems made up of n asynchronous processes that communicate through atomic read-write registers or atomic read-modify-write registers. Given an integer $d \in \{1, \ldots, n-1\}$, two elections problems are considered: *d-election* (at least one and at most d processes are elected) and *exact d-election* (exactly d processes are elected). Full anonymity means that both the processes and the shared registers are anonymous. Memory anonymity means that the processes may disagree on the names of the shared registers. That is, the same register name A can denote different registers for different processes, and the register name A used by a process and the register name B used by another process can address the same shared register. Let n be the number of processes, m the number of atomic read-modify-write registers, and let $M(n,d) = \{k \ : \ \forall \, \ell : 1 < \ell \leq n \ : \ \mathsf{gcd}(\ell, k) \leq d\}$. The following results are presented for solving election in such an adversarial full anonymity context.

- It is possible to solve d-election when participation is not required if and only if $m \in M(n,d)$.
- It is possible to solve exact d-election when participation is required if and only if $\mathsf{gcd}(m,n)$ divides d.
- It is possible to solve d-election when participation is required if and only if $\mathsf{gcd}(m,n) \leq d$.
- Neither d-election nor exact d-election (be participation required or not) can be solved when the processes communicate through read-write registers only.

Keywords: Anonymous processes · Anonymous memory · Distributed computability · Leader election · Process participation · Read-write register · Read-modify-write register · Symmetry-breaking · Tight bounds

1 Introduction

1.1 Leader Election

Leader election is a classic basic problem encountered when processes cooperate and coordinate to solve higher-level distributed computing problems. It consists in designing an algorithm selecting one and only one process from the set of cooperating processes. In classical systems where the processes have distinct identities, leader election

M. Parter (Ed.): SIROCCO 2022, LNCS 13298, pp. 174–190, 2022.
https://doi.org/10.1007/978-3-031-09993-9_10

algorithms usually amount to electing the process with the smallest (or highest) identity. Many textbooks describe such algorithms (e.g., [5,14,20,25]).

This article considers two natural generalizations of the election problem in the presence of both process and memory anonymity, where communication is through shared registers. The first one is *d-election* in which at least one and at most d processes are elected. The second one is *exact d-election* in which exactly d (different) processes are elected.

1.2 System Models

Process Anonymity. Process anonymity means that the processes have no identity, have the same code, and have the same initialization of their local variables. Hence, in a process anonymous system, it is impossible to distinguish a process from another process.

Pioneering work on process anonymity in message-passing systems was presented in [3]. Process anonymity has been studied for a long time in asynchronous shared memory systems (e.g., [4]). It has been more recently addressed in the context of crash-prone asynchronous shared memory systems (e.g., [6,10]).

Assuming a system made up of n anonymous asynchronous processes, we use the notation p_1, ..., p_n to distinguish the processes. The subscript $i \in \{1, \cdots, n\}$ will also be used to identify the local variables of p_i (identified with names written with lower case letters).

Shared Registers. The processes communicate through a shared memory made up of m atomic registers [15] (identified with names written with upper case letters). Hence, the shared memory appears to the processes as an array of registers denoted $R[1..m]$. *Atomic* means that the operations on a register appear as if they have been executed sequentially, each appearing between its start event and its end event [12]. Moreover, for any x, a read of $R[x]$ returns the last value previously written in $R[x]$, where *last* refers to the previous total order on the operations on $R[x]$. (In case $R[x]$ has not been written, the read returns its initial value.) Two communication models are considered in the article.

- Read-write (RW) model. This is the basic model in which a register $R[x]$ can be accessed only by a read or a write operation (as the cells of a Turing machine).
- Read-modify-write (RMW) model. This model is the RW model enriched with a conditional write operation. This conditional write operation atomically reads the register and (according to the value read) possibly modifies it. This conditional write, denoted compare&swap($R[x]$, `old`, `new`), has three parameters, a shared register, and two values. It returns a Boolean value. If $R[x] =$ `old`, it assigns the value `new` to $R[x]$ and returns `true`. If $R[x] \neq$ `old`, $R[x]$ is not modified and the operation returns `false`. An invocation of compare&swap($R[x]$, `old`, `new`) that returns `true` is *successful*.

Memory Anonymity. The notion of an anonymous memory has been introduced in [26]. In a non-anonymous memory, the address $R[x]$ denotes the same register whatever the process that invokes $R[x]$: there is an a priori agreement on the name of each register. In an anonymous memory, there is no such agreement on the names of the shared registers. While the name $R[x]$ used by a given process p_i always denotes the very same register, the same name $R[x]$ used by different processes p_i p_j, p_k ..., may refer to different registers. More precisely, an anonymous memory system is such that:

- For each process p_i, an adversary defined a permutation $f_i()$ over the set $\{1, 2, ..., n\}$ such that, when p_i uses the name $R[x]$, it actually accesses $R[f_i(x)]$,
- No process knows the permutations,
- All the registers are initialized to the same default value.

In an anonymous memory system, ALL the registers are anonymous. Moreover, the size of the anonymous memory is not under the control of the programmer, and it is imposed on her/him. As shown in [2,9] (for non-anonymous processes and anonymous memory), and in this article (for fully anonymous systems) the size of the anonymous memory is a crucial parameter when one has to characterize the pairs $\langle n, m\rangle$ for which election can be solved in fully anonymous n-process systems.

Process Participation. As in previous works on election in anonymous or non-anonymous memory systems [9,24], this article considers two types of assumptions on the behavior of the processes: (1) algorithms that require the participation of all the processes to compete to be leaders, and (2) algorithms that do not (i.e., an arbitrary subset of processes may participate but not necessarily all the processes).

Symmetric Algorithm. Considering a system in which the processes have distinct identifiers, a *symmetric* algorithm is an algorithm where the processes can only compare their identities with equality [24]. So there is no notion of smaller/greater on process identities, and those cannot be used to index entries of arrays, etc. This notion of symmetry associated with process identities is the "last step" before their anonymity. In this article, we will consider algorithms in which the processes (and memories) are anonymous but will also mention symmetric algorithms.

1.3 Related Works on Anonymous Memories

Since its introduction, several problems have been addressed in the context of memory anonymity: mutual exclusion, election, consensus, set-agreement and renaming. We discuss below work on the first two problems that are more related to our work.

Mutual Exclusion. First, we observe that no shared memory-based mutual exclusion algorithm requires the participation of the processes. Let $M(n) = \{k : \forall \ell : 1 < \ell \leq n : \mathsf{gcd}(\ell, k) = 1\}$ (all the integers 2, ..., n are relatively prime with k). The following results have been recently established.

- There is a deadlock-free symmetric mutual exclusion algorithm in the RMW (resp. RW) model made up of m anonymous registers if and only if $m \in M(n)$ (resp. $m \in M(n) \setminus \{1\}$) [2].

- There is a deadlock-free mutual exclusion algorithm in the process anonymous and memory anonymous RMW model made up of m registers if and only if $m \in M(n)$. Moreover, there is no such algorithm in the fully anonymous RW communication model [22].

The conditions relating m an n can be seen as the seed needed to break symmetry despite anonymous memory, and symmetric or anonymous processes, thereby allowing mutual exclusion and election to be solved. A single leader election can be considered one-shot mutual exclusion, where the first process to enter its critical section is elected.

Election in the Symmetric Model. Considering the symmetric process model in which all the processes (unlike in the mutual exclusion problem) are required to participate. The following results are presented in [9] for such a model.

- There is a d-election symmetric algorithm in the memory anonymous RW and RMW communication models made up of m registers if and only if $\gcd(m, n) \leq d$.
- There is an exact d-election symmetric algorithm in the memory anonymous RW and RMW communication models made up of m registers if and only if $\gcd(m, n)$ divides d.

We emphasize that the above results for d-election assume that the processes are symmetric and not that they are anonymous, as done in this article. Finally, fully anonymous agreement problems are investigated in [23].

Remark. While addressing a different problem in a different context, it is worth mentioning the work presented in [8] that addresses the exploration of an m-size anonymous not-oriented ring by a team of n identical, oblivious, asynchronous mobile robots that can view the environment but cannot communicate. Among other results, the authors have shown that there are initial placements of the robots for which gathering is impossible when n and m are not co-prime, i.e., when $\gcd(n, m) \neq 1$. They also show that the problem is always solvable for $\gcd(n, m) = 1$ when $n \geq 17$.

1.4 Motivation and Content

Motivation. The main motivation of this work is theoretical. It investigates a fundamental symmetry-breaking problem (election) in the worst adversarial context, namely asynchronous and fully anonymous systems. Knowing what is possible/impossible, stating computability and complexity lower/upper bounds are at the core of algorithmics [3,11], and trying to find solutions "as simple as possible" is a key if one wants to be able to master the complexity of future applications [1,7]. This article aims to increase our knowledge of what can/cannot be done in the full anonymity context, providing associated necessary and sufficient conditions which enrich our knowledge on the system assumptions under which fundamental problems such as election can be solved.[1]

[1] On an application-oriented side, it has been shown in [16,17,19] that the process of genome-wide epigenetic modifications (which allows cells to utilize the DNA) can be modeled as a fully anonymous shared memory system where, in addition to the shared memory, also the processes (that is proteins modifiers) are anonymous. Hence fully anonymous systems can be useful in the context of biologically-inspired distributed systems [17,19].

When one has to solve a symmetry-breaking problem, the main issue consists in finding the "as weak as possible" initial seed from which the initial system symmetry can be broken. So, considering FULL anonymity, this article complements previously known results on anonymous systems (which were on non-anonymous processes and anonymous memory [2,9]). In all cases, the seed that allows breaking the very strong (adversary) symmetry context defined by FULL anonymity is captured by necessary and sufficient conditions relating the number of anonymous processes and the size of the anonymous memory.

Content of the Article. Let m, n and d be the number of registers, the number of processes and the number of leaders, respectively, and (as previously defined) $M(n, d) = \{k : \forall \ell : 1 < \ell \leq n : \mathsf{gcd}(\ell, k) \leq d\}$. Table 1 summarizes the four main results.

Table 1. Election in the fully anonymous shared memory systems.

Problem	Register type	Participation	Necessary & sufficient condition on $\langle m, n, d\rangle$	Necessary	sufficient	Section
d-election	RMW	not required	$m \in M(n, d)$	Theorem 1	Theorem 2 Algo 1	2
Exact d-election	RMW	required	$\mathsf{gcd}(m, n)$ divides d	Follows from [9]	Theorem 5 Algo 2	3
d-election	RMW	required	$\mathsf{gcd}(m, n) \leq d$	Follows from [9]	Theorem 7	4
d-election and Exact d-election	RW	required or not required	Impossible Corollary 2			5

1. A d-election algorithm in the RMW communication model, which does not require participation of all the processes. It is also shown that the condition $m \in M(n, d)$ is necessary and sufficient for such an algorithm. Notice that $M(n, 1)$ is the set $M(n)$ that appears in the results for fully anonymous mutual exclusion discussed earlier.
2. An exact d-election algorithm for the RMW communication model in which all the processes are required to participate. It is also shown that the necessary and sufficient condition for such an algorithm is $\mathsf{gcd}(m, n)$ divides d.
3. A d-election algorithm (which is based on the previous result) for the RMW communication model in which all the processes are required to participate. It is also shown that $\mathsf{gcd}(m, n) \leq d$ is a necessary and sufficient condition for such an algorithm. (The short algorithm appears in the proof of Theorem 7.)[2]
4. An impossibility result that regardless whether participation is required or not, there is neither d-election nor exact d-election algorithm in the anonymous RW communication model.

Let us notice that, due to the very nature of the anonymous process model, no process can know the "identity" of elected processes. So, at the end of an election algorithm in the anonymous process model, a process only knows if it is or not a leader.

[2] Both the algorithms described in the paper are simple. Their early versions were far from being simple, and simplicity is a first class property. As said by Y. Perlis (the recipient of first Turing Award) "*Simplicity does not precede complexity, but follows it*" [18].

We point out that the leader election problem has several variants, and the most general one, where a process only knows if it is or not a leader is a very common variant [5,21,25].

2 *d*-Election in the RMW Model Where Participation is Not Required

Throughout this section, it is assumed that communication is through RMW anonymous registers and that the processes are not required to participate [13].

2.1 A Necessary Condition for *d*-election

In this subsection, it is further assumed that processes have identities that can only be compared (symmetry constraint). As they are weaker models, it follows that the necessary condition proved below still holds in RMW model where both the processes and the memory are anonymous, and in the model where communication is through anonymous RW registers.

Theorem 1. *There is no symmetric d-election algorithm in the* RMW *communication model for* $n \geq 2$ *processes using* m *anonymous registers if* $m \notin M(n, d)$.

Proof. Let k be an arbitrary positive number such that $1 \leq k \leq n$. Below we examine what must be the relation between k, m and d, when assuming the existence of a symmetric d-election algorithm for n processes using $m \geq 1$ anonymous RMW registers. To simplify the modulo notation, the processes are denoted $p_0, ..., p_{n-1}$.

Let $\gcd(m, k) = \delta$, for some positive number δ. We will construct a run in which exactly k processes participate. Let us partition these k processes into $\delta \geq 1$ disjoint sets, denoted $P_0, ..., P_{\delta-1}$, such that there are exactly k/δ processes in each set. This partitioning is achieved by assigning process p_i (where $i \in \{0, ..., k-1\}$) to the set $P_{i \bmod \delta}$. For example, when $k = 6$ and $\delta = 3$, $P_0 = \{p_0, p_3\}$, $P_1 = \{p_1, p_4\}$, and $P_2 = \{p_2, p_5\}$ (top of Fig. 1). Such a division is possible since, by definition, $\gcd(m, k) = \delta$.

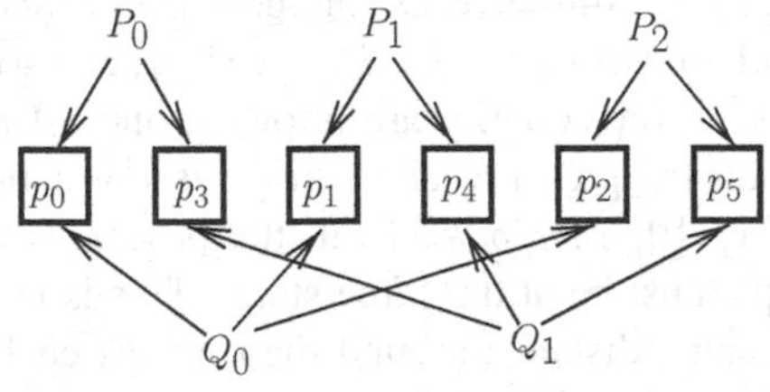

Fig. 1. Illustration of the runs for $k = 6$ and $\delta = 3$

Let us arrange the m registers on a ring with m nodes where each register is placed on a different node. To each one of the δ sets of processes P_i (where $i \in \{0, ..., \delta - 1\}$), let us assign an initial register (namely, the first register that each process in that set accesses) such that for every two sets P_i and its ring successor $P_{(i+1) \bmod \delta}$ the distance between their assigned initial registers is exactly δ when walking on the ring in a clockwise direction. This is possible since $\mathsf{gcd}(m, k) = \delta$.

The lack of global names for the RMW anonymous registers allows us to assign, for each one of the k processes, an initial register and an ordering which determines how the process scans the registers. An execution in which the k processes are running in *lock-steps*, is an execution where we let each process take one step (in the order $p_0, ..., p_{k-1}$), and then let each process take another step, and so on. For a given d-election algorithm A, let us call this execution, in which the processes run in lock-steps, ρ_A. For simplicity, we will omit the subscript A and simply write ρ.

For process p_i and integer j, let $\mathsf{order}(p_i, j)$ denotes the jth new (i.e., not yet assigned) register that p_i accesses during the execution ρ, and assume that we arrange that $\mathsf{order}(p_i, j)$ is the register whose distance from p_i's initial register is exactly $(j-1)$, when walking on the ring in a clockwise direction.

We notice that $\mathsf{order}(p_i, 1)$ is p_i's initial register, $\mathsf{order}(p_i, 2)$ is the next new register that p_i accesses and so on. That is, p_i does not access $\mathsf{order}(p_i, j + 1)$ before accessing $\mathsf{order}(p_i, j)$ at least once, but for every $j\prime \leq j$, p_i may access $\mathsf{order}(p_i, j\prime)$ several times before accessing $\mathsf{order}(p_i, j + 1)$ for the first time. Since the memory is anonymous, when a process accesses a register for the first time, say register $REG[x]$, we may map x to any (physical) register that it hasn't accessed yet. However, when it accesses $REG[x]$ again, it must access the same register it has accessed before when referring to x.

Let us now consider another division of the k processes into sets. We divide the k processes into k/δ disjoint sets, denoted $Q_0, ..., Q_{k/\delta-1}$, such that there are exactly δ processes in each set. This partitioning is achieved by assigning process p_i (where $i \in \{0, ..., k-1\}$) to the set $Q_{\lfloor i/\delta \rfloor}$. For example, when $k = 6$ and $\delta = 3$, $Q_0 = \{p_0, p_1, p_2\}$, and $Q_1 = \{p_3, p_4, p_5\}$. Again, such a partitioning is possible since $\mathsf{gcd}(m, k) = \delta$ (bottom of Fig. 1).

We notice that Q_0 includes the first process to take a step in the execution ρ, in each one of the δ sets, $P_0, ..., P_{\delta-1}$. Similarly, Q_1 includes the second process to take a step in the execution ρ, in each one of the δ sets, $P_0, ..., P_{\delta-1}$, and so on.

Since only comparisons for equality are allowed, and all registers are initialized to the same value –which (to preserve anonymity) is not a process identity– in the execution ρ, for each $i \in \{0, ..., n/\delta - 1\}$, all the processes in the set Q_i that take the same number of steps must be at the same state. (This is because all the processes in Q_i are located at the same distance around the ring. At each lockstep, they invoke the Read/Modify/Write operation into different locations, so because of the symmetry assumption, it is not possible to break the symmetry-between them, (either all or none are elected.) Thus, in the run ρ, it is not possible to break symmetry within a set Q_i ($i \in \{0, ..., k/\delta - 1\}$), which implies that either all the δ processes in the set Q_i will be elected, or no process in Q_i will be elected.

Thus, the number of elected leaders in ρ equals δ times the number of Q_i sets ($i \in \{0, ..., k/\delta - 1\}$) that all their members were elected, and (by definition of d-election) it must be a positive number. That is, the number of elected leaders in ρ equals $a\delta$ for some integer $a \in \{1, ..., k/\delta\}$.

Since in a d-election algorithm at most d leaders are elected in run ρ, it follows from the fact that for some positive integer a, it must be the case that $a\delta \leq d$. Thus, it must be the case that $\text{gcd}(m, k) = \delta \leq d$. Since k was chosen arbitrarily from $\{1, ..., n\}$, it follows that a necessary requirement for a symmetric d-election algorithm for $n \geq 2$ processes using m anonymous RMW registers is that, for every $1 \leq k \leq n$, $\text{gcd}(m, k) \leq d$. $\square_{Theorem 1}$

2.2 A d-election algorithm in RMW fully anonymous systems

Anonymous Memory. The anonymous memory is made up of m RMW registers $R[1..m]$, each initialized to the default value 0. It is assumed that $m \in M(n, d)$ (recall that $M(n, d) = \{k : \forall \ell : 1 < \ell \leq n : \text{gcd}(\ell, k) \leq d\}$).

Local Variables at Each Process p_i. Each process p_i manages the following set of local variables.

- $counter_i$: used to store the number of RMW registers *owned* by p_i. A process *owns* a register when it is the last process that wrote a positive value into this register.
- $myview_i[1..n]$: array of Boolean values, each initialized to `false`. When $myview_i[j]$ is equal to `true`, p_i owns the register $R_i[j]$.
- $round_i$ (initialized to 0): round number (rung number in the ladder metaphor, see below) currently attained by p_i in its competition to be a leader. When $round_i = n - d + 1$, p_i becomes a leader.
- $competitors_i$: maximal number of processes that compete with p_i when it executes a round.

Participation and Output. Any number of processes can invoke the election algorithm. A process exits the algorithm when it invokes return(`res`) where `res` is `leader` or `not leader`.

Description of the of the Algorithm. The code of each anonymous process p_i appears in Fig. 2. When the process p_i invokes elect(), it enters a repeat loop that it will exit at line 11 if it is not elected, and at line 13 if it is elected.

Once in a new round, p_i first writes its new round number in all the registers it owns, those are the registers $R_i[j]$ such that $myview_i[j]$ = `true` (line 4). Then, it strives to own as many registers as possible (without compromising liveness). To this end, it considers all the registers $R_i[j]$ such that $R_i[j] < round_i$ (line 6). If such a register is equal to 0 (i.e., is not currently owned by another process), p_i invoke compare&swap$(R_i[j], 0, round_i)$ to own it (line 7). If it is the case, it accordingly increases $counter_i$ (line 8).

Then p_i computes the maximal number of processes that, at round $round_i$, can compete with them (variable $competitors_i$ at line 9). There are then two cases. If it owns fewer registers than the average number $m/competitors$ (division on real numbers), p_i

resets the registers it owns to their initial value (line 11), and withdraws from the leader competition (line 12). Otherwise, if $round_i < n - d + 1$, p_i re-enters the repeat loop to progress to the next round. If $round_i = n - d + 1$, p_i is one the at most d leaders (line 14).

Let us note that a (successful) assignment of a round number to $R_i[j]$ by a process p_i at line 7 has $R_i[j] = 0$ as pre-condition and $R_i[j] > 0$ as post-condition. Moreover, both the assignment of $R_i[j]$ at lines 4 and 11 have $R_i[j] > 0$ as pre-condition. It follows that, between the lines 3 and 9 $counter_i$ counts the number of registers owned by p_i.

ALGORITHM 1: CODE OF A PROCESS p_i IN THE FULLY ANONYMOUS RMW MODEL (NON MANDATORY PARTICIPATION)

The initial value of all the shared registers is 0.

```
operation elect() is
1  counter_i ← 0; round_i ← 0; for each j ∈ {1,...,m} do myview_i[j] ← false end_do
2  repeat
3    round_i ← round_i + 1                                   // progress to the next round
4    for each j ∈ {1,...,m} do if myview_i[j] then R_i[j] ← round_i fi end_do     // owned
5    for each j ∈ {1,...,m} do                                // try to own more registers
6      while R_i[j] < round_i do               // R[j] < round_i implies myview[j] = false
7        myview_i[j] ← compare&swap(R_i[j], 0, round_i)                // try to own R_i[j]
8        if myview_i[j] then counter_i ← counter_i + 1 fi end_do end_do         // own +1
9    competitors_i ← n − round_i + 1                       // max # of competing processes
10   if counter_i < m/competitors_i then                            // too many competitors
11     for each j ∈ {1,...,m} do if myview_i[j] then R_i[j] ← 0 fi end_do  // free owned
12     return (not leader) fi                                 // withdraw from the election
13 until round_i = n − d + 1 end repeat
14 return (leader).                                                      // p_i is elected
```

Fig. 2. d-election for n anonymous processes and $m \in M(n, d)$ anonymous RMW registers

2.3 Proof of Algorithm 1

Let us say that "process p_i executes round r" when its local variable $round_i = r$. Reminder: $m \in M(n, d)$ where $M(n, d) = \{k \ : \ \forall \ell : 1 < \ell \leq n \ : \ \gcd(\ell, k) \leq d\}$.

Lemma 1. *For every $r \in \{1, ..., n - d + 1\}$, at most $n - r + 1$ processes may execute round r. In particular, at most d processes may execute round $n - d + 1$.*

Proof. The proof is by induction on the number of rounds. The induction base is simple since at most n processes may execute round $r = 1$. Let us assume (induction hypothesis) that the lemma holds for round $r < n - d + 1$ and prove that the lemma also holds (induction step) for round $r + 1$. That is, we need to show that at most $n - r$ processes execute round $r + 1$.

Let P_r be the set of processes that execute round r. If $|P_r| < n-r+1$ then we are done, so let us assume that $|P_r| = n-r+1$. Notice that, since $r < n-d+1$, $|P_r| > d$.

We have to show that at least one process in P_r will not proceed to round $r+1$, i.e., to show that at least one process in P_r will withdraw at line 12. This amounts to show that for at least one process $p_i \in P_r$ the predicate $counter_i < m/competitors_i$ is evaluated to true when p_i executes line 10 during round r.

Assume by contradiction that the predicate $counter_i < m/competitors_i$ in line 13 is evaluated to false for each process $p_i \in P_r$. For each $1 \leq i \leq |P_r|$, let $counter(i)$ denotes the value of $counter_i$ at that time (when the predicate is evaluated to false). Thus, for all $1 \leq i \leq |P_r|$, $counter(i) \geq m/(n-r+1)$. Hence, it follows from the following facts,

1. $counter(1) + \cdots + counter(|P_r|) = m$,
2. $\forall\, 1 \leq i \leq |P_r| \;:\; counter(i) \geq m/(n-r+1)$, and
3. $|P_r| = n-r+1$,

that $\forall\, 1 \leq i \leq |P_r| \;:\; counter(i) = m/(n-r+1)$. Moreover, as

1. $counter(i)$ is a positive integer, we have $\gcd(n-r+1, m) = n-r+1$,
2. $r < n-d+1$ it follows that $n-r+1 > d$,

from which follows that $\gcd(n-r+1, m) > d$, which contradicts the assumption that $m \in M(n,d)$.

Lemma 2. *At most d processes are elected.*

Proof. The proof is an immediate consequence of Lemma 1, which states that at most d processes may execute round $n-d+1$. If they do not withdraw from the competition, each of these processes exits the algorithm at line 14, and considers it is a leader.

$\square_{Lemma\ 2}$

Lemma 3. *For every $r \in \{1, ..., n-d+1\}$, at least one process executes round $r+1$. In particular, at least one process executes round $n-d+1$ at the end of which it claims it is a leader.*

Proof. Considering the (worst) case where the n processes execute round $r = 1$, we show that at least one process attains round 2. To this end, let us assume by contradiction that no process attains round 2. This means that all the processes executed line 10 and found the predicate equal to true (they all withdrew) hence each process p_i is such that $counter_i < m/(n-r+1) = m/n$. Using the notations and the observations of Lemma 1, we have

1. $|P_r| = n$,
2. $counter(1) + \cdots + counter(n) = m$,
3. $\forall\, 1 \leq i \leq n \;:\; counter(i) < m/(n-r+1) = m/n$.

If then follows from the last item that $counter(1) + \cdots + counter(n) < n \times m/n = m$ which contradicts the second item. It follows from this contradiction that there is at least one process for which the predicate of line 10 is false at the end of round 1, and consequently this process progresses to round $r = 2$.

Assuming now by induction that at most $(n - r + 1)$ processes execute round r, we show that at least one process progresses to round $r+1$. The proof follows from the three previous items where $|P_r| = n-r+1$ (item 1), $counter(1)+\cdots+counter(|P_r|) = m$ (item 2), and $\forall\, 1 \leq i \leq |P_r| \,:\, counter(i) < m/(n - r + 1)$ (item 3), from which we conclude $counter(1) + \cdots + counter(|P_r|) < n \times (m - r + 1)(m/(m - r + 1))$, i.e. $m < m$, a contradiction. It follows that at least one process executes the round $r + 1$ during which it finds the predicate of line 10 false and consequently progresses to the next round if $r < n - d + 1$. If $r = n - d + 1$, the process executes line 14 and becomes a leader. $\Box_{Theorem\ 3}$

Theorem 2. *Let m, n and d be such that $m \in M(n, d)$, and assume at least one process invokes* elect(). *Algorithm* 1 *(Fig.* 2*) solves d-election in a fully anonymous system where communication is through* RMW *registers.*

Proof. The proof follows directly from the lemmas 2 and 3. $\Box_{Theorem\ 2}$

3 Exact *d*-election in the RMW Model Where Participation is Required

This section considers the fully anonymous RMW model in which all the processes are required to participate. In such a context, it presents an exact d-election algorithm that assumes that d is a multiple of $\mathsf{gcd}(m, n)$. It also shows that this condition is necessary for exact d-election in such a system model.

3.1 A Necessary Condition for Exact *d*-election

The following theorem, which considers anonymous memory and non-anonymous processes with the symmetry constraint, has been stated and proved in [9].

Theorem 3. (See [9]**).** *There is no symmetric exact d-election algorithm in the* RMW *communication model for $n \geq 2$ processes using m anonymous registers if* $\mathsf{gcd}(m, n)$ *does not divide d.*

As in Sect. 2.1 this impossibility still holds in the RMW model where both the processes and the memory are anonymous, and in the model where communication is through anonymous RW registers.

3.2 An Exact *d*-election algorithm

Anonymous Memory. All the registers of the anonymous memory $R[1..m]$ are RMW registers initialized to 0. Moreover, the size m of the memory is such that $\mathsf{gcd}(m, n)$ divides d.

An anonymous register $R[x]$ will successively contain the values 1, 2, ... where the increases by 1 are produced by successful executions of compare&swap $(R[x], val, val + 1)$ issued by the processes (lines 6 and 8 in Fig. 3). The fact that a process can increase the value of a register to $val + 1$ only if its current value is val is the key of the algorithm.

Underlying Principle. The key idea that governs the algorithm is Bezout's identity, a Diophantine equation that relates any pair of positive integers according to their Greatest Common Divisor[3].

Theorem 4. (Bezout, 1730-1783). *Let m and n be two positive integers and let $d = \mathsf{gcd}(m,n)$. There are two positive integers u and v such that $u \times m = v \times n + d$.*[4]

Consider a rectangle made up of $u \times m$ squares. On one side, this means that u squares are associated with each of the m anonymous registers. On another side, each of the n processes progresses until it has "captured" v squares (from an operational point of view, the capture of a square is a successful invocation of $\mathsf{compare\&swap}(R[x], val, val+1)$.

Then, when $v \times n$ squares have been captured by the processes, each process competes to capture one more square. As it remains only $d = u \times m - v \times n$ squares, the processes that succeed in capturing one more square are the d leaders.

Local variables at each process p_i

- won_i (initialized to 0): number of squares captured by p_i.
- sum_i (initialized to 0): local view of the numbers of squares captured by all the processes.
- $myview_i[1..m]$: local copy (non-atomically obtained) of the anonymous memory $R[1..m]$.

Description of the Algorithm. Assuming d is a multiple of $\mathsf{gcd}(m,n)$ and all the processes participate, Algorithm 2 (described in Fig. 3) solves exact d-election for n anonymous processes and m RMW anonymous registers.

When it invokes $\mathsf{elect}()$, a process p_i enters a repeat loop lines 1–10. Each time it enters the loop, p_i asynchronously reads the anonymous memory non-atomically (line 2) and then counts in sum_i the number of squares that have been captured by all processes as indicated by the previous asynchronously reads (line 3).

If p_i sees a register $R[x]$ that has been captured less than u times (line 4), there are two cases.

- If $won_i < v$, p_i tries to capture one of the u squares of $R[x]$. To this end p_i uses the RMW operation: it invokes $\mathsf{compare\&swap}(R[x], myview_i[x], myview_i[x]+1)$. If it is successful, it increases won_i, the number of squares it has captured so far (line 6).
- If $sum_i \geq v \times n$ (we have then $won_i = v$), p_i strives to capture one more square (line 8). If it is successful, it is elected as of the d leaders.

  In the other case, if $sum_i = u \times m$, all the squares have been captured, so p_i is not a leader (line 11). Otherwise, p_i re-enters the repeat loop.

[3] This principle has already been used in [9] to solve exact d-election with a *symmetric* algorithm in a system where the (non-anonymous) processes cooperate through an anonymous RW registers.

[4] The pair $\langle u, v\rangle$ is not unique. Euclid's $\mathsf{gcd}(m,n)$ algorithm can be used to compute such pairs.

ALGORITHM 2: CODE OF A PROCESS p_i IN THE FULLY ANONYMOUS RMW MODEL (MANDATORY PARTICIPATION)

u and v: smallest positive integers such that $u \times m = v \times n + d$
The initial value of all the shared registers is 0.

operation elect() **is**
1 **repeat**
2 **for each** $j \in \{1, ..., m\}$ **do** $myview_i[j] \leftarrow R[j]$ **end_do** // read of the anony mem.
3 $sum_i \leftarrow myview_i[1] + \cdots + myview_i[m]$ // # successful compare&swap() seen
4 **if** $\exists\, x \in \{1, ..., m\} \; : \; myview_i[x] < u$ **then**
5 **if** $won_i < v$ **then**
6 **if** compare&swap($R[x], myview_i[x], myview_i[x] + 1$) **then** $won_i \leftarrow won_i + 1$ **fi fi**
7 **if** $sum_i \geq v \times n$ **then** // $sum_i \geq v \times n$ implies $won_i = v$
8 **if** compare&swap($R[x], myview_i[x], myview_i[x] + 1$) **then** return (leader) **fi fi**
9 **fi**
10 **until** $sum_i = u \times m$ **end repeat**
11 return (not leader).

Fig. 3. Exact d-election for n anonymous processes and m RMW anonymous registers

Remark. Let α and β be two integers such that $m = \alpha \times \mathsf{gcd}(m, n)$ and $n = \beta \times \mathsf{gcd}(m, n)$. The equations $u \times m = v \times n + d$ and $d = \ell \times \mathsf{gcd}(m, n)$ give rise to the equation $u \times \alpha = v \times \beta + \ell$, which can be used to obtain a more efficient version of the algorithm.

3.3 Proof of Algorithm 2

Theorem 5. *Let m, n and d be such that* $\mathsf{gcd}(m, n)$ *divides d, and assume all the processes invoke* elect(). *Algorithm* 2 *(Fig.* 3*) solves exact d-election in a fully anonymous system where communication is through* RMW *registers.*

Proof. Let us first observe that, due to the atomicity of compare&swap(), if several processes invoke compare&swap($X, v, v+1$) on the very same register X whose value is v, exactly one of them succeeds in writing $v + 1$. It follows that each of the $u \times m$ squares is captured by only one process. Moreover, due to the predicate of line 5, each process eventually captures v squares. Once this occurs, it remains d squares, which are captured by d distinct processes at line 8 (these processes are distinct because, once a process captured such a square, it returns the value leader and stops executing). Moreover, a process can capture one of the d remaining squares only after each process has captured v squares at line 6. It follows that exactly d processes exit the algorithm at line 7 with a successful compare&swap(), and the $(n - d)$ other processes exit the algorithm at line 11. $\Box_{Theorem\ 5}$

4 *d*-Election in the RMW Model Where Participation is Required

This section considers the fully anonymous RMW model in which all the processes are required to participate. In such a context, it presents a d-election algorithm where $\mathsf{gcd}(m, n) \leq d$. It also shows that this condition is necessary for d-election in such a system model.

4.1 A Necessary Condition for *d*-election

The following theorem, which considers anonymous memory and non-anonymous processes with the symmetry constraint, has been stated and proved in [9].

Theorem 6. (See [9]**).** *There is no symmetric d-election algorithm in the* RMW *communication model for* $n \geq 2$ *processes using* m *anonymous registers if* $\mathsf{gcd}(m, n) > d$.

Clearly, this impossibility still holds in the RMW model where both the processes and the memory are anonymous, and in the model where communication is through anonymous RW registers.

4.2 A Necessary and Sufficient Condition for *d*-election

The following corollary is an immediate consequence of Theorem 5.

Corollary 1. *For any pair* $\langle n, m\rangle$*, it is always possible to solve exact* $\mathsf{gcd}(n, m)$*-election in a fully anonymous system where communication is through* RMW *registers.*

Let us also observe that any exact d-election algorithm trivially solves d-election (but then the bound is then not tight). We also have the following theorem.

Theorem 7. *For any pair* $\langle n, m\rangle$*, it is possible to solve d-election in a fully anonymous system where communication is through* RMW *registers if and only if* $\mathsf{gcd}(n, m) \leq d$.

Proof. If direction. For any pair $\langle n, m\rangle$ such that $\mathsf{gcd}(n, m) \leq d$, it is possible to solve d-election by running an exact $\mathsf{gcd}(n, m)$-election algorithm, which exists due to Corollary 1.

As the fully anonymous model is a weaker model than the symmetric model, the "Only if" direction follows from Theorem 6. $\square_{Theorem\ 7}$

5 Impossibility in the RW Communication Model

Theorem 8. *There is neither d-election nor exact d-election algorithms in the process anonymous* RW *non-anonymous communication model.*

Proof. Assuming such an algorithm exists, let us order the participating processes in some fixed order, e.g., p_1, ..., p_x ($x = n$ in the case where full participation is required). Let us consider in such a setting a lock-step execution in which p_1 executes its first (read or write operation on a shared register) operation, then p_2 executes its first operation, etc., until p_x that executes its first (read or write) operation on the shared non-anonymous memory. As all processes have the same code, they all execute the same operation on the same register and are consequently in the same local state after having executed their first operation. The same occurs after they have their (same) second operation, etc. It follows that, whatever the number of steps executed in a lock-step manner by the processes, they all are in the same local state. So, it is impossible to break their anonymity (that would allow us to elect some of them). $\Box_{Theorem\ 8}$

Let us consider an anonymous memory in which the memory adversary associates the same address mapping to all the processes (i.e., $\forall i, j \in \{1, \cdots, n\}$ and $x \in \{1, \cdots, m\}$ we have $f_i(x) = f_j(x)$, see Sect. 1.2). In this case, the model boils down to the process anonymous and non-anonymous memory. The next corollary is then an immediate consequence of the previous theorem.

Corollary 2. *There is neither d-election nor exact d-election algorithms in the fully anonymous* RW *communication model.*

6 Conclusion

This article has investigated the d-election problem in fully anonymous shared memory systems. Namely, systems where not only the processes are anonymous but the shared memory also is anonymous in the sense that there is no global agreement on the names of the shared registers (any register can have different names for distinct processes). Assuming RMW atomic registers, it has shown that both the d-election problem (at least one and at most processes are elected) and the exact d-election problem (exactly d processes are elected) can be solved in such an adversarial context if and only if the three model parameters n (number of processes), m (size of the anonymous memory), and d (number of leaders) satisfy some properties. These necessary and sufficient conditions are:

- $m \in M(n, d)$ for solving d-election when participation is not required,
- $\mathsf{gcd}(m, n)$ divides d for solving exact d-election when participation is required, and
- $\mathsf{gcd}(m, n) \leq d$ for solving d-election when participation is required.

It has also been shown that,

- neither d-election nor exact d-election can be solved in a fully anonymous system where communication is through atomic RW registers.

This work complements previously known research on the symmetry-breaking problem (election) in the context of fully anonymous RW/RMW systems. A very challenging problem remains to be solved: are there other non-trivial functions that can be solved in the fully anonymous RW/RMW setting.

Acknowledgments. The authors want to thank the referees for their constructive comments.

A The Case Where $m = 1$

When the anonymous memory is made up of a single register R, we have $\gcd(1, n) = 1 \leq 1$. In this case there is a very simple d-election algorithm described below, where the single anonymous register is initialized to 0 (Fig. 4).

ALGORITHM 3: CODE OF A PROCESS p_i WHEN $m = 1$
(PARTICIPATION REQUIRED OR NOT REQUIRED)

```
operation elect() is
repeat forever
   myview_i ← R                                   // atomic read the anonymous memory
   if myview_i ≥ d then return (not leader) fi
   if compare&swap(R, myview_i, myview_i + 1) then return (leader) fi
end repeat.
```

Fig. 4. d-election for n anonym. processes when the anonym. memory is a single RMW register

References

1. Aigner M. and Ziegler G., Proofs from THE BOOK (4th edition). Springer, 274 pages, ISBN 978-3-642-00856-6 (2010)
2. Aghazadeh Z., Imbs D., Raynal M., Taubenfeld G., and Woelfel Ph., Optimal memory-anonymous symmetric deadlock-free mutual exclusion. Proc. 38th ACM Symposium on Principles of Distributed Computing (PODC'19), ACM Press, 10 pages (2019)
3. Angluin D., Local and global properties in networks of processes. Proc. 12th Symposium on Theory of Computing (STOC'80), ACM Press, pp. 82–93, (1980)
4. Attiya, H., Gorbach, A., Moran, S.: Computing in totally anonymous asynchronous shared-memory systems. Information and Computation **173**(2), 162–183 (2002)
5. Attiya H. and Welch J.L., Distributed computing: fundamentals, simulations and advanced topics, (2nd Edition), Wiley-Interscience, 414 pages, 2ISBN 0-471-45324-2 (2004)
6. Bouzid, Z., Raynal, M., Sutra, P.: Anonymous obstruction-free (n, k)-set agreement with $(n - k + 1)$ atomic read/write registers. Distributed Computing **31**(2), 99–117 (2018)
7. Dijkstra, E.W.: Some beautiful arguments using mathematical induction. Algorithmica **13**(1), 1–8 (1980)
8. Flocchini, P., Ilcinkas, D., Pelc, A., Santoro, N.: Computing without communicating: ring exploration by asynchronous oblivious robots. Algorithmica **65**(3), 562–583 (2013)
9. Godard E., Imbs D., Raynal M., and Taubenfeld G., From Bezout identity to space-optimal leader election in anonymous memory systems. Proc. 39th ACM Symposium on Principles of Distributed Computing (PODC'20), ACM press, pp. 41–50 (2020)
10. Guerraoui, R., Ruppert, E.: Anonymous and fault-tolerant shared-memory computations. Distributed Computing **20**, 165–177 (2007)
11. Harel D. and Feldman Y., Algorithmics, the spirit of computing. Springer, 572 pages (2012)
12. Herlihy, M.P., Wing, J.M.: Linearizability: a correctness condition for concurrent objects. ACM Transactions on Programming Languages and Systems **12**(3), 463–492 (1990)

13. Imbs, D., Raynal, M., Taubefeld, G.: Election in fully anonymous shared memory systems: tight space bounds and algorithms. ArXiv:2203.02988v1, 17 pages (2022)
14. Kshemkalyani, A.D., Singhal, M.: Distributed computing: principles, algorithms and systems. Cambridge University Press, 736 pages (2008)
15. Lamport, L.: On interprocess communication, Part I: basic formalism. Distrib. Comput. **1**(2), 77–85 (1986)
16. Navlakha, S., Bar-Joseph, Z.: Algorithms in nature: the convergence of systems biology and computational thinking. Molecular Syst. Biol. **7**(546), 1–11 (2011)
17. Navlakha, S., Bar-Joseph, Z.: Distributed information processing in biological and computational systems. Commun. ACM **58**(1), 94–102 (2015)
18. Perlis, A.: Epigrams on programming. ACM SIGPLAN Notices **17**(9), 7–13 (1982)
19. Rashid, S., Taubenfeld, G., Bar-Joseph, Z.: The epigenetic consensus problem. In: Jurdziński, T., Schmid, S. (eds.) SIROCCO 2021. LNCS, vol. 12810, pp. 146–163. Springer, Cham (2021). https://doi.org/10.1007/978-3-030-79527-6_9
20. Raynal, M.: Concurrent programming: algorithms, principles and foundations. Springer, 515 p., ISBN 978-3-642-32026-2 (2013)
21. Raynal, M.: Distributed Algorithms for Message-Passing Systems, 534 p. Springer (2013). ISBN 978-3-642-38122-5 (2013)
22. Raynal, M., Taubenfeld, G.: Mutual exclusion in fully anonymous shared memory systems. Inf. Process. Lett. **158** (2020)
23. Raynal, Michel, Taubenfeld, Gadi: Fully anonymous consensus and set agreement algorithms. In: Georgiou, Chryssis, Majumdar, Rupak (eds.) NETYS 2020. LNCS, vol. 12129, pp. 314–328. Springer, Cham (2021). https://doi.org/10.1007/978-3-030-67087-0_20
24. Styer, E., Peterson, G.L.: Tight bounds for shared memory symmetric mutual exclusion problems. In: Proceedings of 8th ACM Symposium on Principles of Distributed Computing (PODC'89), pp. 177–191. ACM Press (1989)
25. Taubenfeld, G.: Synchronization algorithms and concurrent programming, 423 p., Pearson Education/Prentice Hall, ISBN 0-131-97259-6 (2006)
26. Taubenfeld, G.: Coordination without prior agreement. In: Proceedings of the 36th ACM Symposium on Principles of Distributed Computing (PODC 2017), pp. 325–334. ACM Press (2017)

Dispersion of Mobile Robots on Directed Anonymous Graphs

Giuseppe F. Italiano[1], Debasish Pattanayak[1], and Gokarna Sharma[2(✉)]

[1] LUISS University, Rome, Italy
{gitaliano,dpattanayak}@luiss.it
[2] Kent State University, Kent, OH, USA
gsharma2@kent.edu

Abstract. Given any arbitrary initial configuration of $k \leq n$ robots positioned on the nodes of an n-node anonymous graph, the problem of dispersion is to autonomously reposition the robots such that each node will contain at most one robot. This problem gained significant interest due to its resemblance with several fundamental problems such as exploration, scattering, load balancing, relocation of electric cars to charging stations, etc. The objective is to solve dispersion simultaneously minimizing (or providing a trade-off between) time and memory requirement at each robot. The literature mainly dealt with dispersion on undirected anonymous graphs. In this paper, we initiate the study of dispersion on directed anonymous graphs. We first show that it may not always be possible to solve dispersion when the directed graph is not strongly connected. We then establish some lower bounds on both time and memory requirements at each robot for solving dispersion on a strongly connected directed graph. Finally, we provide two deterministic algorithms solving dispersion on any strongly connected directed graph. Let D be the graph diameter and Δ_{out} be its maximum out-degree. The first algorithm solves dispersion in $O(k^2 \cdot \Delta_{out})$ time with $O(\log(k + \Delta_{out}))$ bits at each robot. The second algorithm solves dispersion in $O(k \cdot D)$ time with $O(k \cdot \log(k + \Delta_{out}))$ bits at each robot, provided that robots in the 1-hop neighborhood can communicate. Both algorithms extend to handle crash faults.

Keywords: Multi-agent systems · Mobile robots · Local and 1-hop communication · Directed graphs · Dispersion · Time and memory complexity

1 Introduction

The dispersion of autonomous mobile robots in a region is a problem of significant interest in distributed robotics [16,17]. Recently, this problem has been formulated by Augustine and Moses Jr. [2] in the context of graphs as follows: *Given any arbitrary initial configuration of $k \leq n$ robots positioned on the nodes of an n-node anonymous graph, the robots reposition autonomously to reach a configuration where each robot is positioned on a distinct node of the graph*, which we call the DISPERSION problem. This problem has many practical applications, e.g., relocating self-driving electric cars (robots) to recharging stations (nodes), assuming that the cars have smart devices to

M. Parter (Ed.): SIROCCO 2022, LNCS 13298, pp. 191–211, 2022.
https://doi.org/10.1007/978-3-031-09993-9_11

communicate with each other to find a free/empty charging station. This problem is also important due to its relationship to many other multi-robot coordination problems, including exploration, scattering, load balancing, and self-deployment [2,19–21].

The objective in DISPERSION is to simultaneously minimize (or provide trade-off between) two fundamental performance metrics, (i) *time* and (ii) *memory* at each robot, to successfully solve the problem. The literature studied DISPERSION mainly on undirected graphs [2,19–22,24,26]. The following question naturally arises: *Is it possible to solve* DISPERSION *on directed anonymous graphs?* The primary motivation behind posing this problem is that the approaches for undirected graphs may not extend to solve dispersion on directed graphs. The existing approaches for undirected graphs rely substantially on visiting a graph edge in both directions. The main challenge for directed graphs is the direction on edges that restricts movement in only one direction. So any technique for directed graphs should take critical care on this aspect. In this paper, we study DISPERSION for the first time on directed graphs.

Contributions. We consider an *anonymous* port-labeled directed graph $G = (V, E)$, $|V| = n$, $|E| = m$, where nodes have no IDs and hence are indistinguishable from each other, but the outgoing ports (leading to outgoing edges and neighbors) at each node are distinguishable. The (outgoing) ports of any node $v \in G$ with out-degree δ^v_{out} have unique labels in $[1, \delta^v_{out}]$. We consider $k \leq n$ robots on graph nodes, initially positioned arbitrarily (at least one node with multiple robots positioned on it; otherwise, the problem is trivially solved). Each robot has a unique ID in $[1, k^{O(1)}]$. The robots have memory, but the graph nodes do not. The setting is *synchronous* – all robots become active and perform their operations simultaneously in synchronized rounds – and time is measured in rounds. Following the literature [2,19,20,22,24,26], we consider *local* and *1-hop* communication models. In the local model, only robots co-located at a graph node can communicate. In the 1-hop model, a robot can communicate with robots positioned on its 1-hop neighbors (in addition to the co-located robots). The robots move along the direction of the edges. The robots are susceptible to crashes at any time, and once crashed a robot stops communicating with other robots.

First, we establish one impossibility and some lower bound results (Sect. 3). We show that it may not always be possible to solve DISPERSION on a directed graph that is not strongly connected; a directed graph is *strongly connected* if there is a path in each direction between each pair of vertices. For strongly connected directed graphs, (i) we observe a time lower bound of $\Omega(k)$ for any $k \leq n$, (ii) we prove a time lower bound of $\Omega(D^2)$ for $k = n$, (iii) we prove a time lower bound of $\Omega(k^2)$ for any DFS traversal based algorithm, and (iv) we prove a memory lower bound of $\Omega(\log k)$ bits per robot. All these lower bounds are deterministic. Second, as our main contribution, we provide two deterministic algorithms solving DISPERSION on any strongly connected directed graph. Specifically, we prove the following theorem.

Theorem 1. *Consider $k \leq n$ robots positioned initially arbitrarily on an n-node strongly connected directed anonymous graph with diameter D and out-degree Δ_{out},* DISPERSION *can be solved in*

a. $O(k^2 \cdot \Delta_{out})$ time with $\Theta(\log(k + \Delta_{out}))$ bits at each robot under the local model with a DFS based algorithm. For $f \leq k$ robot crashes, the time bound becomes $O(f \cdot k^2 \cdot \Delta_{out})$. (Sects. 4 and 6)

b. *$O(k \cdot D)$ time with $O(k \log(k + \Delta_{out}))$ bits at each robot under the 1-hop model. The time bound also applies for the case of $f \leq k$ robots that may experience crash fault. (Sects. 5 and 6)*

To the best of our knowledge, these are the first results for DISPERSION on directed graphs. Theorem 1(a) matches the $\Omega(k^2)$ time lower bound within a $O(\Delta_{out})$ factor, which is asymptotically optimal for $\Delta_{out} = O(1)$, i.e., for constant out-degree directed graphs. Theorem 1(b) shows that, under 1-hop communication, DISPERSION can be solved with $O(D/(k \cdot \Delta_{out}))$ factor improvement on time.

Techniques. The impossibility result is established considering an initial configuration on a not-strongly-connected directed graph so that there is always a node with two robots positioned on it, irrespective of any technique used. The time lower bound $\Omega(k)$ is obtained considering a (strongly connected) directed cycle. The time lower bound $\Omega(D^2)$ is established by constructing a (strongly connected) directed graph G with $k = n$ nodes and diameter (height) $D = \omega(1)$ such that any deterministic algorithm needs $\Omega(D^2)$ rounds to achieve DISPERSION. The time lower bound $\Omega(k^2)$ for a DFS traversal based algorithm is proved considering a strongly connected directed graph where dispersing $k/2$ robots on $k/2$ different nodes needs exactly $k/2$ rounds each, requiring in total at least $\Omega(k^2)$ rounds to disperse all k robots successfully. The memory lower bound $\Omega(\log k)$ bits per robot is established considering the minimum number of bits needed to distinguish IDs of two different robots as per the robot model used.

As our main contribution, the time upper bound $O(k^2 \cdot \Delta_{out})$ is achieved through simulating a *depth first search* (DFS) traversal to visit the nodes of the directed graph, settling a robot on each new node visited. The main difficulty is to backtrack over a directed edge $u \rightarrow v$, i.e., to find a directed path from v to u. The challenge is, if not done carefully, the traversal gets stuck failing to reach u from v (which we call the *local maximum* problem). We avoid the local maximum problem by ranking the nodes (with settled robots) based on the order of the first DFS arrival. During the traversal, if it is found that there is a backtracking path to a lower-ranked node, then the backtracking path is updated to point to that lower-ranked node. We also devise a technique that bounds the length of each backtrack path to $\leq k - 1$ edges, even when DFSs start in parallel from multiple nodes. Furthermore, when a DFS meets another, we subsume one DFS by another based on the unique DFS IDs derived from the robot IDs. Putting these ideas together, visiting $k \cdot \Delta_{out}$ directed edges takes the DFS traversal to at least k different nodes, allowing to settle k robots, in $k \cdot \Delta_{out} \cdot k = O(k^2 \cdot \Delta_{out})$ rounds.

As our another main contribution, the time upper bound $O(k \cdot D)$ is established exploiting the information that can be broadcasted and gathered from the 1-hop communication so that in every D rounds, at least a robot can be settled on a previously empty node. Having 1-hop information is of no use in the DFS based algorithm as it cannot exploit the information collected through 1-hop communication. To bookkeep the information to successfully run broadcast and gather through 1-hop communication, the memory is increased to $O(k \cdot \log(k + \Delta_{out}))$ bits per robot.

For crash faults, we analyze the working principles of the fault-free cases of the above algorithms to see how much extra work (time and memory) is required for each crash. We show that the extra time is proportional to $O(f \cdot k^2 \cdot \Delta_{out})$ for the DFS based algorithm and zero for the second algorithm; the memory bound stays the same as in the fault-free cases in both the algorithms.

Related Work. The literature studied DISPERSION mostly in undirected graphs. Two communication models were considered, local and global.

A significant amount of work in the literature is in the local model which we discuss first. Augustine and Moses Jr. [2] were the first to study DISPERSION assuming $k = n$. They proved a memory lower bound of $\Omega(\log n)$ bits at each robot and a time lower bound of $\Omega(D)$ ($\Omega(n)$ in arbitrary graphs) for any deterministic algorithm. They then provided deterministic algorithms using $O(\log n)$ bits at each robot to solve DISPERSION on lines, rings, and trees in $O(n)$ time. For arbitrary graphs, they provided two algorithms, one using $O(\log n)$ bits at each robot with $O(mn)$ time and another using $O(n \log n)$ bits at each robot with $O(m)$ time, where m is the number of edges in the graph. Kshemkalyani and Ali [19] provided an $\Omega(k)$ time lower bound for arbitrary graphs for $k \leq n$. They then provided three deterministic algorithms on arbitrary graphs: (i) The first algorithm using $O(k \log \Delta)$ bits at each robot with $O(m)$ time, (ii) The second algorithm using $O(D \log \Delta)$ bits at each robot with $O(\Delta^D)$ time, and (iii) The third algorithm using $O(\log(k + \Delta))$ bits at each robot with $O(mk)$ time, where Δ and D, respectively, are the maximum degree and diameter of the graph. Kshemkalyani *et al.* [20] provided an algorithm that runs in $O(\min(m, k\Delta) \cdot \log k)$ time using $O(\log n)$ bits of memory at each robot on arbitrary graphs, given that parameter m, n, k are known to robots. Shintaku *et al.* [32] established the same time bound without robots knowing m, n, k. Recently, Kshemkalyani and Sharma [24] improved the time bound to $O(\min(m, k\Delta))$ keeping memory $O(\log(k + \Delta))$ bits at each robot. For grid graphs, Kshemkalyani *et al.* [22] provided an algorithm that runs in $O(\min(k, \sqrt{n}))$ time using $O(\log k)$ bits at each robot. Randomized algorithms are presented in [26] to solve DISPERSION from rooted initial configurations where the random bits are mainly used to reduce the memory requirement at each robot.

Recently, there is some work in the *global* model which we discuss now. In the global model, the robots on the graph can communicate with each other despite their locations on the graph . Kshemkalyani *et al.* [21] provided two deterministic algorithms on arbitrary graphs: (i) The first algorithm using $O(\log(k + \Delta))$ bits at each robot with $O(\min(m, k\Delta))$ time and (ii) The second algorithm using $O(\log D + \Delta \log k)$ bits at each robot with $O((D + k)\Delta(D + \Delta))$ time. For grid graphs, Kshemkalyani *et al.* [22] provided a $O(\sqrt{k})$ time algorithm with $O(\log k)$ bits at each robot.

DISPERSION in dynamic (undirected) graphs was considered in [23]. Dispersion under crash faults was considered in [29] and under byzantine faults was considered in [27,28]. In this paper, we initiate study of DISPERSION on directed graphs and present results using the local and 1-hop communication models.

One problem that is closely related to DISPERSION is the graph exploration by mobile robots. The exploration problem has been quite heavily studied in the literature for specific as well as arbitrary graphs, e.g., [3,7,9,15,18,25]. The vast majority of works considered undirected graphs. It was shown that a robot can explore an anonymous graph using $\Theta(D \log \Delta)$-bits of memory and the runtime of the algorithm is $O(\Delta^{D+1})$ [15]. In the model where graph nodes also have memory, Cohen *et al.* [7] gave two algorithms: The first algorithm uses $O(1)$-bits at the robot and 2 bits at each node, and the second algorithm uses $O(\log \Delta)$ bits at the robot and 1 bit at each node. The runtime of both algorithms is $O(m)$ with preprocessing time of $O(mD)$. The trade-off between exploration time and number of robots is studied in [25]. The collective exploration by a team of robots is studied in [14] for trees. In directed graphs, two cooperating robots were considered to explore and learn about the anonymous strongly connected directed graph in [6]. The exploration and mapping of directed graphs was also done by using a pebble in [5]. Exploration in directed graphs is also studied for a searcher that tries to minimize the tour required to visit all the nodes in a graph [1] using both randomized and online algorithms [13].

Another problem related to DISPERSION is the scattering of robots in graphs. Scattering has been studied for rings [10,31] and grids [4,30]. Furthermore, DISPERSION is related to the load balancing problem, where a given load at the nodes has to be (re-)distributed among several processors (nodes). This problem has been studied quite heavily in (undirected) graphs, e.g., see [8].

We refer readers to [11,12] for recent developments in the above research topics.

2 Model and Preliminaries

Graph. Let $G = (V, E)$ be an n-node arbitrary, connected, unweighted, directed, port-labeled, anonymous graph with m edges, i.e., $|V| = n$ and $|E| = m$, such that nodes do not have identifiers but, at any node $v_i \in V$, its incident (outgoing) edges are uniquely identified by a *label* (aka port number) in $[1, \delta_{out}^{v_i}]$, where $\delta_{out}^{v_i}$ is the *out-degree* of v_i. The *out-degree* of graph G is $\Delta_{out} = \max_{1 \leq i \leq n} \delta_{out}^{v_i}$, i.e., the maximum $\delta_{out}^{v_i}$ among the nodes in G. There is no bandwidth limitation on the edges, i.e., any number of robots are allowed to traverse an edge at any time following the direction of the arrow. The graph nodes do not have memory.

For any two nodes $u, v \in G$, we denote by path $p(u, v)$ the sequence of consecutive directed edges starting from u and ending at v. The diameter D is the longest shortest directed path $p(u, v)$ between any two nodes $u, v \in G$. A directed graph G is said to be *strongly connected* if for each pair of nodes u, v, there is a directed path $p(u, v)$ reaching to v from u as well as a directed path $p(v, u)$ reaching to u from v.

Robots. Let $\mathcal{R} = \{r_1, r_2, \ldots, r_k\}$ be a set of $k \leq n$ robots residing on the nodes of G. No robot can reside on the edges of G, but one or more robots can occupy the same node. Each robot has a unique $O(\log k)$-bit ID taken from $[1, k^{O(1)}]$. The ID of a robot r_i is denoted by $r_i.ID$ with $r_i.ID = i$. Furthermore, it is assumed that each robot is equipped with memory to store information.

Communication Model. There are two communication models: local and global [21–23]. In the local model, a robot can only communicate with other robots co-located on the same node. In the global model, a robot can communicate with any other robot, irrespective of their positions on graph. We define an intermediate model of 1-hop communication in which a robot can communicate to another robot, if they are located in neighboring nodes. This paper considers the local and 1-hop communication models.

Time Cycle. At any time, a robot $r_i \in \mathcal{R}$ could be active or inactive. When a robot r_i becomes active, it performs the "Communicate-Compute-Move" (CCM) cycle as follows: (i) *Communicate:* For each robot $r_j \in \mathcal{R}$ that is at some node v_i, another robot r_i at v_i (and v_i's neighbors in the 1-hop model) can observe the memory of r_j. Robot r_i can also observe its own memory; (ii) *Compute:* Robot r_i may perform an arbitrary computation using the information observed during the "communicate" portion of the cycle. This includes determining a (possibly) port to use to exit v_i and the information that is communicated to the robot r_j that is at v_i, (iii) *Move:* At the end of the cycle, r_i communicates the information to r_j at v_i (so that r_j will store/update in its memory), and exits v_i using the computed port to reach a neighbor of v_i.

Rooted and General Initial Configurations. An initial configuration is *rooted* if all $k \leq n$ robots are positioned on a single node of G. In a *general* initial configuration, there are robots on $1 < k' < k$ different nodes with at least one node among k' has multiple robots. DISPERSION is trivially solved when $k' = k$.

Crash Faults. A robot is susceptible to crash anytime and once it crashes at time $t \geq 0$, it stops communicating with other robots, i.e., at any time $t' > t$, it appears like it has vanished from the system.

Dispersion. DISPERSION in directed graphs is formally defined as follows. We denote by $f \leq k$ the number of faults. The case of $f = 0$ is a fault-free DISPERSION and the case of $f > 0$ is a faulty DISPERSION.

Definition 1 (DISPERSION). *Given $k \leq n$ robots positioned initially arbitrarily on the nodes of an n-node anonymous directed graph $G = (V, E)$ with $0 \leq f \leq k$ robots may crash at any time, the robots reposition autonomously such that each non-faulty robot is on a distinct node of G and stays stationary thereafter.*

Activation, Time, and Memory. Following previous works [2,19,20,22–24,26], we consider the synchronous setting where every robot is active in every CCM cycle, performing the cycle in synchrony. Time is measured in rounds. Another parameter is memory which we measure as the number of bits.

3 Impossibility and Lower Bounds

We discuss here one impossibility result, three time lower bounds, and one memory lower bound for DISPERSION on directed anonymous graphs. These results collectively show the difficulty in obtaining a solution as well as obtaining fast runtime and low memory when a solution exists.

We first discuss the impossibility result based on the graph type. Consider the case of a directed graph G that is not strongly connected and all the robots are located on a single node $u \in G$. We present the following impossibility result.

Theorem 2. *It may not always be possible to solve* DISPERSION *on a not strongly connected directed graph.*

Proof. By definition, a graph $G = (V, E)$ is strongly connected if and only if, for each pair of vertices $u, v \in V$, there is a directed path $p(u, v)$ from u to v and directed path

$p(v, u)$ from v to u. If G is not strongly connected then there exists at least a pair of vertices $u, v \in V$ such that at least $p(u, v)$ or $p(v, u)$ is not available. Assume that, in graph G, v is not reachable from u, i.e., directed path $p(u, v)$ is not available, but the path $p(v, u)$ is available. Let $U \subset V$ be the set of vertices that are reachable from u. As v is not reachable from u, v must also not be reachable from every vertex of the set U, otherwise there would be a directed path $p(u, v)$ from u to v violating our assumption. Since v is not reachable from u, it cannot be in the set U and it must be the case that $|U| \leq n - 1$. Consider now the rooted initial configuration of n robots located at node u. Since v is not reachable from u (and all the nodes in the set U), at any time there must be at least two robots located at some node in U. □

Remark: Theorem 2 considers a rooted initial configuration of n robots on a node u and a node v that is not reachable from v. If v is reachable from u, the problem has a solution. However, since G is not known a priori, an algorithm may not be designed that is able to achieve such a solution configuration.

We now present three time lower bounds for solving DISPERSION.

Theorem 3. *Any deterministic algorithm for* DISPERSION *on strongly connected directed graphs requires $\Omega(k)$ rounds.*

Proof. Consider a directed cycle G. Note that G is strongly connected since for any two nodes $u, v \in G$, there is a directed path $p(u, v)$ and $p(v, u)$. Consider a rooted initial configuration of $k \leq n$ robots on a single node v_{root} of G. In order for the robots to solve DISPERSION, they need to settle at k distinct nodes of G. To reach a node to settle, some robot must travel $k - 1$ directed edges of G, taking $k - 1$ rounds. □

For $k = n$, we present the following time lower bound for solving DISPERSION.

Theorem 4. *For $k = n$, there exists a strongly connected directed graph G with n nodes and diameter D such that any deterministic algorithm for* DISPERSION *requires $\Omega(D^2)$ rounds.*

We prove the following time lower bound for DISPERSION on directed graphs using a DFS traversal based algorithm, which improves on $\Omega(D^2)$ when $k > O(D)$.

Theorem 5. *Any DFS traversal based algorithm for* DISPERSION *on strongly connected directed graphs requires $\Omega(k^2)$ rounds.*

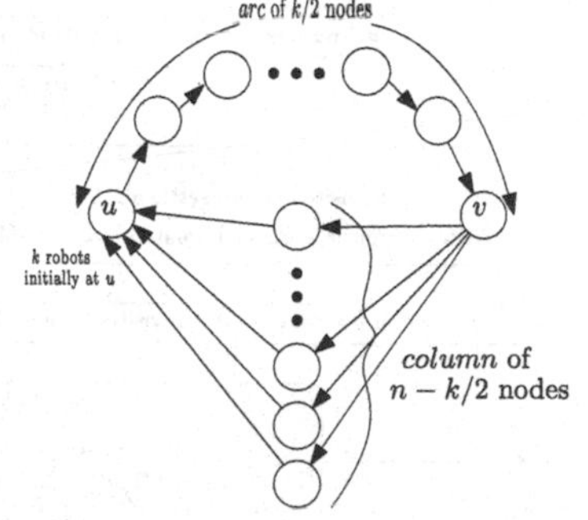

Fig. 1. Lower bound construction for a DFS based DISPERSION algorithm.

Proof. The proof construction is given in Fig. 1. The graph G in Fig. 1 is strongly connected and $k \leq n$ robots start from node u. The proof uses the fact that after $k/2$ robots settle at the top arc in $k/2$ rounds, for the rest $k/2$ robots, it needs $k/2$ rounds each for a robot to settle. This gives in overall $\Omega(k^2)$ rounds lower bound. In the first $k/2 - 1$ rounds, $k/2$ robots settle at $k/2$ nodes on the top arc and a robot settles on a column node in the round $k/2$. After that, each new node in the column is visited in every next $k/2+1$ rounds, after visiting completely the nodes on the top arc. The theorem follows. □

We finally prove $\Omega(\log k)$ bits per robot bound for any deterministic algorithm.

Theorem 6. *Any deterministic algorithm for* DISPERSION *on n-node directed anonymous graphs requires $\Omega(\log k)$ bits at each robot, where $k \leq n$ is the number of robots.*

Proof. Each robot must have its identifier in the range $[1, k^{O(1)}]$ and we count this memory space which must survive across rounds as the space complexity. □

Table 1. Summary of variables used by each robot.

Symbol	Description
$settled$	Indicates a settled robot at a node
$nodeRank$	Indicates the DFS order in which the robot is settled
$backtrackTargetNode$	Indicates the ID of the last settled robot at a node with unvisited ports
$cycleClosingNode$	Indicates whether the node close the current traversal cycle
$cycleClosingNodeRank$	Indicates value in the $nodeRank$ variable of the cycle closing node
$DFSID$	ID of the DFS traversal starting from a node
$recentPort$	Points to the last port via which the unsettled robots exited
$phase$	Indicates $explore$, $retrace1$, or $retrace2$ phase
$status$	Indicates the $visited$ or $fully\text{-}visited$ status of a node
$rootPort = (port, rank)$	Indicates the port and associated rank at a node to reach DFS root

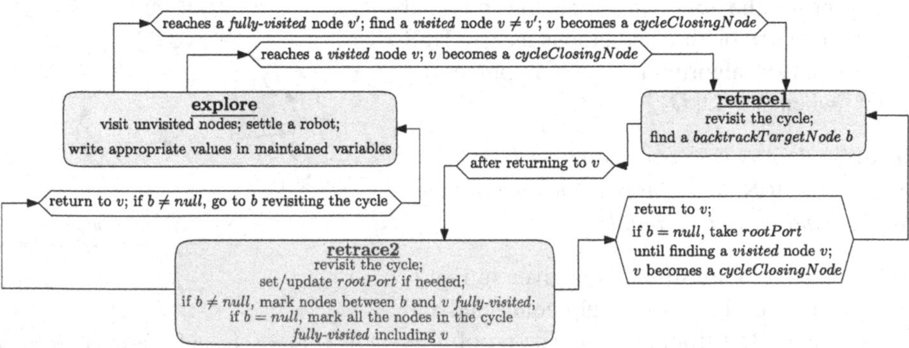

Fig. 2. A state transition diagram depicting each phase of the DFS algorithm and what prompts a transition from one phase to another. Initially, the algorithm starts from the explore phase.

4 DFS Dispersion Algorithm

We present here a deterministic DFS traversal based (or simply DFS) algorithm. We first discuss the variables used, then an overview of the approach, then a detailed discussion, and finally the analysis.

Variables. Each robot maintains the following ten variables (description is given in Table 1). A robot r_i has the variables $r_i.settled$ (initally set to 0), $r_i.recentPort$ (initially set to $null$), $r_i.cycleClosingNode$ (initially set to $null$), $r_i.backtrackTargetNode$ (initially set to $null$), $r_i.phase$ (initially set to $explore$), $r_i.status$ (initially set to $null$), $r_i.nodeRank$ (initially set to $null$), $r_i.cycleClosingNodeRank$ (initially set to $null$), $r_i.rootPort = (port, rank)$ (initially set to $(null, null)$), and $r_i.DFSID$ (initially set to $null$). Variable $r_i.DFSID$ is used only in the general case. The $r_i.settled$ variable indicates the robot r_i that is settled at a node v_i and r_i remains at v_i forever. For a settled robot r_i, $r_i.status \in$ {*visited*, *fully-visited*}; $r_i.status$ becomes $visited$ when v_i is visited by DFS for the first time and becomes $fully\text{-}visited$ after all the outgoing ports at v_i were already visited.

Algorithm 1: DFS algorithm

Input: $k \le n$ robots on $1 \le k' \le k$ nodes of G;
if *r_i is alone at node $v \in G$* **then**
    r_i settles at v writing $r_i.settled \leftarrow 1$ and setting DFS ID i;
else
    each robot on v sets their $DFSID$ i the largest ID among the robots on v and perform DFS i (Algorithm 2);
    if *DFS i meets DFS j* **then**
        if $i < j$ **then**
            all unsettled robots of DFS i traverse DFS j to reach its head node, $head(j)$, to continue the traversal of DFS j;
        else
            DFS i continues its traversal, erasing the values set in all the variables by DFS j and writing all the variables based on DFS i;

Overview of the Algorithm. The algorithm simulates a DFS traversal in three phases: $explore$, $retrace1$, and $retrace2$. Figure 2 depicts how the DFS transitions from one phase to another along with what happens during each phase and what prompts a phase transition. Initially, the DFS is in $explore$ phase. At any time, when the DFS visits an unvisited node, then it settles a robot on it and marks $visited$. When the DFS reaches to a node with a settled robot, say v_x, then backtracking is required. The traversal executes $retrace1$ phase (to find a *backtrack target node*, if any) and $retrace2$ phase (setting backtrack path), one after another, and then it continues $explore$ phase from the backtrack target node. If v_x has *visited* status, then $retrace1$ and $retrace2$ phases start and end at v_x (i.e., v_x becomes the *cycle closing node*). Otherwise, these phases start and end at some other node $v_y \neq v_x$ that has *visited* status (i.e., v_y becomes the cycle closing node). The phases $retrace1$ and $retrace2$ compute a backtrack path setting variable $rootPort$ so that backtracking can be done traversing the $rootPort$ set. After finishing $retrace1$ and $retrace2$ and returning to the cycle closing node, the traversal either (i) moves to the backtrack target node (if not null) to execute the $explore$ phase or (ii) takes $rootPort$ set (if backtrack target node is null) to reach a $visited$ node to continue again $retrace1$ and $retrace2$ phases. This process stops as soon as the robots are settled solving DISPERSION.

Algorithm 2: Algorithm DFS l

---

initially, node rank is set to null for all robots;
let r_i be a robot that belongs to DFS l at node v;
let r_X be the settled robot at node v (if one exists; otherwise DFS l settles r_X);
if $r_i.phase == explore$ **then**
    if *v is an unvisited node* **then**
        settle robot r_X, write $r_X.recentPort \leftarrow r_X.recentPort + 1$, $r_X.status \leftarrow visited$, $r_X.nodeRank \leftarrow r_i.nodeRank + 1$, and exit v via $r_X.recentPort$;
    if *v is a visited node* **then**
        set v (a.k.a., r_X) the cycle closing node and set $r_i.phase \leftarrow retrace1$;
    if *v is a fully-visited node* **then**
        traverse $rootPort$ pointers starting from v until finding a $visited$ node, set that node as the cycle closing node, and set $r_i.phase \leftarrow retrace1$;
if $r_i.phase == retrace1$ **then**
    if *r_i is the smallest ID robot on v* **then**
        r_i revisits the cycle following $recentPort$ pointers and while doing so computes the backtrack target node b that is farthest from v;
        set $r_i.phase \leftarrow retrace2$ after returning to v;
        after returning to v, set $r_i.phase \leftarrow retrace2$;
if $r_i.phase == retrace2$ **then**
    if *r_i is the smallest ID robot on v* **then**
        if $b == null$ **then**
            mark all nodes in the cycle as fully-visited;
            follow $rootPort$ at v until finding a $visited$ node, set v as the cycle closing node, and set $r_i.phase \leftarrow retrace1$;
        else
            r_i revisits the cycle following $recentPort$ pointers and while doing so set the $r_y.rootPort \leftarrow (r_y.recentPort, r_1.cycleClosingNodeRank)$ as well as mark all the nodes after the backtrack target node b and before the cycle closing node $fully\text{-}visited$;
            after returning to v, follow again $recentPort$ pointers to reach b then set $r_i.phase \leftarrow explore$, $r_b.recentPort \leftarrow r_b.recentPort + 1$, and exit via $b.recentPort$ (r_b is the robot settled at b);

We first discuss the rooted case (the pseudocode is in Algorithm 2). After that, we will discuss the general case (the pseudocode is in Algorithms 1 and 2).

Rooted Algorithm. Let $N(v)$ be the set of $k \leq n$ robots $\{r_1, r_2, \ldots, r_k\}$ positioned initially on node $v \in G$ with $r_i.ID = i$, forming a rooted configuration. At round 1, the highest ID robot $r_k \in N(v)$ settles on v setting $r_k.settled \leftarrow 1$, $r_k.status \leftarrow visited$, $r_k.recentPort \leftarrow 1$, and $r_k.nodeRank \leftarrow 1$. Robots $N(v) \backslash \{r_k\}$ exit v via $r_k.recentPort$. The exiting robots carry $r_k.nodeRank$, the rank of robot settled at v.

At round 2, suppose the robots $N(v) \backslash \{r_k\}$ arrive at node w. The robot r_{k-1} settles at w setting $r_{k-1}.settled \leftarrow 1$, $r_{k-1}.status \leftarrow visited$, $r_{k-1}.recentPort \leftarrow 1$, and $r_{k-1}.nodeRank \leftarrow r_k.nodeRank + 1$ (note that $r_k.nodeRank$ was carried from v). The robots in $N(v) \backslash \{r_k, r_{k-1}\}$ exit w via $r_{k-1}.recentPort$. The traversal continues this way. If k different nodes are visited before reaching to a node with an already settled robot, DISPERSION is achieved. However, if the DFS reaches to a node with a settled robot, backtracking is required. In backtracking, the DFS needs to (in sequence)

- find a cycle closing node;
- starting from the cycle closing node, find a backtrack target node b from which the DFS can again continue its *explore* phase settling the robots at new empty nodes;
- create/update backtrack path setting *rootPort* information appropriately; and
- go to backtrack target b from the cycle closing node to continue *explore* phase.

Finding cycle closing and backtrack target nodes happens in *retrace*1 whereas setting backtrack path and going to the cycle closing node happens in *retrace*2 (Fig. 3).

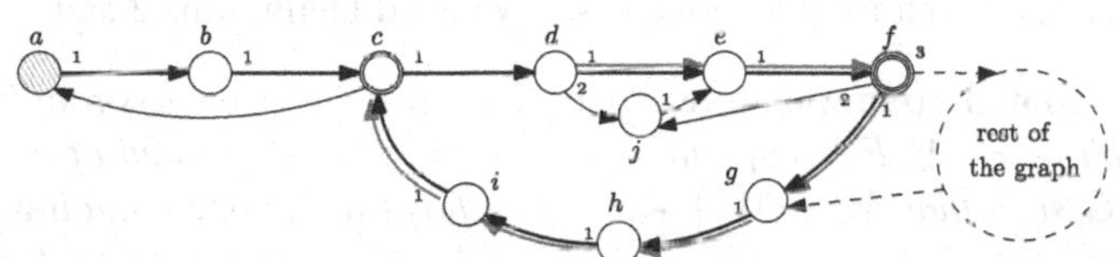

Fig. 3. An illustration of the DFS traversal, starting from node a, after it reaches node c traversing the sequence of edges $a \rightarrow b \rightarrow c \rightarrow d \rightarrow \ldots \rightarrow i \rightarrow c$ (shown in bold). Node c is a *visited* node where $ActivePath(a)$ ends and hence c can serve as a cycle closing node. The node f that is farthest from c in the cycle $c \rightarrow \ldots \rightarrow f \rightarrow \ldots \rightarrow c$ with unvisited out-going ports becomes the backtrack target node during *retrace*1. During *retrace*2, (i) nodes g, h, i have no unvisited out-going ports and marked *fully-visited* and (ii) the current port from d to i are stored in *rootPort* (the corresponding edges are shown in red). (Color figure online)

General Algorithm. Suppose initially robots are positioned on $1 < k' < k$ nodes of G. At round 1, a single robot on a node, if any, settles at that node and the nodes with multiple robots initiate parallel DFSs as in the rooted case (Lines 2–5 of Algorithm 1). The node from which a DFS i starts is called the root of the DFS i and denoted as $root(i)$. If a parallel DFS does not meet another until all robots disperse, we are done. The challenge is how to deal with a traversal meeting another.

A DFS i *meets* DFS j if the robots with DFS ID i arrive at a node x where a robot from DFS ID j is settled. (If robots from DFS ID i and DFS ID j arrive at a node where there is no settled robot, the robot from the DFS with the higher ID settles in that round and the lower ID DFS is said to meet higher ID DFS.) If $i > j$, then we call DFS i *subsuming* otherwise *subsumed.* The *head* of DFS i, denoted as $head(i)$, is the node where the unsettled robots (if any) of that DFS are currently located at (except the robot that is responsible of executing the $retrace1$ and $retrace2$ phases), or else it is the node where the last robot of that DFS settled. That is, while under the $explore$ phase, $head(i)$ has all unsettled robots, however while under $retrace1$ and $retrace2$ phases, the cycle closing node c acts as $head(i)$ and all unsettled robots except the robot performing $retrace1$ and $retrace2$ phases are on c. A DFS is assigned ID i as follows. Suppose a DFS starts initially from a node v_i with a set of $N(v_i)$ robots. The highest ID robot $r_h \in N(v_i)$ serves as ID i. DFS i meeting DFS j at a junction node x is handled as follows (Lines 6–10 of Algorithm 1):

DFS i is Subsuming, i.e., $(i > j)$: DFS i continues its traversal. For each settled robot w belonging to DFS j it visits, it erases the values set in all the variables, except in $w.settled$. It then sets all the variables based on DFS i and continues the traversal.

DFS i is Subsumed, i.e., $(i < j)$: All unsettled robots of DFS i traverse DFS j to reach node $head(j)$. DFS j then continues the traversal from $head(j)$.

4.1 Analysis of the Algorithm

We begin with analyzing the rooted case. We will then analyze the general case building upon the ideas developed for the rooted case. We need Definitions 2 and 3.

Definition 2 (rooted spanning tree). *Consider a node v on a strongly connected directed graph $G = (V, E)$. A spanning tree $\mathcal{T} = (V', E')$ rooted at v is a directed subgraph of G such that $|V| = |V'|$, each node has exactly one outgoing edge except v which has no outgoing edge (i.e., $E' \subseteq E$ with $|E'| = |V| - 1$), and all edges are directed toward v.*

Definition 3 (root path). *Consider the spanning tree $\mathcal{T}$ in Definition 2. A* root path *$RootPath(v', v)$ is a directed path from node $v' \neq v$ to the root v. For each node $w \neq v$, $\mathcal{T}$ contains such a path.*

Time Complexity, Rooted Case, Known $\mathcal{T}$. Let DFS starts from node v which becomes the root. Suppose DFS arrives at node v' for the first time at some round $t > 1$. At that time, v' is marked *visited.* Let v' be the k'-th in the order of the nodes visited by DFS for the very first time. Node v' is assigned $rank(v') := k'$. Initially, v' was marked $null$ (meaning unvisited) and once marked $visited$, v' will be marked *fully-visited* when DFS visits all the outgoing ports of v'. Therefore, after marked visited, v' will never be marked $null$ (unvisited), and after marked fully-visited, it will never be marked visited or $null$ (unvisited).

Definition 4 (active edge). *The edge associated with port $1 \leq port_{out}^{v'} \leq \delta_{out}^{v'}$ of node v' is called* active *if v' is marked* visited *and DFS has exited v' recently through $port_{out}^{v'}$.*

Definition 5 (active path). *An* active path *$ActivePath(v)$ is a directed path that connects nodes marked* visited *through active edges starting from the root v of DFS.*

Lemma 1. *The $RootPath(v', v)$ in $\mathcal{T}$ from any node $v' \neq v$ to the root v always intersects $ActivePath(v)$.*

Lemma 2. *If $ActivePath(v)$ ends at a visited node, then the last edge in it closes a cycle of active edges.*

Lemma 3. *Let $e \rightarrow f$ be the last edge in Lemma 2. We have that $rank(f) < rank(e)$.*

Definition 6 (backtrack target node). *Backtrack target is always on $ActivePath(v)$. Furthermore, let $V(ActivePath(v))$ be the set of nodes in $ActivePath(v)$ such that, for each node $v' \in V(ActivePath(v))$, the port number leading to its active edge is smaller than its degree $\delta_{out}^{v'}$. Node $v' \in V(ActivePath(v))$ is called* backtrack target node *if it is the highest ranked node in $V(ActivePath(v))$.*

Lemma 4. *Backtracking is required when $ActivePath(v)$ points to a visited or fully-visited node.*

Definition 7 (backtracking path). *Consider an active edge $e \rightarrow f$. If f is a fully-visited node, a* backtracking path *$BacktrackPath(f)$ from node f is the following: $RootPath(f, v)$ upto an intersection node c with $ActivePath(v)$ and then $ActivePath(v)$ upto the backtrack target node b (Definition 6). If f is a visited node, $BacktrackPath(f)$ is the segment of $ActivePath(v)$ from f to b.*

Lemma 5 (cycle closing node). *If $ActivePath(v)$ ends at a visited node v', v' becomes a cycle closing node. If $ActivePath(v)$ ends at a fully-visited node v'', let c be the intersection node of $ActivePath(v)$ and $RootPath(v'', v)$. Node c becomes a cycle closing node.*

Lemma 6. *Algorithm 2 solves* DISPERSION *for any rooted initial configuration of $k \leq n$ robots in $O(k^2 \cdot \Delta_{out})$ rounds, given directed rooted spanning tree $\mathcal{T}$.*

Proof. We first prove that Algorithm 2 solves DISPERSION correctly. Notice that, at any time, the cycle closing node as well as the backtrack target node are on $ActivePath(v)$. Moreover, the backtrack target node b is the farthest node from root v in $ActivePath(v)$ with at least one unvisited port left. This property resembles DFS backtrack in undirected graph case. Additionally, the backtrack target node can always be found and reached. Let c be a cycle closing node and f' be a backtrack target node. It is the case that, if f' is in the cycle and $f' \neq c$, $rank(f') > rank(c)$, otherwise c is both the cycle closing node as well as a backtrack target node. If f' is not in the cycle, then $rootPort$ set is taken from c which will take the traversal to a cycle closing node $c' \neq c$ with $rank(c') < rank(c)$ and the process of finding a backtrack target node f' starts from there. Furthermore, for any edge $e \rightarrow f$ on $ActivePath(v)$, e cannot become $fully\text{-}visited$ before f. These properties are sufficient to show that the DFS executes correctly, solving DISPERSION.

We now prove the time bound. Since there are $k \leq n$ robots, any $Backtrack(.)$ path is of length at most k as it visits only the nodes with a robot already settled on each of them. DFS needs to traverse at most $k \cdot \Delta_{out}$ edges before all k robots settle at k different nodes. For each outgoing edge, backtracking is needed at most once. Since backtracking is executed in two phases $retrace1$ and $retrace2$, the total number of rounds to finish these phases is $\leq 2 \cdot k$. In $retrace1$, if DFS visits a $fully\text{-}visited$ node, traversing the $RootPath(.,.)$ is required to find a $visited$ node to make it a cycle closing node. This length is $< k$. In $retrace2$, traversing to the backtrack target node takes additional $< k$ rounds. Therefore, each backtracking finishes in $< 4 \cdot k$ rounds. Therefore, Algorithm 2 needs $< (4 \cdot k) \cdot (k \cdot \Delta_{out}) = O(k^2 \cdot \Delta_{out})$ rounds. □

Time Complexity, Rooted Case, Unknown $\mathcal{T}$. We now remove the assumption of known $\mathcal{T}$ from Lemma 6 and construct it on-the-fly. We start with the following lemmas.

Lemma 7. *For any two visited nodes v', v'' that are eligible to become a backtrack target node, if $rank(v'') < rank(v')$, then v' becomes $fully\text{-}visited$ before v''.*

Lemma 8. *When node v' becomes $fully\text{-}visited$, $RootPath(v', v)$ ends at the smallest ranked node in $ActivePath(v)$ reachable from v'.*

During the execution of DFS, $\mathcal{T}$ may be a forest of many disjoint directed rooted trees. Let $\mathcal{F}$ be the *directed rooted forest* defined as a collection of the disjoint directed rooted trees $\mathcal{T}_1, \ldots, \mathcal{T}_i$.

Lemma 9. *The collection of $RootPath(v', v)$ of the fully-visited nodes v' form a directed rooted forest $\mathcal{F}$.*

Lemma 10. *For any fully-visited node v', $RootPath(v', v)$ of length $\leq k - 1$ can be constructed that leads either to a visited node v'' or to root v with $rank(v) < rank(v'') < rank(v')$.*

Lemma 11. *Algorithm 2 solves* DISPERSION *for any rooted initial configuration of $k \leq n$ robots in $O(k^2 \cdot \Delta_{out})$ rounds, constructing $\mathcal{T}$ on-the-fly.*

Proof. We have from Lemma 10 that the length of $RootPath(v', v)$ from any node $v' \neq v$ to root v is $< k$. Moreover, $RootPath(v', v)$ ends at a $visited$ node. Therefore, a backtrack finishes in $O(k)$ rounds. Furthermore, an edge needs at most one backtrack and there are $k \cdot \Delta_{out}$ edges. The theorem follows. □

Time Complexity, General Case, Unknown $\mathcal{T}$. Since robots are on $1 \leq k' < k$ nodes in any general initial configuration, there will be k' parallel DFSs initiated at round $t = 1$. DFS i does not meet any other DFS j at $t = 1$. If no two DFSs meet until all robots settle, the analysis for the rooted case applies. We analyze here DFS i meeting DFS j and show that $O(k^2 \cdot \Delta_{out})$ time bound can be established. Consider DFS i at some round $t > 1$. Consider the nodes of G that are occupied with robots belonging to DFS i.

Lemma 12. *Let $x \neq head(i)$ be a node of G belonging to DFS i. $head(i)$ is always reachable from x.*

Lemma 13. *If k_i robots belong to DFS i, then $head(i)$ is at distance $\leq k_i - 1$ from any node x in DFS i.*

Since robots have unique IDs, $i \neq j$ for any two DFSs i and j. Suppose DFS i meets DFS j. DFS i is either subsuming ($i > j$) or subsumed ($i < j$). Due to the k' DFSs initiated in parallel, a DFS j may be met by different other DFSs, and DFS j may in turn meet another DFS concurrently. Further, transitive chains of such meetings can occur concurrently. This leads us to formalize a notion of a *meeting graph.*

Definition 8 (meeting graph). *The meeting graph $G_M = (V_M, E_M)$, where V_M is the DFS IDs and there is a directed edge in E_M from i and j if DFS i meets DFS j.*

Nodes in V_M have an arbitrary in-degree but out-degree 1. Moreover, G_M may be composed of multiple connected components. Furthermore, there may be cycles in connected components. We focus on a single connected component C_M in G_M; other components of G_M can be dealt analogously. At round 1, G_M has k' nodes and $E_M = \emptyset$, which are components by themselves. There are multiplicity nodes in at least one component. At round 2 or after, the number of components monotonically decrease and when DISPERSION is achieved, $E_M = \emptyset$ and no multiplicity node in each component.

Observation 1. *A connected component C_M in graph G_M never splits into multiple sub-components.*

Suppose there are M nodes (i.e., DFSs) in C_M. One node (DFS) has the highest ID among the M nodes and that node (DFS) subsumes all the $M - 1$ DFSs in that component. During this process, the number of settled robots monotonically increases. Therefore, C_M never disconnects to form multiple sub-components. Furthermore, C_M may meet another component and they become a single component. This leads us to formalize a notion of a meeting tree for the meeting graph G_M.

Definition 9. (meeting tree). *The k' initial DFSs i form the k' leaf nodes $(i, 0)$ of the meeting tree T_M at level 0. When α nodes $(a_i, h_i), i \in [1, \alpha]$, meet in a component, a node (M, h) is created in T_M as the parent of the child nodes (a_i, h_i), for $i \in [1, \alpha]$, where M is the highest DFS among the α DFSs that met in G_M and $h = 1 + \max_{i \in [1,\alpha]} h_i$.*

When two or more DFSs (and components) meet, one of the DFSs in the formed component subsumes all other DFSs. Therefore, there is a DFS which appears in each level of the meeting tree T_M starting from level 0 upto the highest level in that component. Furthermore, the height h of T_M is $0 \leq h \leq k' - 1$. The maximum height $k' - 1$ represents the sequential meeting of DFS IDs. The height $h < k' - 1$ represents meetings and subsumptions that happen in parallel across different components. Therefore, it will sufficient to bound termination time for sequential cases since it immediately subsumes the time bounds for all other (parallel) cases. The cases we consider are:
(I) DFS l meets DFS $l - 1$, $2 \leq l \leq k'$ (meeting in decreasing order of the DFS IDs),
(II) DFS l meets DFS $l + 1$, $1 \leq l \leq k' - 1$ (meeting in increasing order of the DFS IDs), and
(III) any combination of cases (I) and (II) (meeting sometime in increasing order and sometime in decreasing order of the DFS IDs).

We now discuss the time bound to achieve DISPERSION. Consider node $head(i)$ of DFS i. Consider the situation of $head(i)$ not having any unsettled robot, i.e., all robots belonging to DFS i have been settled. Let the smallest round on which $head(i)$ not having any unsettled robot be t_i. Let DFS j meets DFS i and get subsumed by DFS i $(j < i)$. The unsettled robots of DFS j need to move to $head(i)$ for DFS i to continue its traversal. Let t_j be the round at which the unsettled robots of DFS j reach $head(i)$. The difference $t_j - t_i$ is the wait time for $head(i)$. The total wait time $totalWait(head(i))$ for a DFS i in a component C_M is the sum of the wait times of the $head(i)$ before either it is subsumed or DISPERSION is solved.

Lemma 14. *Consider the highest ID DFS M in component C_M. $totalWait(head(M)) \leq (k_M - 1) \cdot (M - 1)$ for either DFS M to be subsumed or* DISPERSION *is solved, where k_M is the number of robots in C_M.*

Lemma 15. *Consider the highest ID DFS M in component C_M. Either DFS M is subsumed or robots of C_M disperse to k_M nodes in $O((k_M)^2 \cdot \Delta_{out} + k_M \cdot M)$ rounds, where k_M is the number of robots in C_M.*

Lemma 16. *Algorithm 1 solves* DISPERSION *deterministically for any general initial configuration of $k \leq n$ robots in $O(k^2 \cdot \Delta_{out})$ rounds.*

Proof. We first argue that for any DFS j in a component C_M, $totalWait(head(j)) \leq totalWait(head(M))$. Since DFS M is the highest ID DFS in C_M of the meeting graph G_M, it never gets subsumed by any other DFS j in C_M. Therefore, consider Case (I) of DFS M always meeting DFS $l, l < M$. In this case, DFS M has to subsume each DFS met. Since DFS M continues its traversal even after meeting j, $totalWait(head(M)) \geq totalWait(head(j))$. Consider now the Case (II) of DFS l meeting DFS $l+1, 1 \leq l \leq M-1$. In this case, DFS M has to continue its traversal even after DFS $M-1$ reaches $head(M)$, i.e., $totalWait(head(M)) > totalWait(head(j))$. Finally, for Case (III), since DFS M never gets subsumed, $totalWait(head(M)) \geq totalWait(head(j))$ for any other DFS $j \neq M$.

We now prove time bound. Consider the meeting tree T_M of the meeting graph G_M. Each node in T_M is a component (at level 0, a DFS itself is a component) and at the end either there is a single root node in each component tree formed. Therefore, the height of the meeting tree T_M can be at most $\leq k' - 1$ since there are only k' DFS traversals. The total wait time $totalWait(head(M_{\max}))$ of the largest ID DFS $M_{\max}$ is bounded by $k \cdot k' < k^2$. Except the wait time, for k robots, DFS finishes in $O(k^2 \cdot \Delta_{out})$ rounds. Therefore, the total time is $O(k^2 \cdot \Delta_{out}) + totalWait(head(M_{\max})) = O(k^2 \cdot \Delta_{out} + k^2) = O(k^2 \cdot \Delta_{out})$ rounds. □

5 BFS Dispersion Algorithm

We present here a deterministic BFS-type algorithm solving DISPERSION in time $O(k \cdot D)$ rounds with $O(k \log(k + \Delta_{out}))$ bits at each robot satisfying Theorem 1(b), given any general initial configuration. We establish this result under the 1-hop communication model; removing this requirement remains as a future work.

Theorem 7. *To solve* DISPERSION *of $k \leq n$ robots on a n-node strongly-connected directed graph, any deterministic DFS algorithm needs $\Omega(k^2)$ time under the 1-hop communication model.*

Proof. Consider the graph G as in the lower bound of Theorem 6(b) and the rooted initial configuration of all k robots initially at node u (Fig. 1). Starting from u, traversing the upper arc up to v cannot exploit information given by the 1-hop model. From v, even when 1-hop communication tells how many nodes are unvisited, a DFS traversal can only visit one node on the column at a time. After visiting one node, it takes additional $\Omega(k)$ rounds to visit another. Therefore, a DFS algorithm needs $\Omega(k^2)$ rounds. □

We now design a deterministic BFS traversal based algorithm (Algorithm 3) that exploits 1-hop communication and achieves $O(k \cdot D)$ time bound. We note that any deterministic DFS algorithm needs $\Omega(k^2)$ time for DISPERSION even in the 1-hop model (Theorem 7), i.e., 1-hop model is of no benefit for DFS traversal based algorithms.

Suppose initially k robots are on $1 \leq k' < k$ nodes. Denote the node set by $V_{k'}$. Let $k'' \leq k'$ be the multiplicity nodes, i.e., two or more robots positioned on them. Denote the node set by $V_{k''} \subseteq V_{k'}$. Each node $v \in V_{k''}$ is called a *source node*. Let $w \in V_{k'}$ be a node such that it has at least one empty out-going neighbor. Node w is called a *request node*. Let V_R be the set of request nodes. A source node may also be a request node if it has empty out-going neighbbor(s).

Algorithm 3: BFS algorithm for robot r_i at node $v_i \in G$ at some round $t \geq 1$

```
if r_i is alone at node v_i then
    r_i settles at v_i writing r_i.settled <- 1;
if r_i is not alone at v_i and v_i is a source node (no empty outgoing neighbors) then
    if v_i has no settled robot and r_i has the highest ID then
        r_i settles at v_i writing r_i.settled <- 1;
    send broadcast(t, r_i.ID) message to all its outgoing neighbors;
if r_i is not alone at v_i and v_i is a request node (empty outgoing neighbor(s)) then
    if v_i has no settled robot and r_i has the highest ID then
        r_i settles at v_i writing r_i.settled <- 1;
    divide almost equally and send all unsettled robots at v_i to its empty neighbors;
if r_i is alone at v_i and receives a Broadast(.,) message then
    if r_i not a request node then
        r_i forwards it to all its outgoing neighbors if its not duplicate;
    else
        r_i sends Request(t, r_i.ID) to neighbor from which it received Broadcast(.);
if r_i receives a Request(.,) message then
    if r_i not a source node then
        r_i sends to neighbor from which it received Broadcast(.) message;
    else
        r_i sends all the unsettled robots to the request node;
```

The goal in the algorithm is to find a matching between a source node and a request node so that the empty (out-going) neighbors of the request node can be occupied by robots. We exploit 1-hop communication to find such matching in $O(D)$ rounds. Since there are $k \leq n$ robots in total, the algorithm can finish settling all the robots at k different nodes in $O(k \cdot D)$ rounds. Consider a source node $v \in V_{k''}$. The highest ID robot among the robots $R(v)$ on v settles at v. We have two cases.

Case 1 – v is a Source as Well as a Request Node: Let $d(v) \leq \delta^v_{out}$ be the number of empty out-going neighbors of v. Node v sends $\lfloor(|\mathcal{R}(v)| - 1)/d(v)\rfloor$ robots to $d(v) - 1$ empty out-going neighbors and $(|\mathcal{R}(v)| - 1)/d(v) + (|\mathcal{R}(v)| - 1) \mod d(v)$ robots to one empty out-going neighbor.

Case 2 – v is a Source Node But Not a Request Node: The matching is done in three stages, Stages 1–3. Case 2 starts in a round, when v becomes a source node but not a request node. In Stage 1, v sends a *broadcast message* to all its (non-empty) out-going neighbors and the (non-empty) out-going neighbors forward the message to their (non-empty) out-going neighbors, and so on. This message broadcast forms a directed BFS tree $BT(v)$ with v becomes the root and the request nodes become the leaves. As soon as a request node w receives a broadcast message Stage 2 starts in which node w sends a *request message* following the path in $BT(v)$ to reach the source node v. Each internal node u in $BT(v)$ stores a neighbor node from which it received a broadcast message as the parent of u in Stage 1 and this information helps in forwarding the request message to the root. In Stage 3, the source node v sends $|\mathcal{R}(v)| - 1$ robots in the set $\mathcal{R}(v)$ to the request node from which it receives the first request message. At the end of Stage 3, a request node becomes a source as well as a request node and Case 1 applies.

The algorithm terminates with no new source node (i.e., no multiplicity node).

Lemma 17. *A request node can be matched with a source node in $O(D)$ rounds.*

Theorem 8. *Algorithm 3 solves* DISPERSION *deterministically for any initial configuration of $k \leq n$ robots in $O(k \cdot D)$ rounds under the 1-hop communication model.*

6 Extensions to Crash Faults

In this section, we consider that $f \leq k$ robots may experience crash faults. We describe how the DFS and BFS algorithms of Sects. 4 and 5 extend to handle crash faults.

DFS Algorithm. We extend Algorithm 1 to handle crash faults. Crashing of an unsettled robot at any time during Algorithm 1 is not a problem. Therefore, crashing of robots does not affect the $explore$ phase. However, crashing of settled robots is a problem because a robot performing $retrace1$ and $retrace2$ phases should never encounter an empty node. We have two situations: (i) The robot doing the $retrace1$ and $retrace2$ phases crash or (ii) the robots in the path of the robot doing those phases crash.

We deal with the first situation of the possible crash of the robot doing $retrace1$ and $retrace2$ phases by asking all the unsettled robots to perform those phases (not just one robot). In other words, no robot waits at the cycle closing node. All of them perform the $retrace1$ and $retrace2$ phases and the robot settled at the cycle closing node keeps information that it is a cycle closing node. We deal with the second situation of the robot(s) performing $retrace1$ and $retrace2$ encounter an empty node (must be the case that robots at that node was crashed) by starting a new DFS from that crashed node using $DFSID$ variable as a tuple $(roundNo, RID)$, where $roundNo$ is the round at which the DFS is started and RID is the highest ID robot i among the robots on that node. This approach is also used when robots move from the cycle closing node to the backtrack target node to continue the $explore$ phase. At round 1, all k' DFSs initiated in parallel have $roundNo = 1$ and RID is the highest ID robot on each of k' nodes. In the crash-free case, Algorithm 1 works as is even with the tupled ID as subsumption happens based on RID (lexicographical comparison of the $DFSID$s) due to the fact that $roundNo$ remains the same for all DFSs during the traversal. The variable $roundNo > 1$ for a DFS if it is started due to the encounter of a crashed node (robot). A DFS meeting another is handled asking larger $DFSID$ DFS in the lexicographical order to subsume the other.

Lemma 18. *In the same setting of Lemma 16 having $f \leq k$ robot crashes, the DFS algorithm solves* DISPERSION *deterministically in $O(f \cdot k^2 \cdot \Delta_{out})$ rounds.*

Proof of Theorem 1(a): The time bound follows from the proofs of Lemmas 16 and 18. For the memory bound, in the fault-free case, there are total ten different variables maintained at each robot and each variable is of size either $O(1)$ or $O(\log k)$ or $O(\log \Delta_{out})$. Therefore, the total memory bound is $O(\log(k + \Delta_{out}))$ bits per robot. In the crash fault case, the tupled DFSID only adds $O(\log(k + \Delta_{out}))$ factor due to the variable $roundNo$, since $roundNo \leq O(f \cdot k^2 \cdot \Delta_{out}) = O(k^3 \cdot \Delta_{out})$ given that $f \leq k$, and thus can be represented by $O(\log(k + \Delta_{out}))$ bits. □

BFS Algorithm. We extend Algorithm 3 to handle crashes. We again have two cases. Case 1 stays the same. In Case 2, the matching may be interrupted due to the faulty robot(s). Therefore, in Stage 1, a source node v sends broadcast message repetitively in every round until a request message is received by v. If a node receives broadcast message from v multiple times, then it stores the latest copy among them and forwards it to its out-going neighbors. In Stage 2, if a request message can not reach to v, then some intermediate node (robot) in the path to v must have crashed. A non-crashed neighboring robot of the crashed robot on the path to v becomes a request node and sends a request message to v when it receives a broadcast message from v. In Stage 3, if a request node w finds that a node in its path $P(v, w)$ is crashed, the neighboring node becomes a new request node.

Theorem 9. *In the same setting of Theorem 8 having $f \leq k$ robot crashes, the BFS algorithm solves* DISPERSION *deterministically in $O(k \cdot D)$ rounds under the 1-hop communication model.*

Proof of Theorem 1(b): The time bound follows from the proofs of Theorems 8 and 9. For the memory bound, there can be at most $k/2$ source nodes. Therefore, a node may need to keep track of the parent information on the $k/2$ BFS trees that are built by those source nodes, which requires in total $O(k \log(k + \Delta_{out}))$ bits at each robot, combining the $O(\log k)$ bits required to remember the ID of each robot. The broadcast, request, and other messages are of $O(\log k)$ size. Therefore, in total the memory at each robot is $O(k \log(k + \Delta_{out}))$ bits. Even in the faulty case, the memory does not change since for the broadcast messages received from a source node multiple times, the previous records were discarded and only the latest message was stored in memory. □

References

1. Albers, S., Henzinger, M.R.: Exploring unknown environments. In: STOC, pp. 416–425. ACM (1997)
2. Augustine, J., Moses Jr., W.K.: Dispersion of mobile robots: a study of memory-time trade-offs. In: ICDCN, pp. 1:1–1:10 (2018)
3. Bampas, E., Gasieniec, L., Hanusse, N., Ilcinkas, D., Klasing, R., Kosowski, A.: Euler tour lock-in problem in the rotor-router model: I choose pointers and you choose port numbers. In: DISC, pp. 423–435 (2009)
4. Barriere, L., Flocchini, P., Mesa-Barrameda, E., Santoro, N.: Uniform scattering of autonomous mobile robots in a grid. In: IPDPS, pp. 1–8 (2009)
5. Bender, M.A., Fernández, A., Ron, D., Sahai, A., Vadhan, S.P.: The power of a pebble: exploring and mapping directed graphs. In: STOC, pp. 269–278 (1998)
6. Bender, M.A., Slonim, D.K.: The power of team exploration: two robots can learn unlabeled directed graphs. In: FOCS, pp. 75–85 (1994)
7. Cohen, R., Fraigniaud, P., Ilcinkas, D., Korman, A., Peleg, D.: Label-guided graph exploration by a finite automaton. ACM Trans. Algorithms **4**(4), 42:1–42:18 (2008)
8. Cybenko, G.: Dynamic load balancing for distributed memory multiprocessors. J. Parallel Distrib. Comput. **7**(2), 279–301 (1989)
9. Dereniowski, D., Disser, Y., Kosowski, A., Pajak, D., Uznański, P.: Fast collaborative graph exploration. Inf. Comput. **243**(C), 37–49 (2015)
10. Elor, Y., Bruckstein, A.M.: Uniform multi-agent deployment on a ring. Theor. Comput. Sci. **412**(8–10), 783–795 (2011)
11. Flocchini, P., Prencipe, G., Santoro, N.: Distributed Computing by Oblivious Mobile Robots. Morgan & Claypool Publishers, Synthesis Lectures on Distributed Computing Theory (2012)
12. Flocchini, P., Prencipe, G., Santoro, N.: Distributed Computing by Mobile Entities, Theoretical Computer Science and General Issues, vol. 1. Springer (2019). https://doi.org/10.1007/978-3-030-11072-7
13. Foerster, K., Wattenhofer, R.: Lower and upper competitive bounds for online directed graph exploration. Theor. Comput. Sci. **655**, 15–29 (2016)
14. Fraigniaud, P., Gasieniec, L., Kowalski, D.R., Pelc, A.: Collective tree exploration. Networks **48**(3), 166–177 (2006)
15. Fraigniaud, P., Ilcinkas, D., Peer, G., Pelc, A., Peleg, D.: Graph exploration by a finite automaton. Theor. Comput. Sci. **345**(2–3), 331–344 (2005)
16. Hsiang, T.R., Arkin, E.M., Bender, M.A., Fekete, S., Mitchell, J.S.B.: Online dispersion algorithms for swarms of robots. In: SoCG, pp. 382–383 (2003)
17. Hsiang, T., Arkin, E.M., Bender, M.A., Fekete, S.P., Mitchell, J.S.B.: Algorithms for rapidly dispersing robot swarms in unknown environments. In: WAFR, pp. 77–94 (2002)
18. Kshemkalyani, A.D., Ali, F.: Fast graph exploration by a mobile robot. In: AIKE, pp. 115–118 (2018)
19. Kshemkalyani, A.D., Ali, F.: Efficient dispersion of mobile robots on graphs. In: ICDCN, pp. 218–227 (2019)
20. Kshemkalyani, A.D., Molla, A.R., Sharma, G.: Fast dispersion of mobile robots on arbitrary graphs. In: ALGOSENSORS, pp. 23–40 (2019)

21. Kshemkalyani, A.D., Molla, A.R., Sharma, G.: Dispersion of mobile robots in the global communication model. In: ICDCN, pp. 12:1–12:10 (2020)
22. Kshemkalyani, A.D., Molla, A.R., Sharma, G.: Dispersion of mobile robots on grids. In: WALCOM, pp. 183–197 (2020)
23. Kshemkalyani, A.D., Molla, A.R., Sharma, G.: Efficient dispersion of mobile robots on dynamic graphs. In: ICDCS, pp. 732–742 (2020)
24. Kshemkalyani, A.D., Sharma, G.: Near-optimal dispersion on arbitrary anonymous graphs. CoRR (2021)
25. Menc, A., Pajak, D., Uznanski, P.: Time and space optimality of rotor-router graph exploration. Inf. Process. Lett. **127**, 17–20 (2017)
26. Molla, A.R., Jr., W.K.M.: Dispersion of mobile robots: The power of randomness. In: TAMC, pp. 481–500 (2019)
27. Molla, A.R., Mondal, K., Jr., W.K.M.: Efficient dispersion on an anonymous ring in the presence of weak byzantine robots. In: ALGOSENSORS, pp. 154–169 (2020)
28. Molla, A.R., Mondal, K., Jr., W.K.M.: Byzantine dispersion on graphs. In: IPDPS, pp. 942–951. IEEE (2021)
29. Pattanayak, D., Sharma, G., Mandal, P.S.: Dispersion of mobile robots tolerating faults. In: ICDCN, pp. 133–138 (2021)
30. Poudel, P., Sharma, G.: Time-optimal uniform scattering in a grid. In: ICDCN, pp. 228–237 (2019)
31. Shibata, M., Mega, T., Ooshita, F., Kakugawa, H., Masuzawa, T.: Uniform deployment of mobile agents in asynchronous rings. In: PODC, pp. 415–424 (2016)
32. Shintaku, T., Sudo, Y., Kakugawa, H., Masuzawa, T.: Efficient dispersion of mobile agents without global knowledge. In: SSS, pp. 280–294 (2020)

Distributed Interactive Proofs for the Recognition of Some Geometric Intersection Graph Classes

Benjamin Jauregui[1], Pedro Montealegre[2(✉)], and Ivan Rapaport[3]

[1] Departamento de Ingeniería Matemática, Universidad de Chile, Santiago, Chile
bjauregui@dim.uchile.cl

[2] Facultad de Ingeniería y Ciencias, Universidad Adolfo Ibañez, Santiago, Chile
p.montealegre@uai.cl

[3] DIM-CMM (UMI 2807 CNRS), Universidad de Chile, Santiago, Chile
rapaport@dim.uchile.cl

Abstract. A graph $G = (V, E)$ is a geometric intersection graph if every node $v \in V$ is identified with a geometric object of some particular type, and two nodes are adjacent if the corresponding objects intersect. Geometric intersection graph classes have been studied from both the theoretical and practical point of view. On the one hand, many hard problems can be efficiently solved or approximated when the input graph is restricted to a geometric intersection class of graphs. On the other hand, these graphs appear naturally in many applications such as sensor networks, scheduling problems, and others. Recently, in the context of distributed certification and distributed interactive proofs, the recognition of graph classes has started to be intensively studied. Different results related to the recognition of trees, bipartite graphs, bounded diameter graphs, triangle-free graphs, planar graphs, bounded genus graphs, H-minor free graphs, etc., have been obtained.

The goal of the present work is to design efficient distributed protocols for the recognition of relevant geometric intersection graph classes, namely permutation graphs, trapezoid graphs, circle graphs and polygon-circle graphs. More precisely, for the two first classes we give proof labeling schemes recognizing them with logarithmic-sized certificates. For the other two classes, we give three-round distributed interactive protocols that use messages and certificates of size $\mathcal{O}(\log n)$. Finally, we provide logarithmic lower-bounds on the size of the certificates on the proof labeling schemes for the recognition of any of the aforementioned geometric intersection graph classes.

Keywords: Distributed decision · Proof-labeling scheme · Distributed interactive proofs · Intersection Graph Classes

This work was supported by Centro de Modelamiento Matemático (CMM), ACE210010 and FB210005, BASAL funds for centers of excellence from ANID-Chile, FONDECYT 11190482, FONDECYT 1220142 and PAI 77170068.

M. Parter (Ed.): SIROCCO 2022, LNCS 13298, pp. 212–233, 2022.
https://doi.org/10.1007/978-3-031-09993-9_12

1 Introduction

This paper deals with the problem of designing compact distributed certificates and compact distributed interactive proofs for deciding graph properties. In these protocols, the nodes of a connected graph G have to decide, collectively, whether G itself belongs to a particular graph class. As in the centralized case, also in the distributed setting there exists a number of algorithms specially designed to decide whether G belongs to a particular graph class. The specific goal of this work is to decide, through proof-labeling schemes and the more general model of distributed interactive proofs, whether G belongs to relevant intersection graph classes. These classes have applications in topics like biology, ecology, computing, matrix analysis, circuit design, statistics, archaeology, etc. For a nice survey we refer to [35].

1.1 Proof-Labeling Schemes and Distributed Interactive Proofs

In locally decidable algorithms every node is just allowed to send messages to its neighbors, in one round (a less restrictive, but similar scenario, is where the number of rounds is constant, independent of the size of G, see [38]). Some very basic properties can be decided locally (with a local algorithm). For instance, deciding whether the graph G has bounded degree. More generally, if we do not impose bandwidth restrictions, then detecting the existence of any local structure (such as a triangle) can be solved through local algorithms.

In the aforementioned examples, acceptance and rejection are (implicitly) defined as follows. If G satisfies the property, then all nodes must accept; otherwise, at least one node must reject. These very fast local algorithms could be used in distributed fault-tolerant computing, where the nodes, with some regularity, must check whether the current network configuration is in a legal state [29]. Then, if the configuration becomes at some point illegal, the rejecting node(s) raise the alarm or launch a recovery procedure. When there are distributed algorithms designed for particular graph classes, the use of an initial recognizing protocol could avoid the risk of running a distributed protocol for graphs that do not belong to the class for which the protocol was designed.

Of course, many simple properties cannot be decided with one round (or with a constant number of rounds) through local algorithms. In order to overcome this issue, the notion of proof-labeling scheme (PLS) was introduced [29]. PLSs can be seen as a distributed counterpart of the nondeterministic class NP. In fact, in a PLS, a powerful prover gives to every node v a certificate $c(v)$. This provides G with a global distributed proof. Then, every node v performs a local verification using its local information together with $c(v)$.

Incorporating a powerful prover to the model is not just motivated by a purely theoretical interest. In fact, with the rise of the Internet, prover-assisted computing models are ubiquitous. Asymmetric applications –social networks, cloud computing, etc.– where a very powerful central entity stores and process large amounts of data, are already part of our everyday lives. A key issue, which is a central part of the PLS model, is that the devices of the network cannot trust the central entity and are forced to verify the correctness of the distributed proof.

The generalization of the class NP to interactive proof systems, a model where the prover and the verifier are allowed to interact was a breakthrough in computational complexity [3,18,19,32,40]. In the distributed framework, the notion of distributed interactive protocols was introduced in [27] and further studied in [9,14,36,37]. In such protocols, a centralized, untrustable prover with unlimited computation power, named Merlin, exchanges messages with a randomized distributed algorithm, named Arthur.

Let us illustrate the general idea of this model, and also some notation. If we consider, for instance, four interactions, then there are two possible protocols: a dAMAM protocol and dMAMA protocol. In a dAMAM protocol, also denoted dAM[4], the last interaction is performed by Merlin. In a dMAMA protocol, also denoted dMA[4], the last interaction is performed by Arthur. Note that dAM[1]=dM, and we recover exactly the PLS model.

Now, we are going to explain with more detail what happens in a particular case. Let us consider a three interaction, dMAM=dAM[3] protocol. In this case Merlin starts, and he provides a certificate to Arthur (that is, certificates $c(v)$ for every node $v \in V$). Then, a random string (fresh randomness) is generated and made public to both Arthur (the nodes) and Merlin. This is the interaction performed by Arthur, and it should be interpreted as if Arthur was challenging Merlin. Note also that we are considering here the shared randomness setting.

Finally, Merlin replies to the query by sending another distributed certificate. After all these interactions, comes the deterministic distributed verification phase, performed between every node and its neighbors, after which every node decides whether to accept or reject.

We say that an algorithm uses $\mathcal{O}(f(n))$ bits if the messages exchanged between the nodes (in the verification round), and also the certificates sent by the prover Merlin to the nodes, are upper bounded by $\mathcal{O}(f(n))$. We include this bandwidth bound in the notation, which becomes dMA$[k, f(n)]$ and dAM$[k, f(n)]$ for the corresponding protocols.

Interaction may decrease drastically the size of the messages needed to solve some problems. Consider, for instance, the problem symmetry, where the nodes are asked to decide whether the graph G has a non-trivial automorphism (i.e., a non-trivial one-to-one mapping from the set of nodes to itself preserving edges). Any PLS solving the symmetry problem requires certificates of size $\Omega(n^2)$ [21]. Nevertheless, this problem admits distributed interactive protocols with small certificates, and very few interactions. In fact, it can be solved with both a dMAM$[\log n]$ protocol and a dAM$[n \log n]$ protocol [27].

1.2 Geometric Intersection Graph Classes

A graph $G = (V, E)$ is a geometric intersection graph if every node $v \in V$ is identified with a geometric object of some particular type, and two vertices are adjacent if the corresponding objects intersect. The two simplest non-trivial, and arguably two of the most studied geometric intersection graphs are **interval**

graphs and **permutation graphs**. In fact, most of the best-known geometric intersection graph classes are either generalizations of interval graphs or generalizations of permutation graphs. It comes as no surprise that many papers address different algorithmic and structural aspects, simultaneously, in both interval and permutation graph [2,23,30,43].

In both interval and permutation graphs, the intersecting objects are (line) segments, with different restrictions imposed on their positions. In interval graphs, the segments must all lie on the real line. In permutation graphs, the endpoints of the segments must lie on two separate, parallel real lines. In Fig. 1 we show an example of a permutation graph.

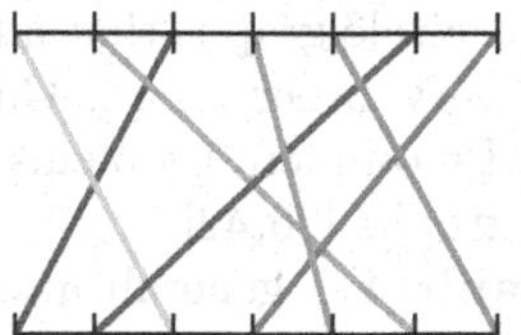
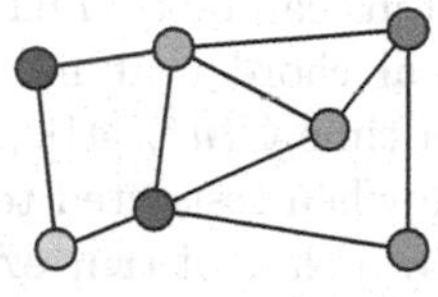

Fig. 1. An example of a permutation graph with its corresponding intersection model.

Although the class of interval graphs is quite restrictive, there are a number of practical applications and specialized algorithms for interval graphs [20,22,28]. Moreover, for several applications, the subclass of unit interval graphs (the situation where all the intervals have the same length) turns out to be extremely useful as well [4,25].

A natural generalization of interval graphs are **circular arc graphs**, where the segments, instead of lying on a line, lie on a circle. More precisely, a circular arc graph is the intersection graph of arcs of a circle. Although circular arc graphs look similar to interval graphs, several combinatorial problems behave very differently on these two classes of graphs. For example, the coloring problem is NP-complete for circular arc graphs while it can be solved in linear time on interval graphs [16]. Recognizing circular-arc graphs can be done in linear-time [24,34].

The class of permutation graphs behaves as the class of interval graphs in the sense that, on one hand, permutation graphs can be recognized in linear time [30] and, on the other hand, many NP-complete problems can be solved efficiently when the input is restricted to permutation graphs [8,31]. A graph G is a **circle graph** if G is the intersection model of a collection of chords in a circle (see Fig. 2).

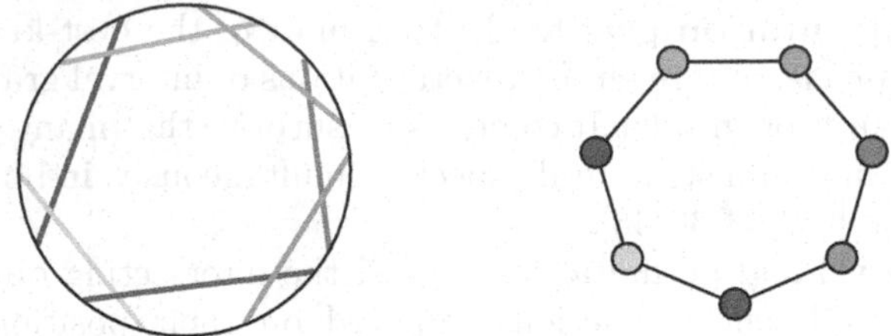

Fig. 2. An example of a circle graph with its corresponding intersection model.

Clearly, circle graphs are a generalization of permutation graphs. In fact, permutation graphs can be characterized as circle graphs that admit an *equator*, i.e., an additional chord that intersects every other chord. Circle graphs can be recognized in time $\mathcal{O}(n^2)$ [41]. Many NP-complete problems can be solve in polynomial time when restricted to circle graphs [26,42].

The well-known class of **trapezoid graphs** is a generalization of both interval graphs and permutation graphs. A trapezoid graph is defined as the intersection graph of trapezoids between two horizontal lines (see Fig. 3). Ma and Spinrad [33] showed that trapezoid graphs can be recognized in $\mathcal{O}(n^2)$ time. Trapezoid graphs were applied in various contexts such as VLSI design [10] and bioinformatics [1]. Note that trapezoid and circle graphs are incomparable: the trapezoid graph of Fig. 3 is not a circle graph, while the circle graph of Fig. 2 is not a trapezoid graph.

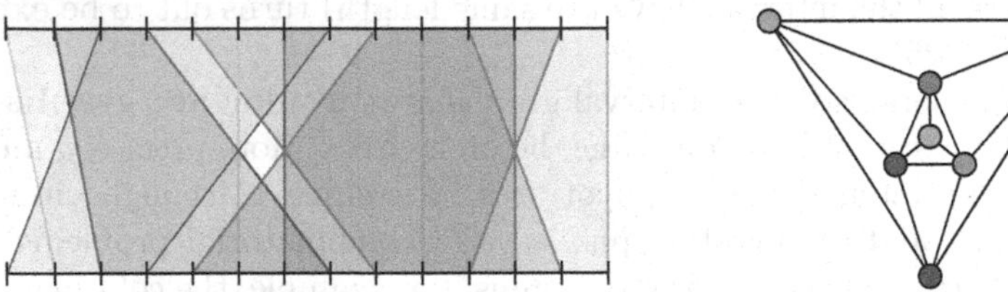

Fig. 3. An example of a permutation graph with its corresponding intersection model.

Recall that the way permutation graphs were generalized to circular graphs is by placing the ends of the segments in a circle (as chords) instead of placing the ends of the segments in two parallel lines. The same approach is used to generalize trapezoidal graphs and thus introducing polygon circle graphs.

More precisely, a **polygon circle graph** is the intersection graph of convex polygons of k sides, all of whose vertices lie on a circle. In this case we refer to a k-polygon circle graphs. In Fig. 4 we show an example of a 3-polygon circle graph. Both trapezoid graphs and circle graphs are proper subclasses of polygon circle graphs. Note that the polygon circle graph of Fig. 4 is neither a trapezoid graph nor a circle graph.

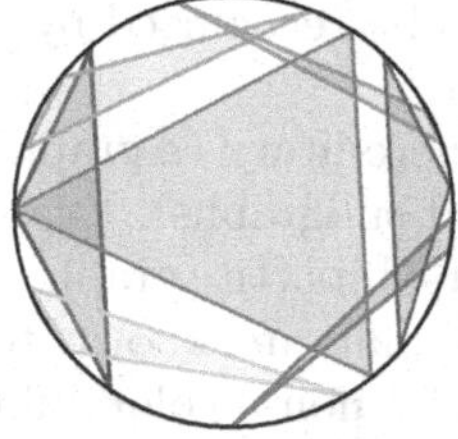 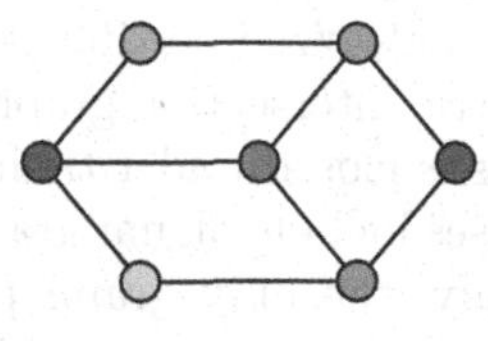

Fig. 4. An example of a 3-polygon circle graph with its corresponding intersection model.

The problem of recognizing whether a graph is a k-polygon circle graphs, for any $k \geq 3$, is NP-complete [39]. Nevertheless, many NP-complete problems have polynomial time algorithms when restricted to polygon circle graphs [16,17]. In an unpublished result, M. Fellows proved that the class of polygon circle graphs is closed under taking induced minors.

1.3 Our Results

In Sect. 3 we recall, explain and develop some tools, building blocks for the rest of the paper. In Sect. 4 we prove that trapezoid graphs can be recognized by PLSs with certificates of size $\mathcal{O}(\log n)$, and then we obtain the result for permutation graphs as a corollary. Then, in Sect. 5 we prove that k-polygon circle graphs can be recognized with a three round, dMAM protocol woth proof-size $\mathcal{O}(\log n)$, and we obtain the result for circle graphs as a particular case when $k = 2$. Finally, in Sect. 6, we prove that any PLS for recognizing permutation graphs, trapezoid graphs, circle graphs or polygon circle graphs requires certificates of size $\Omega(\log n)$.

1.4 Related Work

We already know the existence of PLSs (with logarithmic size certificates) for the recognition of many graph classes such as acyclic graphs [29], planar graphs [13], graphs with bounded genus [12], graph classes defined by a finite set of forbidden minors [5], etc.

The distributed interactive proof model is clearly more powerful than the PLS model. Some problems, for which any PLS requires huge certificates, can be solved with a distributed interactive protocol with small certificates and, in fact, with very few interactions. This is the case of problem symmetry, where the system must decide whether a graphs has a non-trivial automorphism. Problem symmetry is in $\mathsf{dMAM}[\log n]$ and also in $\mathsf{dAM}[n \log n]$, while the size of any certificate in a PLS must be of size at least $\Omega(n^2)$ [27].

It is always important to keep in mind the compiler defined in [37] which turns, automatically, any problem solved in NP in time $\tau(n)$ into a dMAM protocol that uses bandwidth $\tau(n) \log n/n$. Therefore, any class of sparse graphs that can be recognized in linear time, can also be recognized by a dMAM protocol with logarithmic-sized certificates.

Any geometric intersection graph class is hereditary (a graph class is hereditary if the class is closed under taking induced subgraphs). Examples of hereditary graph classes include planar graphs, forests, bipartite graphs, perfect graphs, etc. Interestingly, the only graph properties that are known to require PLSs with large certificates (e.g. small diameter [7], non-3-colorability [21], having a non-trivial automorphism [21]), are non-hereditary. This raises the question of whether we can improve the result of this paper and design a PLS for the recognition of any geometric intersection graph.

2 Preliminaries

Let G be a simple connected n-node graph, let $I : V(G) \to \{0,1\}^*$ be an input function assigning labels to the nodes of G, where the size of all inputs is polynomially bounded on n. Let $\mathsf{id} : V(G) \to \{1, \ldots, \text{poly}(n)\}$ be a one-to-one function assigning identifiers to the nodes. A *distributed language* $\mathcal{L}$ is a (Turing-decidable) collection of triples (G, id, I), called *network configurations*.

A distributed interactive protocol consists of a constant series of interactions between a *prover* called Merlin, and a *verifier* called Arthur. The prover Merlin is centralized, has unlimited computing power and knows the complete configuration (G, id, I). However, he cannot be trusted. On the other hand, the verifier Arthur is distributed, represented by the nodes in G, and has limited knowledge. In fact, at each node v, Arthur is initially aware only of his identity $\mathsf{id}(v)$, and his label $I(v)$. He does not know the exact value of n, but he knows that there exists a constant c such that $\mathsf{id}(v) \leq n^c$. Therefore, for instance, if one node v wants to communicate $\mathsf{id}(v)$ to its neighbors, then the message is of size $\mathcal{O}(\log n)$.

Given any network configuration (G, id, I), the nodes of G must collectively decide whether (G, id, I) belongs to some distributed language $\mathcal{L}$. If this is indeed the case, then all nodes must accept; otherwise, at least one node must reject (with certain probabilities, depending on the precise specifications we are considering).

Interactive protocols have two phases: an interactive phase and a verification phase. If Arthur is the party that starts the interactive phase, he picks a random string r_1 (known to all nodes of G because we are considering the shared randomness setting) and send it to Merlin. Merlin receives r_1 and provides every node v with a certificate $c_1(v)$ that is a function of v, r_1 and (G, id, I). Then again Arthur picks a random string r_2 and sends r_2 to Merlin, who, in his turn, provides every node v with a certificate $c_2(v)$ that is a function of v, r_1, r_2 and (G, id, I). This process continues for a fixed number of rounds. If Merlin is the party that starts the interactive phase, then he provides at the beginning every node v with a certificate $c_0(v)$ that is a function of v and (G, id, I), and the interactive process continues as explained before. In the last interaction round, the verification phase begins. This phase is a one-round deterministic algorithm executed at each node. More precisely, every node v broadcasts a message M_v to its neighbors. This message may depend on $\mathsf{id}(v)$, $I(v)$, all random strings generated by Arthur, and all certificates received by v from Merlin. Finally, based on all the knowledge accumulated by v (i.e., its identity, its input label, the generated random strings, the certificates received from Merlin, and all the messages received from its neighbors), the protocol either accepts or rejects at node v. Note that Merlin knows the messages each node broadcasts to its neighbors because there is no randomness in this last verification round.

Definition 1. *Let $\mathcal{V}$ be a verifier and $\mathcal{M}$ a prover of a distributed interactive proof protocol for languages over graphs of n nodes. If $(\mathcal{V}, \mathcal{M})$ corresponds to an Arthur-Merlin k-round, $\mathcal{O}(f(n))$ bandwidth protocol, we write $(\mathcal{V}, \mathcal{M}) \in \mathsf{dAM_{prot}}[k, f(n)]$.*

Definition 2. *Let $\varepsilon \leq 1/3$. The class $\mathsf{dAM}_\varepsilon[k, f(n)]$ is the class of languages $\mathcal{L}$ over graphs of n nodes for which there exists a verifier $\mathcal{V}$ such that, for every configuration (G, id, I) of size n, the two following conditions are satisfied.*

Completeness. *If $(G, \mathsf{id}, I) \in \mathcal{L}$ then, there exists a prover $\mathcal{M}$ such that $(\mathcal{V}, \mathcal{M}) \in \mathsf{dAM_{prot}}[k, f(n)]$ and*

$$\mathbf{Pr}\Big[\mathcal{V} \text{ accepts } (G, \mathsf{id}, I) \text{ in every node given } \mathcal{M}\Big] \geq 1 - \varepsilon.$$

Soundness. *If $(G, \mathsf{id}, I) \notin \mathcal{L}$ then, for every prover $\mathcal{M}$ such that $(\mathcal{V}, \mathcal{M}) \in \mathsf{dAM_{prot}}[k, f(n)]$,*

$$\mathbf{Pr}\Big[\mathcal{V} \text{ rejects } (G, \mathsf{id}, I) \text{ in at least one nodes given } \mathcal{M}\Big] \geq 1 - \varepsilon.$$

We also denote $\mathsf{dAM}[k, f(n)] = \mathsf{dAM}_{1/3}[k, f(n)]$, and omit the subindex ε when its value is obvious from the context.

For small values of k, instead of writing $\mathsf{dAM}[k, f(n)]$, we alternate M's and A's. For instance: $\mathsf{dMAM}[f(n)] = \mathsf{dAM}[3, f(n)]$. In particular $\mathsf{dAM}[f(n)] = \mathsf{dAM}[2, f(n)]$. Moreover, we denote $\mathsf{dM}[f(n)]$ the model where only Merlin provides a certificate, and no randomness is allowed (in other words, the model dM is the PLS model).

In this paper, we are interested mainly in the languages of graphs that are permutation, trapezoid, circle, polygon-circle and unit square. Formally,

Permutation-Recognition $= \{\langle G, \mathsf{id}\rangle$ s.t. G is a permutation graph$\}$.
Trapezoid-Recognition $= \{\langle G, \mathsf{id}\rangle$ s.t. G is a trapezoid graph$\}$.
Circle-Recognition $= \{\langle G, \mathsf{id}\rangle$ s.t. G is a circle graph$\}$.
k-Polygon-Circle-Recognition $= \{\langle G, \mathsf{id}\rangle$ s.t. G is a k-polygon-circle graph$\}$.

We denote by $[n]$ the set $\{0, \ldots, n-1\}$ and S_n the set of permutations of $[n]$. In the following, all graphs $G = (V, E)$ are simple and undirected. When the nodes of an n-node graph are enumerated with unique values in $[n]$, we denote $G = ([n], E)$. In a distributed problem, we always assume that the input graph is connected. We use the standard definitions and notations for (induced) subgraph, neighborhood, path, cycle, tree, clique, etc. For more details we refer to the textbook of Diestel [11].

3 Toolbox

In our results, we use some previously defined protocols as subroutines. In some cases, we consider protocols that solve problems which are more general than just decision problems (as, for instance, the construction of a spanning tree).

3.1 Spanning Tree and Related Problems

The construction of a spanning tree is an important building block for several protocols in the PLS model. Given a network configuration $\langle G, \mathsf{id}\rangle$, the Spanning-Tree problem asks to construct a spanning tree T of G, where each node has to end up knowing which of its incident edges belong to T.

Proposition 1. *There is a 1-round protocol for* Spanning-Tree *with certificates of size* $\mathcal{O}(\log n)$.

From the protocol of Proposition 1 it is easy to construct another one for problem Size, where the nodes, given the input graph $G = (V, E)$, have to verify the precise value of $|V|$ (recall that we are assuming that the nodes are only aware of a polynomial upper bound on $n = |V|$).

Proposition 2 [29]. *There is a 1-round protocol for* Size *with certificates of size* $\mathcal{O}(\log n)$.

Finally, for two fixed nodes $s, t \in V$, problem $s, t -$ Path is defined in the usual way: given a network configuration $\langle G, \mathsf{id}\rangle$, the output is a path P that goes from s to t. In other words, each node must end up knowing whether it belongs to P, and, in the positive cases, which of its neighbors are its predecessor and successor in P.

Proposition 3 [29]. *There is a 1-round protocol for* $s, t -$ Path *with certificates of size* $\mathcal{O}(\log n)$.

3.2 Problems Equality and Permutation

A second important building block, this time for interactive protocols, is a protocol for solving problem Equality, which is defined as follows. Given G a connected n-node graph, each node v receives two natural numbers $a(v)$ and $b(v)$, both of them encoded with $\mathcal{O}(log(n))$ bits. The problem Equality consists of verifying whether the multi-sets $\mathcal{A} = \{a(v)\}_{v \in V}$ and $\mathcal{B} = \{b(v)\}_{v \in V}$ are equal.

Proposition 4 *[37]. Problem* EQUALITY *belongs to* $\mathsf{dAM}_{1/3}[\log n]$.

A closely related problem is PERMUTATION, where some function π is given as input, and the nodes must verify whether π is indeed a permutation (a bijective function from V to $[n]$). Note that the input is given in a distributed way, by given $\pi(v)$ to each node $v \in V$. Using the protocol for EQUALITY as subroutine, it is possible to solve PERMUTATION with certificates of size $\mathcal{O}(\log n)$.

Proposition 5 *[37]. Problem* PERMUTATION *belongs to* $\mathsf{dMAM}_{1/3}[\log n]$.

We now introduce a new problem called CORRESPONDING ORDER, which is defined for inputs of the form $\langle G = (V, E), \mathsf{id}, (x, \pi)\rangle$, where the nodes must verify that: (i) π is a bijection from V to $[n]$; (ii) x is an injective function from V to $[N]$, where $N \geq n$; and (iii) for every $u, v \in V$, $\pi(u) \geq \pi(v) \iff x(u) \geq x(v)$.

Proposition 6. *Problem* CORRESPONDING ORDER *belongs to the class* $\mathsf{dMAM}[\log N]$.

Proof. The following is a protocol for CORRESPONDING ORDER. In the first round, each node v receives from the prover:

- The certification for the size of V and that π is an injective function.
- $a(v) = (x(v), \pi(v))$ and $b(v) = (y(v), \pi(v) + 1 \mod n)$, with $y(v) \in [N]$.

Suppose that π is an injective function and that n is known by all the nodes. Observe that, if $\{a(v)\}_{v \in V}$ and $\{b(v)\}_{v \in V}$ are equal, then $y(v) = x(u)$, where u is the successor of v, i.e. $\pi(u) = \pi(v) + 1$. Then, $\langle G = (V, E), \mathsf{id}, (x, \pi)\rangle$ is a yes-instance of CORRESPONDING ORDER if and only if π is an injective function, $\{a(v)\}_{v \in V} = \{b(v)\}_{v \in V}$, and $x(v) \leq y(v)$ for each node v such that $\pi(v) < n-1$.

Then, in the two remaining rounds, the nodes interact with the prover in order to prove that π is an injective function and that $\{a(v)\}_{v \in V}$, $\{b(v)\}_{v \in V}$ are equal multi-sets, using the protocols for PERMUTATION and EQUALITY, respectively. The nodes also check that $x(v) \leq y(v)$ (except for the node v such that $\pi(v) = n-1$). The communication bounds, as well as the correctness and soundness of the protocol follows from the ones of protocols for SPANNING-TREE, SIZE, EQUALITY and PERMUTATION. □

Note that our protocol for CORRESPONDING ORDER can be easily extended to the case when the range of function x is a set S of size N that admits a total order.

4 Permutation and Trapezoid Graphs

To recognize trapezoid graphs, first we present a useful characterization of them that are used to build a compact one-round PLS for problem TRAPEZOID-RECOGNITION, from which we derive a protocol for problem PERMUTATION.

Remember that in a model of a trapezoid graph, there are two parallel lines $\mathcal{L}_t$ and $\mathcal{L}_b$. We denote this lines the *top and bottom lines*, respectively. Each trapezoid has sides contained in each line, and then defined by four vertices, two in the top line, and two in the bottom line. Formally, each trapezoid T is defined by the set $T = \{t_1, t_2, b_1, b_2\}$, where $t_1 < t_2$ and $b_1 < b_2$, with $t_1, t_2 \in \mathcal{L}_t$ and $b_1, b_2 \in \mathcal{L}_b$ (see Fig. 5).

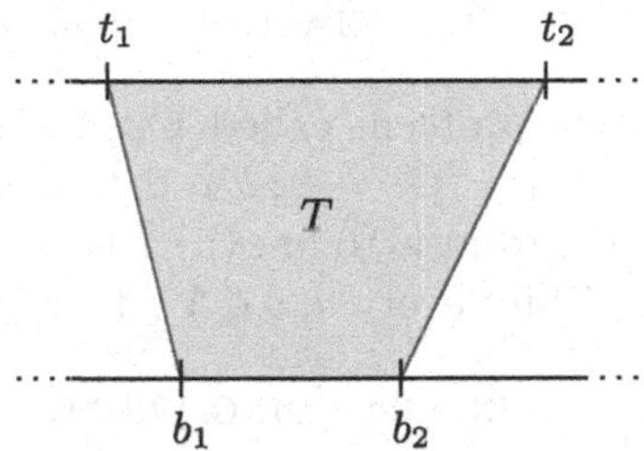

Fig. 5. Each trapezoid T is defined by the set $T = \{b_1, b_2, t_1, t_2\}$.

The definition of a trapezoid graph can be restated as follows (see [6]): A trapezoid graph $G = (V, E)$ is the intersection graph of a set of trapezoids $\{T_v\}_{v \in V}$ satisfying the following conditions. The vertices of each trapezoid have values in $[2n]$, two corresponding to the upper line and the other to the bottom line. The vertices defining the set $\{T_v\}_{v \in V}$, are all different, i.e., no pair of trapezoids share vertices. Therefore, in both the top and the bottom lines, each element in $[2n]$ correspond to a vertex of some trapezoid. The trapezoid model in the example of reffig:Extrapezoid satisfies these conditions.

For $v \in V$, we call $\{t_1(v), t_2(v), b_1(v), b_2(v)\}$ the vertices of T_v. Moreover, we say that $\{t_1(v), t_2(v), b_1(v), b_2(v)\}$ are the *vertices* of node v. In the following, a trapezoid model satisfying the abode conditions is called a *proper trapezoid model* for G. Given a graph $G = (V, E)$ (that is not necessarily a trapezoid graph), a *semi-proper trapezoid model* for G is a set of trapezoids $\{T_v\}_{v \in V}$ satisfying previous conditions, such that, for every $\{u, v\} \in E$, the trapezoids T_v and T_u have nonempty intersection. The difference between a proper and a semi-proper model is that in the first we also ask every pair of non-adjacent edges have non-intersecting trapezoids.

Given a trapezoid graph $G = (V, E)$ and a proper trapezoid model $\{T_v\}_{v \in V}$, we define the following sets for each $v \in V$:

$$F_t(v) = \{i \in [2n] \mid i < t_1(v) \text{ and } i \in \{t_1(w), t_2(w)\} \text{ for some } w \notin N(v)\}$$
$$F_b(v) = \{i \in [2n] \mid i < b_1(v) \text{ and } i \in \{b_1(w), b_2(w)\} \text{ for some } w \notin N(v)\}$$

We also call $f_t(v) = |F_t(v)|$ and $f_b(v) = |F_b(v)|$. The following lemmas characterize trapezoid graphs.

Lemma 1. *Let $G = (V, E)$ an connected trapezoid a graph with n nodes. Then each proper trapezoid model $\{T_v\}_{v \in V}$ of G satisfies for every $v \in V$:*

$$b_1(v) - f_b(v) = t_1(v) - f_t(v)$$

Proof. Let $\{T_v\}_{v \in V}$ be a proper trapezoid model of G. Then, given a node $v \in V$, all the coordinates in $F_t(v)$ are vertices of some $w \neq N(v)$. Such trapezoids T_w have their two upper vertices in the set $[t_1(v)]$ and their two lower vertices in $[b_1(v)]$, as otherwise T_w and T_v would intersect. Then, the cardinality of the set $[t_1(v)] \setminus F_t(v)$ is even, and the same holds for $[b_1(v)] \setminus F_b(v)$. Moreover, the cardinality of the set $[t_1(v)] \setminus F_t(v)$ equals the cardinality of $[b_1(v)] \setminus F_b(v)$, as every position in $[2n]$ corresponds to a vertex of some trapezoid's node. We deduce that

$$t_1(v) - f_t(v) = |\{1, \ldots, t_1(v)\} \setminus F_t(v)| = |\{1, \ldots, b_1(v)\} \setminus F_b(v)| = b_1(v) - f_b(v).$$

Lemma 2. *Let $G = (V, E)$ be a n-node graph that is not a trapezoid graph. Then, for every semi-proper trapezoid model $\{T_v\}_{v \in V}$ of G, at least one of the following conditions is true:*

1. *$\exists v \in V$ such that some value in $\{b_1(v), \ldots, b_2(v)\}$ or $\{t_1(v), \ldots, t_2(v)\}$ is a vertex of $\omega \notin N(v)$.*
2. *$\exists v \in V$ such that $b_1(v) - f_b(v) \neq t_1(v) - f_t(v)$.*

Proof. Let G be a graph that is not a trapezoid graph and $\{T_v\}_{v \in V}$ a semi-proper trapezoid model. As G is not a permutation graph, by definition necessarily there exist a pair $\{v, \omega\} \notin E$ such that $T_v \cap T_\omega \neq \emptyset$. We distinguish two possible cases (see Fig. 6):

1. $[b_1(v), b_2(v)]_{\mathbb{N}} \cap [b_1(\omega), b_2(\omega)]_{\mathbb{N}} \neq \emptyset$ or $[t_1(v), t_2(v)]_{\mathbb{N}} \cap [t_1(\omega), t_2(\omega)]_{\mathbb{N}} \neq \emptyset$.
2. $[b_1(v), b_2(v)]_{\mathbb{N}} \cap [b_1(\omega), b_2(\omega)]_{\mathbb{N}} = \emptyset$ and $[t_1(v), t_2(v)]_{\mathbb{N}} \cap [t_1(\omega), t_2(\omega)]_{\mathbb{N}} = \emptyset$.

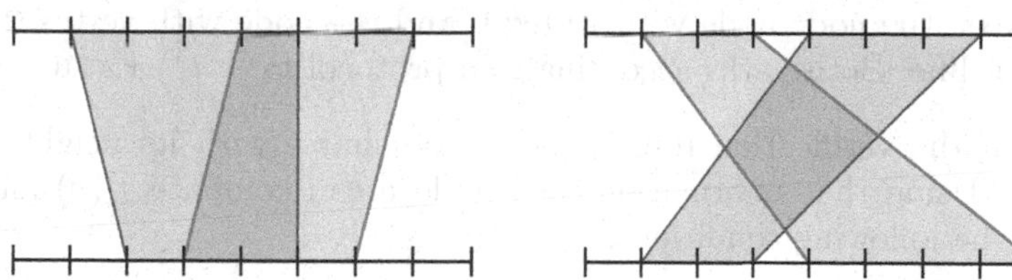

Fig. 6. A representation of the two possible cases. In the first case, depicted in left, at least one vertex of a trapezoid is contained in the other. In the second case, in the right hand, the trapezoids intersect, but not in the vertices.

Clearly if the first case holds, then condition 1 is satisfied. Suppose then that there is no pair $\{v, \omega\} \notin E$ such that $T_v \cap T_\omega \neq \emptyset$ satisfying the first case. Then necessarily the second case holds. Let u be a node for which exists $\omega \in V \setminus N(u)$ such that $T_u \cap T_\omega \neq \emptyset$. For all possible choices of u, let us pick the one such that $b_1(u)$ is minimum. Then u satisfies the following conditions:

(a) Exists a node $\omega \in V$ such that $\omega \notin N(v)$ and $T_u \cap T_\omega \neq \emptyset$
(b) All nodes $\omega \in V$ such that $\omega \notin N(v)$ and $T_u \cap T_\omega \neq \emptyset$ satisfy that $t_2(\omega) < t_1(u)$ and $b_2(u) < b_1(\omega)$
(c) None of the positions in $\{1, \ldots, b_1(u)\}$ is occupied by a vertex of a node ω such that $\{u, \omega\} \notin E$ and $T_u \cap T_\omega \neq \emptyset$.

Observe that conditions (a) and (b) imply that $t_1(u) - f_t(u) > 0$, while condition (c) implies that $b_1(u) - f_b(u) = 0$. We deduce that condition 2 holds by u. □

We are now ready to define our protocol and main result regarding TRAPEZOID-RECOGNITION.

Theorem 1. *There is a 1-round proof labelling scheme for* TRAPEZOID-RECOGNITION *with certificates of size* $\mathcal{O}(\log n)$.

Proof. The following is a one-round PLS for TRAPEZOID-RECOGNITION

Given an instance $\langle G = (V, E), \mathsf{id}\rangle$, the certificate provided by the prover to node $v \in V$ is interpreted as follows.

1. The certification of the total number of nodes n, according to some protocol for SIZE.
2. Values $b_1(v), b_2(v), t_1(v), t_2(v) \in [2n]$, such that $b_1(v) < b_2(v)$ and $t_1(v) < t_2(v)$, representing the vertices of a trapezoid T_v.
3. Value p_v corresponding to the minimum position in the upper line greater that $t_1(v)$ that is not a vertex of a neighbor of v.
4. Value q_v corresponding minimum position in the lower line grater than $b_1(v)$ that is not a vertex of a neighbor of v.
5. The certification of a path P_t between the node with vertex 0 and the node with vertex $2n-1$ in the upper line (respecting assignment in 2.) and a path P_b between the node node with vertex 0 and the node with vertex $2n-1$ in the lower line. Both paths according to a protocol for $s, t-$ PATH.

Then, in the verification round, each node shares with its neighbors their certificates. Using that information each node v can compute $f_t(v)$ and $f_b(v)$, and check the following conditions:

a. The correctness of the value of n, according to some protocol for SIZE.
b. The correctness of the paths P_b and P_t, according to a protocol for $s, t-$PATH.
c. The vertices of the trapezoid of v are in $[2n]$.
d. $T_v \cap T_\omega \neq \emptyset$ for all $\omega \in N(v)$.
e. All values in $\{t_1(v)+1, \ldots, t_2(v)-1\}$ and $\{b_1(v)+1, \ldots, b_2(v)-1\}$ are a vertex of some neighbor of v.
f. $t_2(v) < p_v$ and $b_2(v) < q_v$.
g. If $\omega \in N(v)$ and $p_\omega < t_2(v)$, then v verifies that p_ω is a vertex of some other neighbor.

h. If $\omega \in N(v)$ and $q_\omega < b_2(v)$, then v verifies that q_ω is a vertex of some other neighbor.
i. $b_1(v) - f_b(v) = t_1(v) - f_t(v)$.

We now analyze the soundness and completeness of our protocol.

Completeness: Suppose that G is a trapezoid graph. An honest prover just has to send the real number of nodes n, a trapezoid model $\{T_v\}_{v \in V}$ of G and valid paths P_b and P_t according the trapezoid model. Then, the nodes will verify a, b by the completeness of the protocols for SIZE and s,t – PATH. Conditions c, d, e ,f, g and h are verified by the correctness of the model $\{T_v\}_{v \in V}$. Condition i is also verified, by Lemma 1.

Soundness: Suppose G is not a trapezoid graph. If a dishonest prover provides a wrong value of n, or wrong paths P_t or P_b, then at least one node will reject verifying a or b. Then, we assume that the prover cannot cheat on these values.

Suppose that the prover gives values $\{T_v\}_{v \in V}$ such that is fulfilled $\bigcup_{v \in V}\{t_1(v), t_2(v)\} \neq [2n]$. If some vertex of a node is not in the set $[2n]$, then that node fails to verify condition c and rejects. Without loss of generality, we can assume that there exists a $j \in [2n]$ such that $t_1(v), t_2(v) \neq j$, for every $v \in V$. If a node ω satisfies that $t_1(\omega) < j < t_2(\omega)$, then node ω fails to verify condition e and rejects. Then j is not contained in any trapezoid. As P_t is correct, j must be different than 1 and $2n$. Also by the correctness of P_t, there exist a pair of adjacent nodes $u, v \in V$ such that $t_2(u) < j < t_1(v)$. From all possible choices for u and v, we pick the one such that $t_2(u)$ is maximum. We claim that v fails to check condition g. Since j is not a vertex of any node, then $p_u \leq j$. If v verifies condition g, then necessarily $p_u < j$. Then, there must exist a node $\omega \in N(v)$ such that $p_u = t_1(\omega)$. But since we are assuming that j is not contained in any trapezoid, we have that $t_2(\omega) < j$, contradicting the choice of u.

Therefore, if conditions a - h are verified, we can assume that the nodes are given a semi-proper trapezoid model of G. Since we are assuming that G is not a trapezoid graph, by Lemma 2 we deduce that condition i cannot be satisfied and some node rejects.

We now analyze the communication complexity of the protocol: the certification for SIZE and s,t – PATH is $\mathcal{O}(\log n)$, given by refprop:sizeofG and refprop:stpath. On the other hand, for each $v \in V$, the values $b_1(v)$, $b_2(v)$, $t_1(v)$, $t_2(v)$, p_v, q_v are computable in $\mathcal{O}(\log n)$ space as they are numbers in $[2n]$. Overall the total communication is $\mathcal{O}(\log n)$. □

Observe that permutation graphs are exactly the trapezoid graphs that admit a proper model where $t_2(v) = t_1(v) + 1$ and $b_2(v) = b_1(v) + 1$, for every $v \in V$. Then, previous protocol can be adapted to solve PERMUTATION-RECOGNITION, simply asking the nodes to accept only the models that satisfy previous condition. We conclude that the problem PERMUTATION-RECOGNITION admits a PLS with certificates of size $\mathcal{O}(\log n)$.

5 Circle and Polygon Circle Graphs

In this section, we give a three-round protocol for the recognition of polygon-circle graph. This extension is based in a non-trivial extension of the properties of circle graphs.

Remember that a n-node graph $G = (V, E)$ is a k-polygon-circle graph if and only if G is the intersection model of a set of n polygons of k vertices inscribed in a circle, namely $\{P_v\}_{v \in V}$. Further, every k-polygon-circle graph admits a model satisfying the following conditions [6]: (1) for each $v \in V$, the polygon P_v is represented as a set of k vertices $\{p_0(v), \dots, p_{k-1}(v)\}$ such that, for each $i \in [k-1]$, $1 \leq p_i(v) < p_{i+1}(v) \leq n \cdot k$, and (2) $\bigcup_{v \in V} \bigcup_{i \in [k]} \{p_i(v)\} = [n \cdot k]$. In other words, each value in $[k \cdot n]$ corresponds to a unique vertice of some polygon. A set of polygons satisfying conditions (1) and (2) are called a *proper polygon model* for G. Similar to previous cases, when we just ask that adjacent nodes have intersecting polygons (but not necessarily the reciprocal) we say that the model is a *semi-proper polygon model* for G.

Let G be a graph and $\{P_v\}$ be a semi-proper model for G. For each $v \in V$. Let us call $\alpha(v)$ the set of points in $\{1, \dots, p_1(v)\} \cup \{p_k(v), \dots, kn\}$ that do not correspond to a neighbor of v. For each $i \in [k]$, we also call $\beta_i(v)$ the set of nodes $w \notin N(v)$ such that $p_i(w) \in \alpha(v)$. Formally,

$$\alpha(v) = \{i \in [0, p_1(v)]_{\mathbb{N}} \cup [p_k(v), kn - 1]_{\mathbb{N}} : \forall u \in N(v), i \notin P_u\}$$
$$\beta_i(v) = \{w \in V : p_i(w) \in \alpha(v)\}$$

Lemma 3. *Let $G = (V, E)$ be a graph, and let $\{P_v\}_{v \in V}$ a semi-proper model for G. Then $\{P_v\}_{v \in V}$ is a proper model for G if and only if $|\alpha(v)| = k|\beta_1(v)|$ for every $v \in V$.*

Proof. Let us suppose first that $\{P_v\}_{v \in V}$ is a proper model for G and v be an arbitrary node. If $\alpha(v) = \emptyset$ the result is direct. Then, let us suppose that $\alpha(v) \neq \emptyset$, and let us pick $q \in \alpha(v)$. Then necessarily there exists $i \in [k]$ such that q belongs to $\beta_i(v)$. Observe that for each node w in $\beta_i(v)$, all the vertices of the polygon P_w are contained α_v. Otherwise, the polygons P_w and P_v would have non-empty intersection, which contradicts the fact that $\{P_v\}_{v \in V}$ is a proper model. This implies that $|\alpha(v)| = k|\beta_i(v)|$, for every $i \in [k]$. In particular $|\alpha(v)| = k|\beta_1(v)|$.

Let us suppose now that $\{P_v\}_{v \in V}$ is not a proper model for G. Let us define the set C of vertices having non-neighbor with intersecting polygons, formally

$$C = \{v \in V : \exists w \in V, \{v, w\} \notin E \text{ and } P_v \cap P_w \neq \emptyset\}.$$

Now pick $v \in C$ such that $p_1(v)$ is maximum, and call C_v the set of non-neighbors of v whose polygons intersect with P_v. Let w be an arbitrary node in C_v. By the maximality of $p_1(v)$, we know that $p_1(w) \in \beta_1(v)$. But since $P_w \cap P_v \neq \emptyset$, there must exist $i \in [k]$ such that $p_i(w) \notin \beta_i(v)$. This implies that $|\beta_1(v)| \geq |\beta_i(v)|$ for every $i \in [k]$, and at least one of these inequalities is strict. Since $|\alpha(v)| = \sum_{i \in [k]} |\beta_i(v)|$, we deduce that $k|\beta_1(v)| > |\alpha(v)|$. □

Let $G = (V, E)$ be a graph, and $\{P_v\}_{v \in V}$ be a semi-proper polygon model for G. For each $i \in [k]$ and $v \in V$, we denote by $\pi_i(v)$ the cardinality of the set $\{u \in V : p_i(u) < p_i(v)\}$, and denote by $\sigma_i(v)$ the cardinality of the set $\{q < p_i(v) : \exists u \in V : p_1(u) = q \vee p_k(u) = q\}$. For a node v, we denote $N_{1,k}(v)$ the number vertices of polygons corresponding to neighbors of v, that are contained $[0, p(v_1)]_{\mathbb{N}} \cup [p(v_k), kn]_{\mathbb{N}}$. Formally,

$$N_{1,k}(v) = |\{q \in [0, p(v_1)]_{\mathbb{N}} \cup [p(v_k), kn]_{\mathbb{N}} : \exists w \in N(v), q \in P_w\}|$$

Lemma 4. *Let $G = (V, E)$ be a graph, and $\{P_v\}_{v \in V}$ be a semi-proper polygon model for G. Then, $|\alpha(v)| = kn - p_k(v) + p_1(v) - 1 - N_{1,k}(v)$, and $|\beta_1(v)| = n - \sigma_k(v) + \pi_k(v) + \pi_1(v)$.*

Proof. Let $\{P_v\}_{v \in V}$ be a semi-proper polygon model for G and v be an arbitrary node. First, observe that there are $p_1(v)$ integer positions for vertices in $[p_1(v)]$ and $(kn-1) - p_k$ positions for vertices in $[p_k, kn-1]$. Then, there are $kn - p_k + p_1(v) - 1$ available integer positions in $[p_1(v)] \cup [p_k(v), kn-1]_{\mathbb{N}}$. Since $N_{1,k}(v)$ of these positions are occupied by a polygon corresponding some neighbor of v, we deduce that $\alpha(v) = kn - p_k + p_1(v) - 1 - N_{1,k}(v)$.

Second, observe that the set $\{p_1(u), p_k(u)\}_{u \in V}$ uses $2n$ of the kn possible positions. Then, there are $2n - \sigma_k(v)$ positions used the elements of $\bigcup_{u \in V}\{p_1(u), p_k(u)\} \cap [p_k(v)+1, kn-1]_{\mathbb{N}}$. On the other hand, there are $n - \pi_k$ positions used by vertices in $\bigcup_{u \in V}\{p_k(u)\} \cap [p_k(v)+1, kn-1]_{\mathbb{N}}$. Therefore, there are $n - \sigma_k(v) + \pi_k(v)$ positions used by $\bigcup_{u \in V}\{p_1(u)\} \cap [p_k(v)+1, kn-1]_{\mathbb{N}}$. Finally, noticing that there are $\pi_1(v)$ positions used by $\bigcup_{u \in V}\{p_1(u)\} \cap [0, p_1(v)-1]_{\mathbb{N}}$, we deduce that $|\beta_1(v)| = n - \sigma_k(v) + \pi_k(v) + \pi_1(v)$. □

We are now ready to give the main result of this section.

Theorem 2. *k-POLYGON-CIRCLE-RECOGNITION belongs to* $\mathsf{dMAM}[\log n]$.

Proof. Consider the protocols for SIZE, PERMUTATION and CORRESPONDING ORDERof Propositions 2, 5 and 6. Given an instance $\langle G, \mathsf{id}\rangle$, consider the following three round protocol. In the first round, the prover provides each node v with the following information:

1. The certification of the total number of nodes n, according to the protocol for SIZE.
2. The vertices of the polygon P_v, denoted $V(P_v) = \{p_1(v), \ldots, p_k(v)\}$.
3. The values of $\pi_1(v)$, $\pi_k(v)$ and $\sigma(v)$.
4. The certification of $\bigcup_v V(P_v) = [k \cdot n]$ according to the protocol for PERMUTATION.
5. The certification of the correctness of $\{(p_1(v), \pi_1(v))\}_{v \in V}$ according to the protocol for CORRESPONDING ORDER.
6. The certification of the correctness of $\{(p_k(v), \pi_k(v))\}_{v \in V}$ according to the protocol for CORRESPONDING ORDER.
7. The certification of the correctness of $\{(p_1(v), \sigma_1(v)\}_{v \in V}$ and the collection $\{p_k(v), \sigma_k(v)\}_{v \in V}$ according to the protocol for CORRESPONDING ORDER.

Then, in the second and third round the nodes perform the remaining interactions of the protocols for PERMUTATION and CORRESPONDING ORDER. In the verification round, the nodes first check the correctness of 1-7 according to the verification rounds for SIZE, PERMUTATION and CORRESPONDING ORDER.

Remark: in order to check 7, each node has to play the role of two different nodes v', v'', one to verify $\{(p_1(v), \sigma(v')\}_{v \in V}$, and the other one to verifies $\{(p_k(v), \sigma(v'')\}_{v \in V}$, where $\sigma(v') = \sigma_1(v)$ and $\sigma(v'') = \sigma_k(v)$. To do so, Merlin gives v the certificates of v' and v'', and v answers with the random bits as if they would be generated by v' and v''. Obviously, this increases the communication cost by a factor of 2.

Then, each node v computes $|\beta(v)|$ and $|\alpha(v)|$ according to the expressions of Lemma 4, and checks the following conditions:

a. $\forall u \in N(v)$, $P_u \cap P_v \neq \emptyset$.
b. $|\alpha(v)| = k|\beta_1(v)|$.

We now analyze completeness and soundness.

Completeness: Suppose that input graph G is a k-polygon-circle graph. Then Merlin gives a proper polygon model $\{P_v\}_{v \in V}$ for G. Merlin also provides the correct number of nodes n, correct orders $\{\pi_1(v)\}_{v \in V}$ and $\{\pi_k(v)\}_{v \in V}$, $\{\sigma_1(v)\}_{v \in V}$ and $\{\sigma_k(v)\}_{v \in V}$, and the certificates required in the corresponding sub-routines. Then, the nodes verify correctness of 1-7 with probability greater than 2/3, by the correctness of the protocols for SIZE, PERMUTATION and CORRESPONDING ORDER. Finally, condition a is verified by definition of a proper model, and condition b is verified by Lemma 3. We deduce that every node accepts with probability greater than 2/3.

Soundness: Suppose now that G is not a k-polygon-circle graph. By the soundness of the protocols for SIZE, PERMUTATION and CORRESPONDING ORDER, we now that at least one node rejects the certificates not satisfying 1-7, with probability greater than 2/3. Suppose then that conditions 1-7 are verified. Observe that every set of polygons satisfying condition a form a semi-proper polygon model for G. Since G is not a k-polygon-circle graph, by Lemma 3 we deduce that at least one node fails to verify a or b. All together, we deduce that at least one node rejects with probability greater than 2/3.

We now analyze the communication complexity of the protocol: the certification for SIZE, PERMUTATION and CORRESPONDING ORDER is $\mathcal{O}(\log n)$, given by Proposition 2, 5 and 6. On the other hand, for each $v \in V$, the values $\pi_1(v), \pi_k(v), \sigma_1(v), \sigma_2(v), V(P_v)$ can be encoded in $\mathcal{O}(\log n)$ as they are numbers in $[n]$, $[2n]$ or $[kn]$. Overall the total communication is $\mathcal{O}(\log n)$.

Since circle graphs are 2-polygon-circle graphs, we deduce that problem CIRCLE-RECOGNITION is in $\mathsf{dMAM}[\log n]$.

6 Lower Bounds

In this section we give logarithmic lower-bounds in the certificate sizes of any PLS that recognizes the class of permutation, trapezoid, circle or polygon-circle graphs. In order to so, we use a technique given by Fraigniaud et al. [15], called *crossing edge*, and which we detail as follows. Let $G = (V, E)$ be a graph and let $H_1 = (V_1, E_1)$ and $H_2 = (V_2, E_2)$ be two subgraphs of G. We say that H_1 and H_2 are independent if and only if $V_1 \cap V_2 = \emptyset$ and $E \cap (V_1 \times V_2) = \emptyset$.

Definition 3 ([15]). *Let $G = (V, E)$ be a graph and let $H_1 = (V_1, E_1)$ and $H_2 = (V_2, E_2)$ be two independent isomorphic subgraphs of G with isomorphism $\sigma\colon V_1 \to V_2$. The* CROSSING *of G induced by σ, denoted by $\sigma_{\bowtie}(G)$, is the graph obtained from G by replacing every pair of edges $\{u, v\} \in E_1$ and $\{\sigma(u), \sigma(v)\} \in E_2$, by the pair $\{u, \sigma(v)\}$ and $\{\sigma(u), v\}$.*

Then, the tool that we use to build our lower-bounds is the following.

Theorem 3 ([15]). *Let $\mathcal{F}$ be a family of network configurations, and let $\mathcal{P}$ be a boolean predicate over $\mathcal{F}$. Suppose that there is a configuration $G_s \in \mathcal{F}$ satisfying that (1) G contains as subgraphs r pairwise independent isomorphic copies $H_1, ..., H_r$ with s edges each, and (2) there exists r port-preserving isomorphisms $\sigma_i\colon V(H_1) \to V(H_i)$ such that for every $i \neq j$, the isomorphism $\sigma^{ij} = \sigma_i \circ \sigma_j^{-1}$ satisfies $\mathcal{P}(G_s) \neq \mathcal{P}(\sigma^{ij}_{\bowtie}(G)_s)$. Then, the verification complexity of any PLS for $\mathcal{P}$ and $\mathcal{F}$ is $\Omega\left(\frac{log(r)}{s}\right)$.*

Let us consider first permutation and trapezoid graphs. Let $\mathcal{F}$ the family of instances of PERMUTATION-RECOGNITION, induced by the family of graphs $\{Q_n\}_{n>0}$. Each graph Q_n consists of $5n$ nodes forming a path $\{v_1, \ldots, v_{5n}\}$ where we add the edge $\{v_{5i-3}, v_{5i-1}\}$, for each $i \in [n]$. It is easy to see that for each $n > 0$, Q_n is a permutation graph (and then also a trapezoid graph). In Fig. 7 is depicted the graph Q_3 and its corresponding model.

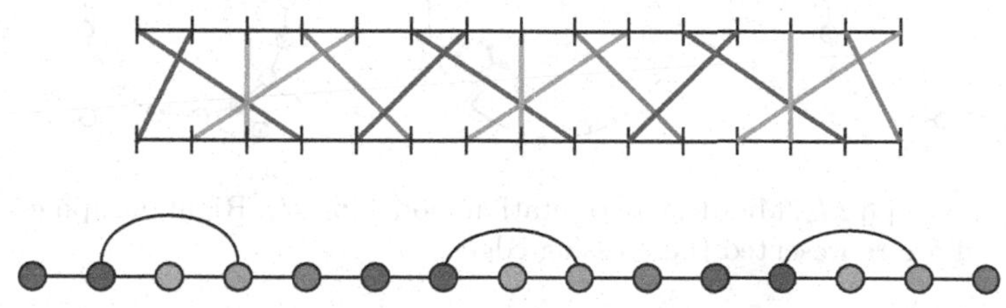

Fig. 7. Graph Q_3 and a permutation model for Q_3.

Given Q_n defined above, consider the subgraphs $H_i = \{v_{5i-2}, v_{5i-1}\}$, for each $i \in [n]$, and the isomorphism $\sigma_i\colon V(H_1) \to V(H_i)$ such that $\sigma_i(v_3) = v_{5i-2}$ and $\sigma_i(v_4) = v_{5i-1}$.

Lemma 5. *For each $i \neq j$, the graph $\sigma_{\bowtie}^{ij}(Q_n)$ it is not a trapezoid graph.*

Proof. Given $i < j$, observe that in $\sigma^{ij}: V(H_j) \to V(H_i)$, the nodes v_{5j-3}, v_{5j-2}, v_{5i-1}, v_{5i-3}, v_{5i-2}, v_{5j-1} form an induced cycle of length 6 (see Fig. 8 for an example).

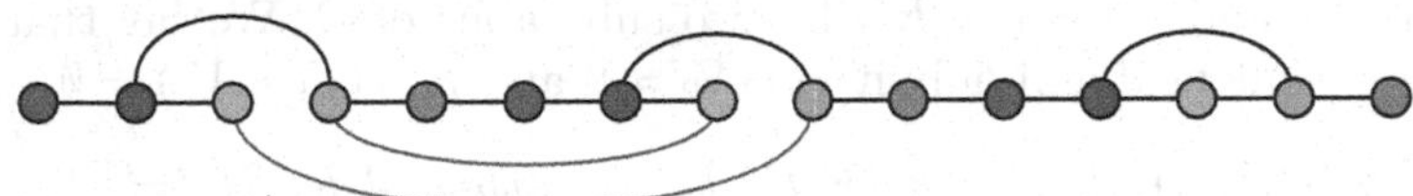

Fig. 8. Graph $\sigma_{\bowtie}^{12}(Q_3)$, where in red are represented the crossing edges. Observe that this graph is not a trapezoid graph, as it contains an induced cycle of length 6.

As a trapezoid graph have induced cycles of length at most 6, we deduce that $\sigma_{\bowtie}^{ij}(Q_n)$ is not a trapezoid graph.

By Theorem 3 and the abode result, the lower bound result is direct.

Theorem 4. *Any PLS for* Permutation-Recognition *or* Trapezoid-Recognition *has proof-size of $\Omega(\log n)$ bits.*

We now tackle the lower-bound for circle and polygon-circle graphs. Let $\mathcal{G}$ the family of instances of Circle-Recognition, defined by the family of graphs $\{M_n\}_{n>2}$. Each graph M_n consists of $6n$ nodes, where $4n$ nodes form a path $\{v_1, \ldots, v_{4n}\}$ where we add, for each $i \in [n]$, the edges $\{v_{4i-3}, v_{4n+i}\}$, $\{v_{4i-2}, v_{5n+i}\}$ and $\{v_{4n+i}, v_{5n+i}\}$. It is easy to see that for each $n > 0$, M_n is a circle graph (and then also a polygon-circle graph). In Fig. 9 is depicted the graph M_4 and its corresponding model.

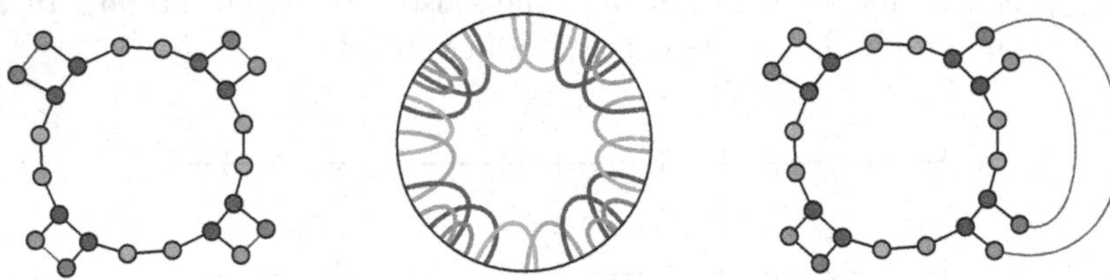

Fig. 9. Left: Graph M_4. Middle: a permutation model for M_4. Right: Graph $\sigma_{\bowtie}^{i,i+1}(M_4)$, where in red are represented the crossing edges.

Given M_n defined above, consider the subgraphs $H_i = \{v_{4n+i}, v_{5n+i}\}$, for each $i \in [n]$, and the isomorphism $\sigma_i: V(H_1) \to V(H_i)$ such that $\sigma_i(4n+1) = v_{5n+i}$ and $\sigma_i(5n+1) = v_{4n+i}$.

Lemma 6. *For every $k > 0$ and each $i \neq j$, the graph $\sigma_{\bowtie}^{ij}(M_n)$ it is not a k-polygon-circle graph.*

Proof. First, observe that in $\sigma_{\bowtie}^{ij}(M_n)$ we have two induced cycles defined by $C_1 = v_1, \ldots, v_{4n}$, and $C_2 = v_{4j-3}, v_{4n+j}, v_{5n+i}, v_{4i-2}, v_{4i-3}, v_{4n+i}, v_{5n+j}$. Moreover $|V(C_1) \cap V(C_2)| = 4$, $|V(C_1) - V(C_2)| = 4n - 4$ and $|V(C_2) - V(C_1)| = 4$. See Fig. 9 for a representation of $\sigma_{\bowtie}^{i,i+1}(M_4)$.

Claim. Every graph G consisting in two graphs C_1 and C_2 such that $|V(C_1) \cap V(C_2)| \geq 4$, $|V(C_1) - V(C_2)| \geq 2$ and $|V(C_2) - V(C_1)| \geq 2$ is not a k-polygon-circle graph, for every $k > 0$.

Let us denote by v_i and v_f two special nodes with degree 3, connecting the two cycles. Suppose there exists a k-polygon-circle model for G. Observe that, if we delete all polygons corresponding to nodes of $V(C_2) - V(C_1)$, we obtain a polygon model for C_1. However, the cycle C_1 has at least 4 nodes, because $|V(C_1) - V(C_2)| \geq 2$ and $|V(C_1) \cap V(C_2)| \geq 4$. Then, there is no way to add the removed polygons corresponding to $V(C_2) - V(C_1)$, without intersecting a polygon of $V(C_1) - V(C_2)$.

Then, by Sect. 6, we deduce that the graph induced by $C_1 \cup C_2$ is not a k-polygon-cycle graph. Since the class of polygon-circle graphs is hereditary, we deduce that $\sigma_{\bowtie}^{ij}(M_n)$ it is not a k-polygon-circle graph. □

Direct by Theorem 3 and Lemma 6 we deduce the following result.

Theorem 5. *Any PLS for* Circle-Recognition *or* k-Polygon-Circle-Recognition *has proof-size of* $\Omega(\log n)$ *bits.*

References

1. Abouelhoda, M.I., Ohlebusch, E.: Chaining algorithms for multiple genome comparison. J. Discrete Algorithms **3**(2–4), 321–341 (2005)
2. Asdre, K., Ioannidou, K., Nikolopoulos, S.D.: The harmonious coloring problem is np-complete for interval and permutation graphs. Discrete Appl. Math. **155**(17), 2377–2382 (2007)
3. Babai, L., Moran, S.: Arthur-merlin games: a randomized proof system, and a hierarchy of complexity classes. J. Comput. Syst. Sci. **36**(2), 254–276 (1988)
4. Beeri, C., Fagin, R., Maier, D., Yannakakis, M.: On the desirability of acyclic database schemes. J. ACM (JACM) **30**(3), 479–513 (1983)
5. Bousquet, N., Feuilloley, L., Pierron, T.: Local certification of graph decompositions and applications to minor-free classes. arXiv preprint arXiv:2108.00059 (2021)
6. Brandstädt, A., Van Le, B., Spinrad, J.P.: Graph Classes: A Survey. Society for Industrial and Applied Mathematics, January 1999
7. Censor-Hillel, K., Paz, A., Perry, M.: Approximate proof-labeling schemes. Theoret. Comput. Sci. **811**, 112–124 (2020)
8. Chao, H.S., Hsu, F.-R., Lee, R.C.T.: An optimal algorithm for finding the minimum cardinality dominating set on permutation graphs. Discrete Appl. Math. **102**(3), 159–173 (2000)
9. Crescenzi, P., Fraigniaud, P., Paz, A.: Trade-offs in distributed interactive proofs. In: 33rd International Symposium on Distributed Computing (DISC 2019). Schloss Dagstuhl-Leibniz-Zentrum fuer Informatik (2019)

10. Dagan, I., Golumbic, M.C., Pinter, R.Y.: Trapezoid graphs and their coloring. Discrete Appl. Math. **21**(1), 35–46, 1988
11. Diestel, R.: Graph theory, 3rd ed. Graduate texts in mathematics, 173 (2005)
12. Feuilloley, L., Fraigniaud, P., Montealegre, P., Rapaport, I., Rémila, É., Todinca, I.: Local certification of graphs with bounded genus. arXiv preprint arXiv:2007.08084 (2020)
13. Feuilloley, L., Fraigniaud, P., Montealegre, P., Rapaport, I., Rémila, É., Todinca, I.: Compact distributed certification of planar graphs. Algorithmica, pp. 1–30 (2021)
14. Fraigniaud, P., Montealegre, P., Oshman, R., Rapaport, I., Todinca, I.: On distributed Merlin-Arthur decision protocols. In: Censor-Hillel, K., Flammini, M. (eds.) SIROCCO 2019. LNCS, vol. 11639, pp. 230–245. Springer, Cham (2019). https://doi.org/10.1007/978-3-030-24922-9_16
15. Fraigniaud, P., Patt-Shamir, B., Perry, M.: Randomized proof-labeling schemes. Distrib. Comput. **32**(3), 217–234 (2019)
16. Garey, M.R., Johnson, D.S., Miller, G.L., Papadimitriou, C.H.: The complexity of coloring circular arcs and chords. SIAM J. Algebraic Discrete Methods **1**(2), 216–227 (1980)
17. Gavril, F.: Maximum weight independent sets and cliques in intersection graphs of filaments. Inf. Process. Lett. **73**(5–6), 181–188 (2000)
18. Goldreich, O., Micali, S., Wigderson, A.: Proofs that yield nothing but their validity or all languages in np have zero-knowledge proof systems. J. ACM (JACM) **38**(3), 690–728 (1991)
19. Goldwasser, S., Micali, S., Rackoff, C.: The knowledge complexity of interactive proof systems. SIAM J. Comput. **18**(1), 186–208 (1989)
20. Golumbic, M.C.: Algorithmic graph theory and perfect graphs. Elsevier (2004)
21. Göös, M., Suomela, J.: Locally checkable proofs in distributed computing. Theory Comput. **12**(1), 1–33 (2016)
22. Halldórsson, M.M., Konrad, C.: Improved distributed algorithms for coloring interval graphs with application to multicoloring trees. Theoretical Comput. Sci. **811**, 29–41 (2020)
23. Kanté, M.M., Limouzy, V., Mary, A., Nourine, L., Uno, T.: On the enumeration and counting of minimal dominating sets in interval and permutation graphs. In: Cai, L., Cheng, S.-W., Lam, T.-W. (eds.) ISAAC 2013. LNCS, vol. 8283, pp. 339–349. Springer, Heidelberg (2013). https://doi.org/10.1007/978-3-642-45030-3_32
24. Kaplan, H., Nussbaum, Y.: A simpler linear-time recognition of circular-arc graphs. In: Arge, L., Freivalds, R. (eds.) SWAT 2006. LNCS, vol. 4059, pp. 41–52. Springer, Heidelberg (2006). https://doi.org/10.1007/11785293_7
25. Kaplan, H., Shamir, R.: Pathwidth, bandwidth, and completion problems to proper interval graphs with small cliques. SIAM J. Comput. **25**(3), 540–561 (1996)
26. Ton Kloks. Treewidth of circle graphs. In International Symposium on Algorithms and Computation, pages 108–117. Springer, 1993
27. Gillat Kol, Rotem Oshman, and Raghuvansh R Saxena. Interactive distributed proofs. In ACM Symposium on Principles of Distributed Computing, pages 255–264. ACM, 2018
28. Konrad, C., Zamaraev, V.: Distributed minimum vertex coloring and maximum independent set in chordal graphs. In 44th International Symposium on Mathematical Foundations of Computer Science (MFCS 2019). Schloss Dagstuhl-Leibniz-Zentrum fuer Informatik (2019)
29. Korman, A., Kutten, S., Peleg, D.: Proof labeling schemes. Distrib. Comput. **22**(4), 215–233 (2010)

30. Kratsch, D., McConnell, R.M., Mehlhorn, K., Spinrad, J.P.: Certifying algorithms for recognizing interval graphs and permutation graphs. SIAM J. Comput. **36**(2), 326–353 (2006)
31. Lappas, E., Nikolopoulos, S.D., Palios, L.: An o (n)-time algorithm for the paired domination problem on permutation graphs. Eur. J. Combinatorics **34**(3), 593–608 (2013)
32. Lund, C., Fortnow, L., Karloff, H., Nisan, N.: Algebraic methods for interactive proof systems. J. ACM (JACM) **39**(4), 859–868 (1992)
33. Ma, T.-H., Spinrad, J.P.: On the 2-chain subgraph cover and related problems. J. Algorithms **17**(2), 251–268 (1994)
34. McConnell, R.M.: Linear-time recognition of circular-arc graphs. Algorithmica **37**(2), 93–147 (2003)
35. McKee, T.A., McMorris, F.R.: Topics in Intersection Graph Theory. Society for Industrial and Applied Mathematics, January 1999. https://doi.org/10.1137/1.9780898719802
36. Montealegre, P., Ramírez-Romero, D., Rapaport, I.: Shared vs private randomness in distributed interactive proofs. LIPIcs, vol. 181, pp. 51:1–51:13. Schloss Dagstuhl - Leibniz-Zentrum für Informatik (2020)
37. Naor, M., Parter, M., Yogev, E.: The power of distributed verifiers in interactive proofs. In: ACM-SIAM Symposium on Discrete Algorithms, pp. 1096–115. SIAM (2020)
38. Naor, M., Stockmeyer, L.: What can be computed locally? SIAM J. Comput. **24**(6), 1259–1277 (1995)
39. Pergel, M.: Recognition of Polygon-Circle Graphs and Graphs of Interval Filaments Is NP-Complete. In: Brandstädt, A., Kratsch, D., Müller, H. (eds.) WG 2007. LNCS, vol. 4769, pp. 238–247. Springer, Heidelberg (2007). https://doi.org/10.1007/978-3-540-74839-7_23
40. Shamir, A.: Ip= pspace. J. ACM (JACM) **39**(4), 869–877 (1992)
41. Spinrad, J.: Recognition of circle graphs. J. Algorithms **16**(2), 264–282 (1994)
42. Tiskin, A.: Fast distance multiplication of unit-monge matrices. Algorithmica **71**(4), 859–888 (2015)
43. Yamazaki, K., Saitoh, T., Kiyomi, M., Uehara, R.: Enumeration of nonisomorphic interval graphs and nonisomorphic permutation graphs. Theoret. Comput. Sci. **806**, 310–322 (2020)

Exactly Optimal Deterministic Radio Broadcasting with Collision Detection

Koko Nanahji(✉)

University of Toronto, Toronto, ON M5S1A4, Canada
koko.nanahji@gmail.com

Abstract. We consider the broadcast problem in synchronous radio broadcast models with collision detection. One node of the network is given a message that must be learned by all nodes in the network. We provide a deterministic algorithm that works on the beeping model, which is a restricted version of the radio broadcast model with collision detection. This algorithm improves on the round complexity of previous algorithms. We prove an exactly matching lower bound in the radio broadcast model with collision detection. This shows that the extra power provided by the radio broadcast model with collision detection does not help improve the round complexity.

Keywords: Broadcast · Distributed · Radio network · Collision Detection · Beep Model · Encoding

1 Introduction

Broadcast is a fundamental problem in distributed computing that is frequently used as a building block in other problems. We study the broadcast problem in synchronous radio networks, modeled as undirected connected graphs, where nodes do not have identifiers and, initially, have no knowledge about the network. In this problem, one node of the network is given a message that must be learned by all nodes in the network.

In the radio broadcast model with collision detection, a listening node receives the entire message of its neighbour if that neighbour is its only neighbour which transmits that round. If more than one of its neighbours transmit, then it receives a collision signal. The beeping model is a restricted version in which nodes can only transmit the collision signal.

Chlebus, Gąsieniec, Gibbons, Pelc, and Rytter [6], provide a deterministic broadcasting algorithm that works on the beeping model. This algorithm takes $\Theta(D \cdot \log_2 \mu)$ rounds to broadcast any message $\mu \in \mathbb{Z}^+$, where D is the source eccentricity of the network. Czumaj and Davies [12] and, independently, Hounkanli and Pelc [18], provide an asymptotically optimal deterministic broadcasting algorithm in the beeping model that takes $D + 6 \cdot \lceil \log_2 \mu \rceil + 11$ rounds to broadcast any message $\mu \in \mathbb{Z}^+$. They also prove a lower bound of $\Omega(D + \log_2 \mu)$ rounds.

M. Parter (Ed.): SIROCCO 2022, LNCS 13298, pp. 234–252, 2022.
https://doi.org/10.1007/978-3-031-09993-9_13

In this paper, we provide a deterministic algorithm to broadcast any value from a predefined set of values that has exactly optimal round complexity. In particular, in Sect. 4, we provide a deterministic algorithm to broadcast any value from set $\{1, \ldots, m\}$ that works on the beeping model and takes $D + r(m)$ rounds, where $r(m) \leq 2 \cdot \lceil \log_2 m \rceil + 2$. In Sect. 5, we prove that $D + r(m)$ is a lower bound in the radio broadcast model with collision detection. This shows that allowing nodes to send arbitrarily long messages instead of just a collision signal does not help improve the round complexity.

2 Model

There are a number of variants of the synchronous radio broadcast model that are considered. The three most relevant to this paper are the following:

- **Radio broadcast model without collision detection.** In this model, the nodes communicate in synchronous rounds. At each round, a node can idle, listen, or transmit the same value to all its neighbours. If a node transmits or idles, it gets no feedback from the environment at that round. If a node listens and none of its neighbors transmit or at least two of its neighbors transmit, it receives nothing. If a node listens and exactly one of its neighbors transmits, then it receives the value transmitted by that neighbour.
- **Radio broadcast model with collision detection.** This model is same as the previous model except that, if at least two neighbors of a node transmit, it receives a collision signal. Thus, a node that listens can distinguish between none of its neighbours transmitting and at least two of its neighbours transmitting.
- **Beeping model.** This is a special case of the radio broadcast model with collision detection, in which the only message a node can transmit is the collision signal (also called a beep). Thus, in the beeping model, when a node listens, it receives a beep if at least one of its neighbours transmits at that round.

We assume that, initially, the nodes are indistinguishable and have no knowledge about the network. Each node starts by listening (without recording any information) up to and including the first round in which at least one of its neighbours transmits. We say that a node *wakes up* at some round t, if at least one of its neighbours transmits for the first time at round t. A node that wakes up at round t starts executing the given algorithm at round $t + 1$.

In the broadcast problem, an external source wakes up one of the nodes in the network by giving it a value from a finite set of possible values. This node is called the source node. We define round 1 of an execution to be the round immediately following the round in which the source node is woken up. The nodes at distance ℓ from the source node are said to be at level ℓ. The maximum level of any node in the network is called the source eccentricity.

3 Related Work

Early approaches to solve the broadcast problem used randomization. Bar-Yehuda, Goldreich, and Itai [2] were first to provide a randomized algorithm for the broadcasting problem in the radio broadcast model without collision detection. They assumed that nodes are anonymous (i.e., they do not have any identifiers) and they are given the size of the network and the maximum number of neighbours of any node in the network. Their algorithm takes $\mathcal{O}(D \cdot \log n + \log^2 n)$ rounds with high probability, where n is the size of the network and D is the source eccentricity of the network. Later, Czumaj and Rytter [13] and, independently, Kowalski and Pelc [19] provided a randomized broadcasting algorithm where nodes have distinct identifiers and they are given the size of the network. Their algorithm takes expected $\mathcal{O}(D \log \frac{n}{D} + \log^2 n)$ rounds. Kushilevitz and Mansour [21] proved an expected $\Omega(D \log \frac{n}{D})$ round lower bound, when nodes only know the size and the diameter of the network. Alon, Bar-Noy, Linial, and Peleg [1] proved an $\Omega(\log^2 n)$ round lower bound, even if nodes have distinct identifiers and every node knows the entire network. Thus, the expected round complexity of the broadcast problem without collision detection is in $\Theta(D \log \frac{n}{D} + \log^2 n)$ when nodes have distinct identifiers and are given the size of the network.

Ghaffari, Haeupler, Khabbazian [17] studied randomization in the radio broadcast model with collision detection. They provided a broadcasting algorithm that takes $\mathcal{O}(D + \log^6 n)$ rounds, with high probability, in which the nodes are given the size of the network and the source eccentricity of the network. They also presented a broadcasting algorithm that takes $\mathcal{O}(D + \log^2 n)$ rounds, with high probability, in which the nodes know the entire network topology.

Deterministic approaches have also been studied. Most of the papers about the deterministic broadcast problem are for the radio broadcast model without collision detection. Initially, researchers considered the case in which each node has a distinct identifier. Some papers assume that nodes are not given any information about the network, except possibly the size of the network n, the maximum number of neighbours of any node in the network Δ, and the source eccentricity of the network D. There are many broadcasting algorithms for this setting [6–9]. The fastest such algorithm was provided by Czumaj and Davies [11], which takes $\mathcal{O}(n \log D \log \log \frac{D\Delta}{n})$ rounds.

Other papers consider algorithms designed with knowledge of the entire network topology. They also assume that nodes have distinct identifiers. Chlamtac [5] provided a deterministic algorithm for the broadcast problem that takes $\mathcal{O}(D \cdot \log^2 n)$ rounds. Kowalski and Pelc [20] provided an optimal deterministic algorithm, which takes $\mathcal{O}(D + \log^2 n)$ rounds.

Another approach was to carefully assign short labels to the nodes of the network, instead of assuming they have distinct identifiers. Ellen, Gorain, Miller, and Pelc [15] showed that broadcast can be done in any radio broadcast network after assigning 2-bit labels to each node. Their algorithm takes $\Theta(n)$ rounds. They also showed that 1-bit labels were sufficient for a class of networks. Gewu, Potop-Butucaru, Rabie [3] proved that 1-bit labels are sufficient to solve the broadcast problem for a larger class of networks. Ellen and Gilbert [14] provide an algorithm that uses 3-bit labels and takes $\mathcal{O}(D \log^2 n)$ rounds and an algorithm that uses 4-bit labels and takes $\mathcal{O}(\sqrt{Dn})$ rounds. Ellen, Gorain, Miller, and Pelc [15] observed that, without identifiers and without collision detection, it is impossible to solve the broadcast problem deterministically in some radio networks because it is impossible to break symmetry.

The known broadcasting algorithms for the beeping model [6,12,18] do not assume that nodes have identifiers. Some of the techniques used in these algorithms are closely related to techniques used in our paper, so we describe them in more detail.

In the algorithm by Chlebus, Gąsieniec, Gibbons, Pelc, and Rytter [6], all bits of the message are transmitted to nodes at the same level before any bits of this message are transmitted to nodes that are at higher levels. The algorithm to broadcast message μ is executed in phases of $\mathcal{O}(\log_2 \mu)$ rounds. At each phase $1 \leq i \leq D$, the nodes at level $i-1$ transmit the message (bit by bit) to the nodes at level i. Thus, this algorithm takes $(D \cdot \log_2 \mu)$ rounds.

The algorithms presented by Czumaj and Davies [12] and Hounkanli and Pelc [18] are very similar. Both rely on the *Beep Waves* method, introduced by Ghaffari and Haeupler [16], to relay a message using beeps. The idea is to broadcast the bits of the message level by level in a pipelined fashion. At every third round, starting with round 1, the source node announces a bit of the message. The source node beeps at that round if and only if the bit is 1. When any other node learns a bit of the message, it conveys the value of that bit to its neighbours at the next level during the following round. Then it idles for one round, ignoring the information those neighbours send. In particular, nodes at level $\ell \geq 1$ learn the ith bit of the message at round $\ell + 3(i-1)$. During the execution of Beep Waves, a beep of the source node propagates like a wave throughout the network, spreading to the next level at each round.

In all three algorithms, a non-source node learns the message bit by bit over several rounds and, hence, it needs a mechanism to detect the end of the message. If the set of messages is $\{1, \ldots, m\}$, it suffices for the source node to send the $\lceil \log_2 m \rceil$ bits of a binary encoding of the message. However, if the message can be any positive integer, the encoding of the message must be self-delimiting. For example, Czumaj and Davies [12] start the encoding with 10, then duplicate each bit in the binary representation of the message, and end the encoding with 10. Hence, binary representation of 5 becomes 1011001110 in their encoding.

4 Algorithm

First, we describe an algorithm in the beeping model in which the source node can broadcast any value from a prefix-free set of binary strings. The algorithm uses a variant of the *Beep Waves* method. However, the pipelining operates slightly differently, where transmitting a 0 takes fewer rounds. Later, we design a new encoding scheme which takes this into account. The combination of the algorithm and the encoding scheme is shown to be optimal in the next section.

Given a binary string x, let $|x|$ denote the length of x, let x_i denote the ith character of x for $1 \leq i \leq |x|$, and let $pre_j(x)$ be the prefix of length j of x, where $0 \leq j \leq |x|$.

Let S be a prefix-free set of binary strings (i.e. no element of S is a prefix of another element of S). First, we provide a detailed description of the algorithm to broadcast any string $s \in S$.

Algorithm 1: Algorithm for source node to broadcast s after waking up

```
Beep
Listen
if received nothing then Terminate
Idle
for i from 1 to |s| − 1 do
    if s_i = 1
    then Beep
         Idle
         Idle
    else Idle
if s_|s| = 1 then Beep
Terminate
```

Algorithm 2: Algorithm for a non-source node v after being woken up

```
acc_v ⟵ ϵ
Beep
Listen
if received nothing then terminal_v ⟵ True else terminal_v ⟵ False
while acc_v ∉ S do
    Listen
    if received signal
    then acc_v ⟵ acc_v · 1
         if acc_v ∉ S
         then Beep
              Idle
         else if terminal_v = False then Beep
    else acc_v ⟵ acc_v · 0
Terminate
```

The pseudocode for the source node is presented in Algorithm 1. In the first round after it is woken up, the source node beeps. In the second round, the source node listens. If it receives nothing, then it terminates. Otherwise, it idles for one round and then, for $1 \leq i \leq |s| - 1$, it does one of the following. If $s_i = 0$, then the source node idles. If $s_i = 1$, then the source node beeps and idles for two rounds. Finally, if the last character of s is 1, then the source node beeps.

The pseudocode for an arbitrary non-source node, v, is presented in Algorithm 2. Its local variable acc_v is used to accumulate the characters of s and is initially set to the empty string. Once v is woken up, it beeps in the next round and listens in the following round. It receives a beep in this round if and only if it is connected to a node at a higher level. This information is recorded in its local variable $terminal_v$.

While node v has not yet received the entire message, it repeatedly executes phases. Each phase starts by a listen round, in which it learns the next bit of the message. Specifically, receiving nothing means this bit is 0 and receiving a beep means this bit is 1. In the latter case, it beeps at the next round to relay this information to the nodes at the next level and then idles for one round. In the last phase, immediately after it has learned that the last bit of the message is 1, a node that is not connected to a node at a higher level can immediately terminate. If it is connected to a node at a higher level, then it can terminate after beeping.

Now, we prove the correctness and bound the round complexity of this algorithm when broadcasting message s. Since each node beeps after waking up, the nodes at level ℓ wake up at round ℓ. Moreover, if there are nodes at level 1, then the source node receives a signal at round 2 and, hence, continues with the rest of its algorithm. Likewise, a non-source node v sets $terminal_v$ to *True* if and only if it is connected to a node at a higher level.

Let $C(x)$ be the number of 0's in the binary string x plus three times the number of 1's in x. In particular, $C(x) = 0$, if x is the empty string.

Observe that, the source node executes 3 rounds before executing the first iteration of the loop. If $s_i = 0$, then the ith iteration of the loop takes $C(s_i) = 1$ round. If $s_i = 1$, then the ith iteration of the loop takes $C(s_i) = 3$ rounds. Therefore, $3 + C(pre_{i-1}(s)) + 1$ is the first round of the ith iteration, for $1 \leq i \leq |s| - 1$.

Observation 1. *Suppose the source node does not terminate at the end of round 2. Then, for all* $1 \leq i \leq |s|$, *the source node beeps at round* $3 + C(pre_{i-1}(s)) + 1$ *if and only if* $s_i = 1$. *Furthermore, the source node terminates at round* $3 + C(pre_{|s|-1}(s)) + s_{|s|}$.

We will show that when $D > 0$, all nodes terminate by the end of round $D + C(pre_{|s|-1}(s)) + 3$.

In the next two results, we identify the rounds in which each node learns each bit of the message. First, we identify the round in which each node learns the first bit of s. Then, we generalize it to all bits of s.

Lemma 1. *For all* $1 \leq \ell \leq D$, *each node* v *at level* ℓ *appends a character to* acc_v *for the first time at round* $\ell + 3$ *and the appended character is* s_1.

Proof. The source node beeps at round 4 if and only if $s_1 = 1$. Each node u at level 1 wakes up at round 1 and appends a character to acc_u for the first time at round 4. The nodes at level 2 wake up at round 2 and listen at round 4. Therefore, nodes at level 1 receive a signal at round 4 if and only if $s_1 = 1$. Since u appends 1 to acc_u if it receives a signal and appends 0 if it receives nothing, u appends s_1 to acc_u at round 4.

Let $2 \leq \ell \leq D$ and assume the claim is true for $\ell - 1$. In particular, each node w at level $\ell - 1$ appends s_1 to acc_w at round $\ell + 2$. From the pseudocode, if $terminal_w = False$, then w beeps at round $\ell + 3$ if and only if $s_1 = 1$. Since nodes at level ℓ wake up at round ℓ, each node v at level ℓ listens and appends 0 or 1 to acc_v for the first time at round $\ell + 3$. If $\ell < D$, the nodes at level $\ell + 1$ wake up at round $\ell + 1$ and listen at round $\ell + 3$. Therefore, v receives a signal at round $\ell + 3$ if and only if $s_1 = 1$. Since v appends 1 to acc_v if it receives a signal and appends 0 if it receives nothing, v appends s_1 to acc_v at round $\ell + 3$.

Lemma 2. *For all $1 \leq j \leq |s|$, each node v at level $1 \leq \ell \leq D$ appends a character to acc_v for the jth time at round $\ell + C(pre_{j-1}(s)) + 3$ and the appended character is s_j.*

Proof. We will prove the claim by induction on $1 \leq j \leq |s|$ and $1 \leq \ell \leq D$. Note that, if $j = 1$, then $C(pre_{j-1}(s)) = 0$. Thus, by Lemma 1, the claim holds for all $1 \leq \ell \leq D$ when $j = 1$.

Let $2 \leq j' \leq |s|$ and $1 \leq \ell' \leq D$. Assume the claim holds for $1 \leq \ell \leq D$ when $j = j' - 1$ and for $1 \leq \ell \leq \ell' - 1$ when $j = j'$.

Let v be a node at level ℓ'. It follows from the induction hypothesis that v appends $s_{j'-1}$ to acc_v at round $\ell' + C(pre_{j'-2}(s)) + 3$ and the value of acc_v becomes $pre_{j'-1}(s)$. Since S is prefix-free, $pre_{j'-1}(s) \notin S$ at the end of round $\ell' + C(pre_{j'-2}(s)) + 3$. From the pseudocode, if $s_{j'-1} = 0$, then $C(s_{j'-1}) = 1$ round later, v listens and appends a character to acc_v. If $s_{j'-1} = 1$, then $C(s_{j'-1}) = 3$ rounds later, v next listens and appends a character to acc_v. Thus, v appends the next character to acc_v at round $\ell' + C(pre_{j'-2}(s)) + 3 + C(s_{j'-1}) = \ell' + C(pre_{j'-1}(s)) + 3$.

Let u be a node at level $\ell' + 1$. By the induction hypothesis, u appends $s_{j'-1}$ to acc_u at round $\ell' + 1 + C(pre_{j'-2}(s)) + 3$. Thus, if $s_{j'-1} = 0$, then u listens at round $\ell' + 1 + C(pre_{j'-2}(s)) + 3 = \ell' + C(pre_{j'-1}(s)) + 3$. If $s_{j'-1} = 1$, then, u idles at round $2 + \ell' + 1 + C(pre_{j'-2}(s)) + 3 = \ell' + C(pre_{j'-1}(s)) + 3$. Hence, nodes at level $\ell' + 1$ do not beep at round $\ell' + C(pre_{j'-1}(s)) + 3$.

If $\ell' = 1$, then, by Observation 1, the source node beeps at round $3 + C(pre_{j'-1}(s)) + 1 = \ell' + C(pre_{j'-1}(s)) + 3$ if and only if $s_{j'} = 1$. So, suppose $\ell' \geq 2$. By the induction hypothesis, each node w at level $\ell' - 1$ appends $s_{j'}$ to acc_w at round $\ell' - 1 + C(pre_{j'-1}(s)) + 3$. From the pseudocode, if $terminal_w = False$, w beeps at round $\ell' + C(pre_{j'-1}(s)) + 3$ if and only if $s_{j'} = 1$. Therefore, v receives a signal at round $\ell' + C(pre_{j'-1}(s)) + 3$ if and only if $s_{j'} = 1$. Since v appends 1 to acc_u if it receives a signal and appends 0 if it receives nothing, v appends $s_{j'}$ to acc_v at round $\ell' + C(pre_{j'-1}(s)) + 3$.

Now, we show that all nodes learn s and terminate by the end of round $D + C(pre_{|s|-1}(s)) + 3$.

Lemma 3. *For all $1 \leq \ell \leq D$, nodes at level ℓ learn the value of s at round $\ell + C(pre_{j-1}(s)) + 3$. Each node v at level ℓ terminates at round $\ell + C(pre_{|s|-1]}(s)) + 3 + s_{|s|}$ if $terminal_w = False$ and terminates at round $\ell + C(pre_{|s|-1}(s)) + 3$ if $terminal_w = True$.*

Proof. Let $1 \leq \ell \leq D$ and let v be a node at level ℓ. By Lemma 2, $acc_v = s$ at the end of round $\ell + C(pre_{j-1}(s)) + 3$. If $s_{|s|} = 0$ or $terminal_v = True$, then v terminates at the end of this round. Otherwise, v terminates one round later.

Let $m \geq 2$. We now explain how to modify the algorithm to broadcast any value from $\{1, \ldots, m\}$. Since each value of an arbitrary set of size m can be mapped to a value in $\{1, \ldots, m\}$, this method can be used to broadcast a value from an arbitrary set of size m.

First, we recursively construct a prefix-free set of binary strings W_i such that, for all $i \geq 3$ and all $w \in W_i$, $C(pre_{|w|-1}(w)) \leq i - 3$. Let $W_0 = \{0\}$, $W_1 = W_2 = \{1\}$, $W_3 = \{0, 1\}$, $W_4 = \{00, 01, 1\}$ and $W_5 = \{000, 001, 01, 1\}$. For all $i \geq 6$, W_i consists of the strings in W_{i-1} prepended by 0 (denoted by $0 \cdot W_{i-1}$) and the strings in W_{i-3} prepended by 1 (denoted by $1 \cdot W_{i-3}$). For all $i \geq 0$, let $w_i = |W_i|$. Then, $w_0 = w_1 = w_2 = 1$ and, for all $i \geq 3$, $w_i = w_{i-1} + w_{i-3}$. Note that, $w_0, w_1, w_2, \ldots$ is Narayana's cows sequence [23]. We show that W_i is prefix-free.

Lemma 4. *For all $i \geq 0$, there is no string in W_i that is a prefix of any other string in w_i.*

Proof. By inspection, the claim holds for $0 \leq i \leq 5$. Let $i \geq 6$ and assume the claim is true for W_{i-1} and W_{i-3}. Hence, no string in $0 \cdot W_{i-1}$ is a prefix of any other string in $0 \cdot W_{i-1}$. Similarly, no string in $1 \cdot W_{i-3}$ is a prefix of any other string in $1 \cdot W_{i-3}$. Since $W_i = 0 \cdot W_{i-1} \cup 1 \cdot W_{i-3}$, no string in W_i is a prefix of any other string in W_i.

Next, we describe the relationship between $w \in W_i$ and the function C for all $i \geq 3$.

Lemma 5. *For all $i \geq 3$, all $w \in W_i$, $C(pre_{|w|-1}(w)) \leq i - 3$.*

Proof. By inspection, the claim holds for $3 \leq i \leq 5$. Let $i \geq 6$ and assume the claim is true for W_{i-1} and W_{i-3}. Let $w \in W_i$ be arbitrary. If $w = 0 \cdot w'$, then $w' \in W_{i-1}$. By the induction hypothesis, $C(pre_{|w'|-1}(w')) \leq i - 4$. Hence, $C(pre_{|w|-1}(w)) \leq i - 3$. If $w = 1 \cdot w'$, then $w' \in W_{i-3}$. By the induction hypothesis, $C(pre_{|w'|-1}(w')) \leq i - 6$. Hence, $C(pre_{|w|-1}(w)) \leq i - 3$.

Next, we describe how to modify the algorithm to broadcast a value from $\{1, \ldots, m\}$. Let r be the smallest value such that $w_r \geq m$. Since $m \geq 2$, it follows that $r \geq 3$. Let $S = W_r = \{s_1, s_2, \ldots, s_{w_r}\}$. To broadcast $\mu \in \{1, \ldots, m\}$, the source node sets $s = \mathcal{M}(\mu)$ and performs the steps in Algorithm 1. Each non-source node performs steps in Algorithm 2, but when it terminates it decodes the message as $\mathcal{M}^{-1}(acc_v)$. By observation 1 and Lemma 3, all nodes terminate within $D + C(pre_{|\mathcal{M}(\mu)|-1]}(\mathcal{M}(\mu))) + 3$ rounds. Since $r \geq 3$, Lemma 5 implies that $C(pre_{|\mathcal{M}(\mu)|-1]}(\mathcal{M}(\mu))) \leq r - 3$. Thus, all nodes learn μ and terminate within $D + r$ rounds.

It is known that $w_i = \lfloor dc^i + \frac{1}{2} \rfloor$, where $c \approx 1.4656$ is the real root of $x^3 - x^2 - 1$ and $d \approx 0.6115$ is the real root of $31 \cdot x^3 - 31 \cdot x^2 + 9 \cdot x - 1$ [10]. Thus $\lfloor dc^{r-1} \rfloor = w_{r-1} < m$. Since m is an integer, $dc^{r-1} < m$, so $r < \log_c m - \log_c d + 1 < 2 \log_2 m + 3$. Since r is an integer, $r \leq 2\lceil \log_2 m \rceil + 2$.

Theorem 1. *Let $m \geq 2$ and let $r \geq 3$ be the smallest value such that $w_r \geq m$. Consider the algorithm in Algorithms 1 and 2 that enables the source node to broadcast any message from $\{1, \ldots, m\}$. For all $D \geq 1$ and all $\mu \in \{1, \ldots, m\}$, all nodes learn μ and terminate within $D + r \leq D + 2\lceil \log_2 m \rceil + 2$ rounds during the execution of the algorithm with message μ on every graph of source eccentricity D.*

5 Lower Bound

It is convenient for the proof of the lower bound to strengthen the model by assuming that each node knows its level and whether that level is the final level of the graph. We also assume that the source node is fixed. Since a node that listens can simulate idling by throwing away any information it receives, we assume that nodes either listen or transmit at each round.

We define a family of graphs used in the proof. For all $D \geq 2$, let E_D be the graph with a source node and two nodes at each level 1 through D, such that, for all $0 \leq \ell \leq D - 1$, each node at level ℓ is connected to both nodes at level $\ell + 1$. Figure 1 shows graph E_3.

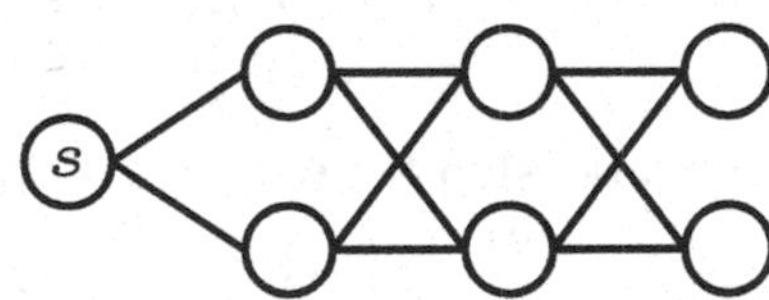

Fig. 1. The graph E_3 with source node s.

Algorithms executing on E_D have some nice properties. In the full version of this paper [22], we consider an arbitrary algorithm $\mathcal{A}$ that does not necessarily perform broadcast, but does have the property that all nodes eventually wake up during the execution of $\mathcal{A}$ with message μ on E_D, for all $D \geq 12$ and all messages $\mu \in M$. We prove that, all nodes at the same level, ℓ, wake up at the same round, $t_{\mathcal{A}}(D, \mu, \ell)$, during the execution of $\mathcal{A}$ with message μ on E_D, for each $D \geq 12$ and each $\mu \in M$. Then we show that, for each level $\ell \geq 2$, all nodes at level ℓ transition to the same state, $\mathbf{c}_{\mathcal{A}}(\ell, 0)$, when they wake up during the execution of $\mathcal{A}$ with message μ on E_D, for every $\mu \in M$ and every $D \geq 12$, provided ℓ is not the last level of E_D.

The main result in this section is the following theorem, where $w_0, w_1, w_2, \ldots$ is Narayana's cows sequence. It proves that the algorithm in 1 and 2 is optimal in the radio broadcast model with collision detection.

Theorem 2. *Let $m \geq 2$ and let $r \geq 3$ be the smallest value such that $w_r \geq m$. Consider any algorithm $\mathcal{B}$ that enables the source node to broadcast any message from $\{1, \ldots, m\}$. Then there exist $D \geq 12$, a message $\mu \in \{1, \ldots, m\}$, and a non-source node that uses at least $D + r$ rounds during the execution of $\mathcal{B}$ with message μ on E_D.*

Let $m \geq 2$ and let $r \geq 3$ be the smallest value such that $w_r \geq m$. Since $w_{r-1} < m \leq w_r$, an algorithm that enables the source node to broadcast any message from $\{1, \ldots, m\}$ also enables the source node to broadcast any message from $\{1, \ldots, w_{r-1} + 1\}$. Therefore, it suffices to assume that $m = w_{r-1} + 1$.

To obtain a contradiction, assume there exists an algorithm $\mathcal{B}$ that enables the source node to broadcast any message from $\{1, \ldots, w_{r-1} + 1\}$ such that, for all $D \geq 12$ and all $\mu \in \{1, \ldots, w_{r-1} + 1\}$, all non-source nodes terminate within $D + r - 1$ rounds during the execution of $\mathcal{B}$ with message μ on E_D.

From algorithm $\mathcal{B}$, we construct an algorithm $\mathcal{B}'$, which still terminates within $D + r - 1$ rounds, but the nodes at each level wake up as early as possible. This reduces the number of cases that will have to be considered. In particular, nodes at level D wake up at round D of $\mathcal{B}'$ and, hence, distinguish between all $w_{r-1} + 1$ messages within $r - 1$ rounds after waking up. We show that, for all $2 \leq \ell \leq D - 1$, during $\mathcal{B}'$, all nodes at level ℓ also distinguish between all source messages within $r - 1$ rounds after waking up. These properties are formalised in the following definition.

Definition 1. *An algorithm $\mathcal{A}$ has property $P(k)$ if, for all $D \geq 12$,*

1. *for all $\mu \in \{1, \ldots, w_{k-1} + 1\}$ and for all $1 \leq \ell \leq D$, the nodes at level ℓ wake up at round ℓ during the execution of $\mathcal{A}$ with message μ on E_D and,*
2. *for all distinct $\mu, \mu' \in \{1, \ldots, w_{k-1} + 1\}$ and for all $2 \leq \ell \leq D - 1$, there exists $0 \leq t_\ell \leq k - 1$ such that the nodes at level ℓ are in different states at the end of round $\ell + t_\ell$ during the executions of $\mathcal{A}$ with messages μ and μ' on E_D.*

First, we will explain why algorithm $\mathcal{B}'$ has property $P(r)$. The proof of this result appears in the full version of this paper [22]. Then, we inductively prove that, for all $k \geq 3$, no algorithm has property $P(k)$. Since algorithm $\mathcal{B}'$ has property $P(r)$, this contradicts the existence of algorithm $\mathcal{B}'$ and, hence, algorithm $\mathcal{B}$ does not exist.

First, we construct algorithm $\mathcal{B}'$ and show that it satisfies condition 1 of property $P(r)$.

Lemma 6. *There exists an algorithm $\mathcal{B}'$ that enables the source node to broadcast any message from $\{1, \ldots, w_{r-1}+1\}$, such that, for all $D \geq 12$, all $1 \leq \ell \leq D$ and all $\mu \in \{1, \ldots, w_{r-1} + 1\}$, the nodes at level ℓ wake up at round ℓ and all non-source nodes terminate by the end of round $D + r - 1$ during the execution of $\mathcal{B}'$ with message μ on E_D.*

Proof Sketch. We show that there are finitely many levels $\ell \geq 2$ such that nodes in state $\mathbf{c}_{\mathcal{B}}(\ell, 0)$ listen. Let $\ell_0 \geq 9$ be such that nodes in state $\mathbf{c}_{\mathcal{B}}(\ell, 0)$ transmit for all $\ell \geq \ell_0$. Since all non-source nodes terminate within $\ell_0 + D + r - 1$ rounds during the execution of $\mathcal{B}$ on E_{ℓ_0+D} and nodes at level $\ell_0 + \ell$ wake up at or after round $\ell_0 + \ell$, the nodes at level $\ell_0 + \ell$ terminate within $D - \ell + r - 1$ rounds after waking up. We also show that, for all $D \geq 12$ and all $0 \leq \ell \leq D$, the nodes at level $\ell_0 + \ell$ wake up at round $t_{\mathcal{B}}(\ell_0 + D, \mu, \ell_0 + \ell) = t_{\mathcal{B}}(\ell_0 + 1, \mu, \ell_0) + \ell$ during the execution of $\mathcal{B}$ with message μ on E_{ℓ_0+D}.

We construct algorithm $\mathcal{B}'$ so that, for all $D \geq 12$, every execution of $\mathcal{B}'$ with a message $\mu \in \{1, \ldots, w_{r-1} + 1\}$ on E_D simulates the execution of $\mathcal{B}$ with the same message on E_{ℓ_0+D}: During the execution of $\mathcal{B}'$ on E_D, the source node simulates levels 0 through ℓ_0 in E_{ℓ_0+D} and, each non-source node, from when it wakes up, simulates a node ℓ_0 levels higher in E_{ℓ_0+D}, from when it wakes up. In $\mathcal{B}'$, each node transmits the round after it wakes up and, hence, the nodes at level ℓ wake up at round ℓ for $1 \leq \ell \leq D$. More generally, during the execution of $\mathcal{B}'$ with message μ on E_D, for all $t \geq 0$, at the end of round $\ell + t$, the nodes at level ℓ are in the same state as the nodes at level $\ell_0 + \ell$ at the end of round $t_{\mathcal{B}}(\ell_0 + 1, \mu, \ell_0) + \ell + t$ in the execution of $\mathcal{B}$ with message μ on E_{ℓ_0+D}. Since the nodes at level $\ell_0 + \ell$ terminate within $D - \ell + r - 1$ rounds after waking up during the execution of $\mathcal{B}$ with message μ on E_{ℓ_0+D}, the nodes at level ℓ terminate within $D - \ell + r - 1$ rounds after waking up (at round ℓ) during the execution of $\mathcal{B}'$ with message μ on E_D.

Next, we show that algorithm $\mathcal{B}'$ also satisfies condition 2 of property $P(r)$. In particular, for any pair of messages in $\{1, \ldots, w_{r-1} + 1\}$, the nodes at level $2 \leq \ell \leq D - 1$ must be able to distinguish between them within $r - 1$ rounds after waking up.

Lemma 7. *For all distinct $\mu, \mu' \in \{1, \ldots, w_{r-1} + 1\}$, all $D \geq 12$ and all $2 \leq \ell \leq D - 1$, there exists $0 \leq t_\ell \leq r - 1$ such that, the nodes at level ℓ are in different states at the end of round $\ell + t_\ell$ during the executions of $\mathcal{B}'$ with messages μ and μ' on E_D.*

Proof Sketch. To obtain a contradiction, suppose there exist distinct $\mu, \mu' \in \{1, \ldots, w_{r-1} + 1\}$, $D \geq 12$ and a level $2 \leq \ell \leq D - 1$ such that, for all $0 \leq t \leq r - 1$, the nodes at level ℓ are in the same states at the end of round $\ell + t$ of the execution with message μ on E_D as they are at the end of round $\ell + t$ of the execution with message μ' on E_D. We show that, for all $0 \leq t \leq r - 1$, the nodes at level D are in the same states at the end of round $D + t$ of the execution of $\mathcal{B}'$ with message μ on E_D as they are at the end of round $D + t$ of the execution of $\mathcal{B}'$ with message μ' on E_D. Hence, the nodes at level D cannot distinguish between μ and μ' at the end of first $D + r - 1$ rounds of the execution of $\mathcal{B}'$ with message μ on E_D. This is a contradiction because the nodes at level D terminate within $D + r - 1$ rounds.

Let $k \geq 3$ be the smallest value such that there exists an algorithm $\mathcal{X}$ with property $P(k)$. First, we show that, if algorithm $\mathcal{X}$ has some additional properties, then we can construct another algorithm $\mathcal{X}''$ from $\mathcal{X}$ that satisfies property

$P(k')$, where $k' < k$. We present a high level description of the construction of algorithm $\mathcal{X}''$. A more detailed description appears in the full version of this paper [22].

Lemma 8. *Suppose that there exist a level $2 \leq \ell_0 \leq 4$, a round $3 \leq t_0 \leq 5$, and a set S of size $w_{k-t_0+1}+1$ such that, for all $0 \leq t \leq t_0$, the nodes at level ℓ_0 are in the same state at the end of round $\ell_0 + t$ during the execution of $\mathcal{X}$ with all messages $\mu \in S$ on E_{12}. Then, there exists an algorithm $\mathcal{X}''$ that has property $P(k - t_0 + 2)$.*

Proof Sketch. For all $D \geq 12$, every execution of $\mathcal{X}''$ with message $\mu \in S$ on E_D simulates the execution of $\mathcal{X}$ with the same message on E_{ℓ_0+D+6}.

In $\mathcal{X}''$, each node transmits the round after it wakes up and, hence, the nodes at level ℓ wake up at round ℓ for $1 \leq \ell \leq D$. Thus, $\mathcal{X}''$ satisfies the first condition of property $P(k - t_0 + 2)$.

The source node listens for three rounds after first transmitting. At round 5, the source node performs the same operation (either transmitting or listening) as the nodes at level ℓ_0 at round $\ell_0 + t_0 + 3$ in the execution of $\mathcal{X}$ with message μ on E_{ℓ_0+D+6}. From then on, the source node simulates the nodes at levels 0 through ℓ_0 in $\mathcal{X}$ starting at round $\ell_0 + t_0 + 4$, communicating with the nodes at level 1 during $\mathcal{X}''$ in the same way as the nodes at level ℓ_0 communicate with the nodes at level $\ell_0 + 1$ during $\mathcal{X}$.

For $1 \leq \ell \leq \ell_0 + D - 1$, let $\mathbf{s}(\ell, t)$ denote the state of the nodes at level ℓ at the end of round t of during the execution of $\mathcal{X}$ with the message μ on E_{ℓ_0+D+6}. Nodes at level 1 listen for two rounds after first transmitting and, then, at the end of round 4, they transition to state $\mathbf{s}(\ell_0 + 1, \ell_0 + t_0 + 2)$. Thus, at round 5, nodes at level 1 perform the same operation (transmitting or listening) as the nodes at level $\ell_0 + 1$ at round $\ell_0 + t_0 + 3$. Starting at round 6, the nodes at level 1 simulate the nodes at level $\ell_0 + 1$ starting at round $\ell_0 + t_0 + 4$.

For $2 \leq \ell \leq D - 1$, at round $\ell + 2$, the nodes at level ℓ perform the same operation (transmitting or listening) as the nodes at level $\ell_0+\ell$ at round $\ell_0+\ell+t_0$. At the end of round $\ell+2$, the nodes at level ℓ transition to state $\mathbf{s}(\ell_0+\ell, \ell_0+\ell+t_0)$. Starting at round $\ell+3$, each node at level ℓ simulates a node at level $\ell_0+\ell$ starting at round $\ell_0+\ell+t_0+1$. So, for all $t \geq 0$, at the end of round $\ell+2+t$, the nodes at level ℓ are in state $\mathbf{s}(\ell_0 + \ell, \ell_0 + \ell + t_0 + t)$. Table 1 shows some of the states of nodes at levels 1 through 4 in $\mathcal{X}''$ at the end of rounds 4, 5, and 6.

Table 1. Some of the states of nodes at levels 1 through 4 at rounds 4,5, and 6.

Round	Level 1	Level 2	Level 3	Level 4
4	$\mathbf{s}(\ell_0 + 1, \ell_0 + t_0 + 2)$	$\mathbf{s}(\ell_0 + 2, \ell_0 + t_0 + 2)$		
5	$\mathbf{s}(\ell_0 + 1, \ell_0 + t_0 + 3)$	$\mathbf{s}(\ell_0 + 2, \ell_0 + t_0 + 3)$	$\mathbf{s}(\ell_0 + 3, \ell_0 + t_0 + 3)$	
6	$\mathbf{s}(\ell_0 + 1, \ell_0 + t_0 + 4)$	$\mathbf{s}(\ell_0 + 2, \ell_0 + t_0 + 4)$	$\mathbf{s}(\ell_0 + 3, \ell_0 + t_0 + 4)$	$\mathbf{s}(\ell_0 + 4, \ell_0 + t_0 + 4)$

Starting at round $D+2$, nodes at level D simulate the nodes at levels $\ell_0 + D$ through ℓ_0+D+6 starting at round ℓ_0+D+t_0, communicating with the nodes at level $D-1$ during $\mathcal{X}''$ in the same way as the nodes at level ℓ_0+D communicate with the nodes at level ℓ_0+D-1 during $\mathcal{X}$.

Since $\mathcal{X}$ has property $P(k)$, such that, it follows that, for all distinct $\mu, \mu' \in S$, all $D \geq 12$ and all $2 \leq \ell \leq D-1$, there exists $0 \leq t_{\ell_0+\ell} \leq r-1$ such that, the nodes at level $\ell_0+\ell$ are in different states at the end of round $\ell_0+\ell+t_{\ell_0+\ell}$ during the executions of $\mathcal{X}$ with messages μ and μ' on E_{ℓ_0+D+6}. We show that the nodes at level $\ell_0+\ell$ are in the same states at the end of the first $\ell_0+\ell+t_0$ rounds during the execution of $\mathcal{X}$ with all messages $\mu \in S$ on E_{ℓ_0+D+6}. Hence $t_{\ell_0+\ell} > t_0$. The nodes at level ℓ, at the end of round $\ell+t_{\ell_0+\ell}-t_0+2 \geq \ell+2$ during the execution of $\mathcal{X}''$ on E_D, are in the same state as the nodes at level $\ell_0+\ell$ at the end of round $\ell_0+\ell+t_{\ell_0+\ell}$ during the execution of $\mathcal{X}$ with the same message on E_{ℓ_0+D+6}. Thus, the nodes at level ℓ are in different states at the end of round $\ell+t_{\ell_0+\ell}-t_0+2$ during the executions of $\mathcal{X}''$ with messages μ and μ' on E_D. Since $t_{\ell_0+\ell}-t_0+2 \leq k-t_0+1$ and $|S| = w_{k-t_0+1}+1$, it follows that $\mathcal{X}''$ satisfies the second condition of property $P(k-t_0+2)$.

The first condition of property $P(k)$ implies that, for all $\ell \geq 2$, nodes in state $\mathbf{c}_{\mathcal{X}}(\ell, 0)$ transmit and transition to the same state. We call this state $\mathbf{c}_{\mathcal{X}}(\ell, 1)$. If nodes in state $\mathbf{c}_{\mathcal{X}}(\ell, 1)$ transmit, let $\mathbf{c}_{\mathcal{X}}(\ell, 2)$ be the state to which they transition after transmitting. Otherwise, let $\mathbf{c}_{\mathcal{X}}(\ell, 2)$ be the state to which they transition after receiving a collision signal. If $\ell+1$ is not the last level in the graph, the nodes at level $\ell+1$ transmit at round $\ell+2$, so the nodes at level ℓ receive a collision signal at round $\ell+2$ if they are listening. Thus, if $\ell \leq D-2$, the nodes at level ℓ are in state $\mathbf{c}_{\mathcal{X}}(\ell, 2)$ at the end of round $\ell+2$ during the execution of $\mathcal{X}$ on E_D.

Lemma 9. *For all $D \geq 12$, all $2 \leq \ell \leq D-2$ and all $0 \leq t \leq 2$, the nodes at level ℓ are in the same states at the end of round $\ell+t$ during the execution of $\mathcal{X}$ with all messages $\mu \in \{1, \ldots, w_{k-1}+1\}$ on E_D.*

Hence, $\mathcal{X}$ does not satisfy the second condition of property $P(3)$. Since $\mathcal{X}$ has property $P(k)$, it follows that $k \geq 4$.

We focus on nodes at levels 2, 3, and 4 and get an upper bound on the number of messages that they cannot distinguish from one another three rounds after they wake up.

Lemma 10. *For $2 \leq \ell \leq 4$, there are at most w_{k-2} different messages $\mu \in \{1, \ldots, w_{k-1}+1\}$ such that, during the execution of $\mathcal{X}$ with message μ on E_{12}, the nodes at level ℓ are in the same state at the end of round $\ell+3$.*

Proof. To get a contradiction, suppose there exist a level $2 \leq \ell' \leq 4$, a set $S \subseteq \{1, \ldots, w_{k-1}+1\}$ with $|S| \geq w_{k-2}+1$, and a state q such that, for all $\mu \in S$, the nodes at level ℓ' are in state q at the end of round $\ell'+3$ during the execution of $\mathcal{X}$ with message μ on E_{12}.

During the execution of $\mathcal{X}$ with all messages $\mu \in S$ on E_{12}, the nodes at level ℓ' are in the same state at the end of round $\ell'+t$ for $0 \leq t \leq 2$ and they are in state q at the end of round $\ell'+3$. Thus, the precondition of Lemma 8 holds for $\ell_0 = \ell'$ and $t_0 = 3$. Hence, there is an algorithm with property $P(k-1)$. By definition of k, no such algorithm exists. Thus, $|S| \leq w_{k-2}$.

Recall that, for $2 \leq \ell \leq 4$, the nodes at level ℓ are in state $\mathbf{c}_{\mathcal{X}}(\ell, 2)$ at the end of round $\ell+2$ during the execution of $\mathcal{X}$ for every message $\mu \in \{1, \ldots, w_{k-1}+1\}$ on E_{12}. If nodes in state $\mathbf{c}_{\mathcal{X}}(\ell, 2)$ transmit, nodes at level ℓ transition to the same state at the end of round $\ell+3$. By Lemma 10, there are at most $w_{k-2} < w_{k-1}+1$ different messages such that the nodes at level ℓ are in the same state at the end of round $\ell + 3$. Thus, nodes in state $\mathbf{c}_{\mathcal{X}}(\ell, 2)$ listen. Nodes at levels 2 and greater can receive only a collision signal or nothing while listening during any execution on E_{12}. Thus, there are two states that are reachable by nodes at level ℓ from state $\mathbf{c}_{\mathcal{X}}(\ell, 2)$, we call these states $\mathbf{c}_{\mathcal{X}}(\ell, 3)$ and $\mathbf{c}'_{\mathcal{X}}(\ell, 3)$. Next, we show that nodes in state $\mathbf{c}_{\mathcal{X}}(\ell, 3)$ perform different operation than nodes in state $\mathbf{c}'_{\mathcal{X}}(\ell, 3)$, for $\ell = 2, 3$.

Lemma 11. *For $\ell = 2, 3$, nodes in exactly one of the states $\mathbf{c}_{\mathcal{X}}(\ell, 3)$ and $\mathbf{c}'_{\mathcal{X}}(\ell, 3)$ transmit.*

Proof. To obtain a contradiction, assume that for $\ell = 2$ or $\ell = 3$, nodes in both $\mathbf{c}_{\mathcal{X}}(\ell, 3)$ and $\mathbf{c}'_{\mathcal{X}}(\ell, 3)$ transmit or nodes in both these states listen. Let $\mu, \mu' \in \{1, \ldots, w_{k-1}+1\}$. Let α_μ and $\alpha_{\mu'}$ be the executions of $\mathcal{X}$ with messages μ and μ' on E_{12}, respectively. At the end of round $\ell+3$ of α_μ and $\alpha_{\mu'}$, the nodes at level ℓ are in one of the states $\mathbf{c}_{\mathcal{X}}(\ell, 3)$ and $\mathbf{c}'_{\mathcal{X}}(\ell, 3)$, the nodes at level $\ell+1$ are in state $\mathbf{c}_{\mathcal{X}}(\ell+1, 2)$, and the nodes at level $\ell+2$ are in state $\mathbf{c}_{\mathcal{X}}(\ell+2, 1)$. Since nodes in both of the states $\mathbf{c}_{\mathcal{X}}(\ell, 3)$ and $\mathbf{c}'_{\mathcal{X}}(\ell, 3)$ transmit or nodes in both of these states listen, the nodes at level $\ell+1$ cannot distinguish between executions α_μ and $\alpha_{\mu'}$ at the end of round $\ell+4$. Hence, the nodes at level $\ell+1$ are in the same state at the end of round $\ell+4$ of α_μ and $\alpha_{\mu'}$. Therefore, for all $\mu \in \{1, \ldots, w_{k-1}+1\}$, during the execution of $\mathcal{X}$ with message μ on E_{12}, the nodes at level $\ell+1$ are in the same state at the end of round $(\ell+1)+3$. However, this contradicts Lemma 10 because $w_{k-1}+1 > w_{k-2}$.

For $\ell = 2, 3$, suppose nodes in state $\mathbf{c}_{\mathcal{X}}(\ell, 3)$ transmit and nodes in state $\mathbf{c}'_{\mathcal{X}}(\ell, 3)$ listen. Let $U \subseteq \{1, \ldots, w_{k-1}+1\}$ be the set of messages such that nodes at level 2 are in state $\mathbf{c}_{\mathcal{X}}(2, 3)$ at the end of round 5 during the executions of $\mathcal{X}$ with message $\mu \in U$ on E_{12} and let $U' = \{1, \ldots, w_{k-1}+1\} \setminus U$. By Lemma 10, $|U| \leq w_{k-2}$. Since $|U|+|U'| = w_{k-1}+1$, it follows that $|U'| \geq w_{k-1}+1-w_{k-2} = w_{k-4}+1$. Similarly, $|U'| \leq w_{k-2}$ and $|U| \geq w_{k-4}+1$.

By assumption, nodes in state $\mathbf{c}_{\mathcal{X}}(2, 3)$ transmit and, hence, there can be only one state to which nodes in state $\mathbf{c}_{\mathcal{X}}(2, 3)$ transition. Thus, the nodes at level 2 cannot distinguish between the values in U at the end of the first 4 rounds after they wake up on E_{12}.

Lemma 12. *For all $0 \leq t \leq 4$, the nodes at level 2 are in the same states at the end of round $2+t$ during the execution of $\mathcal{X}$ with all messages $\mu \in U$ on E_{12}.*

Hence, $\mathcal{X}$ does not satisfy the second condition of property $P(5)$. Since $\mathcal{X}$ has property $P(k)$, it follows that $k \geq 6$.

Now, we show that, during the execution of $\mathcal{X}$ on E_{12} with message $\mu \in U$, the nodes at level 3 listen at round 7.

Lemma 13. *Nodes in state* $\mathbf{c}_{\mathcal{X}}(3,2)$ *transition to state* $\mathbf{c}'_{\mathcal{X}}(3,3)$ *after receiving a collision signal.*

Proof. To obtain a contradiction, assume that nodes in state $\mathbf{c}_{\mathcal{X}}(3,2)$ transition to state $\mathbf{c}_{\mathcal{X}}(3,3)$ after receiving a collision signal. We will construct an algorithm that has property $P(k-3)$. By definition of k, no such algorithm exists. Thus, we obtain a contradiction.

For all $\mu \in U$, during the execution of $\mathcal{X}$ with message μ on E_{12}, at the end of round 5, the nodes at level 2 are in state $\mathbf{c}_{\mathcal{X}}(2,3)$ and the nodes at level 3 are in state $\mathbf{c}_{\mathcal{X}}(3,2)$. Nodes in state $\mathbf{c}_{\mathcal{X}}(2,3)$ transmit and nodes in state $\mathbf{c}_{\mathcal{X}}(3,2)$ listen. Thus, by assumption, the nodes at level 3 are in state $\mathbf{c}_{\mathcal{X}}(3,3)$ at the end of round 6 and, hence, they transmit at round 7.

Let $\mathbf{c}_{\mathcal{X}}(2,4)$ be the state to which nodes in state $\mathbf{c}_{\mathcal{X}}(2,3)$ transition after transmitting. If nodes in state $\mathbf{c}_{\mathcal{X}}(2,4)$ transmit, then let $\mathbf{c}_{\mathcal{X}}(2,5)$ be the state to which they transition after transmitting. If nodes in state $\mathbf{c}_{\mathcal{X}}(2,4)$ listen, then let $\mathbf{c}_{\mathcal{X}}(2,5)$ be the state to which they transition after receiving a collision signal. For all $\mu \in U$, during the execution of $\mathcal{X}$ with message μ on E_{12}, since the nodes at level 3 transmit at round 7, the nodes at level 2 receive a collision signal at round 7 if they are listening and, hence, they are in state $\mathbf{c}_{\mathcal{X}}(2,5)$ at the end of round 7, whether they listen or transmit at round 7. Since nodes at level 2 are in state $\mathbf{c}_{\mathcal{X}}(2,t)$ at the end of round $2+t$ for $0 \le t \le 2$, during the execution of $\mathcal{X}$ with any message $\mu \in \{1, \ldots, w_{k-1}+1\}$ on E_{12}, it follows that, for all $0 \le t \le 5$, the nodes at level 2 are in state $\mathbf{c}_{\mathcal{X}}(2,t)$ at the end of round $2+t$ during the execution of $\mathcal{X}$ with any message $\mu \in U$ on E_{12}. Thus, the precondition of Lemma 8 holds for $\ell_0 = 2$ and $t_0 = 5$. Hence, there is an algorithm with property $P(k-3)$.

We know that at the end of round 5 nodes at level 4 are in state $\mathbf{c}_{\mathcal{X}}(4,1)$ and at the end of round 6 they are in state $\mathbf{c}_{\mathcal{X}}(4,2)$. At round 6, either they listen and receive a collision signal or they transmit. Now we show that they never transmit at round 6.

Lemma 14. *Nodes in state* $\mathbf{c}_{\mathcal{X}}(4,1)$ *listen.*

Proof. To obtain a contradiction, suppose nodes in state $\mathbf{c}_{\mathcal{X}}(4,1)$ transmit. For all $\mu \in \{1, \ldots, w_{k-1}+1\}$, at the end of round 5 during the execution of $\mathcal{X}$ with message μ on E_{12}, the nodes at level 3 are in state $\mathbf{c}_{\mathcal{X}}(3,2)$ and the nodes at level 4 are in state $\mathbf{c}_{\mathcal{X}}(4,1)$. Since nodes in state $\mathbf{c}_{\mathcal{X}}(4,1)$ transmit, the nodes at level 4 transmit at round 6 and the nodes at level 3 receive a collision signal. Therefore, for all $\mu \in \{1, \ldots, w_{k-1}+1\}$, during the execution of $\mathcal{X}$ with message μ on E_{12}, the nodes at level 3 are in state $\mathbf{c}'_{\mathcal{X}}(3,3)$ at the end of round 6. However, this contradicts Lemma 10 because $w_{k-1}+1 > w_{k-2}$.

By definition of U', for all $\mu \in U'$, at the end of round 5 during the execution of $\mathcal{X}$ with message μ on E_{12}, the nodes at level 2 are in state $\mathbf{c}'_{\mathcal{X}}(2,3)$, the nodes at level 3 are in state $\mathbf{c}_{\mathcal{X}}(3,2)$, and the nodes at level 4 are in state $\mathbf{c}_{\mathcal{X}}(4,1)$. Since nodes in state $\mathbf{c}'_{\mathcal{X}}(2,3)$ listen, the nodes at level 2 listen at round 6. By Lemma 14, the nodes at level 4 listen at round 6. Thus, the nodes at level 3 listen in state $\mathbf{c}_{\mathcal{X}}(3,2)$ at round 6 and receive nothing. By Lemma 13, nodes in state $\mathbf{c}_{\mathcal{X}}(3,2)$ transition to state $\mathbf{c}'_{\mathcal{X}}(3,3)$ after receiving a collision signal and transition to state $\mathbf{c}_{\mathcal{X}}(3,3)$ after receiving nothing. Therefore, we have the following observation.

Observation 2. *For all $\mu \in U'$, the nodes at level 3 are in state $\mathbf{c}_{\mathcal{X}}(3,3)$ at the end of round 6 during the execution of $\mathcal{X}$ with message μ on E_{12}.*

Now we improve the upper bound on the cardinality of U and U'.

Lemma 15. $|U|, |U'| \leq w_{k-3}$.

Proof. To obtain a contradiction, suppose $|U| \geq w_{k-3}+1$ or $|U'| \geq w_{k-3}+1$. By definition of U, for all $\mu \in U$, the nodes at level 2 are in state $\mathbf{c}_{\mathcal{X}}(2,3)$ at the end of round 5 during the execution of $\mathcal{X}$ with message μ on E_{12}. By Observation 2, for all $\mu \in U'$, the nodes at level 3 are in state $\mathbf{c}_{\mathcal{X}}(3,3)$ at the end of round 6 during the execution of $\mathcal{X}$ with message μ on E_{12}.

Let S be one of the sets U and U'. If $S = U$, then let $\ell' = 2$. Otherwise, let $\ell' = 3$. Since nodes in state $\mathbf{c}_{\mathcal{X}}(\ell',3)$ transmit, there can be only one state to which nodes in state $\mathbf{c}_{\mathcal{X}}(\ell',3)$ transition after transmitting. We call this state $\mathbf{c}_{\mathcal{X}}(\ell',4)$. Since nodes at level ℓ' are in state $\mathbf{c}_{\mathcal{X}}(\ell',t)$ at the end of round $\ell'+t$ for $0 \leq t \leq 2$, during the execution of $\mathcal{X}$ with any message $\mu \in \{1,\ldots,w_{k-1}+1\}$ on E_{12}, it follows that, for all $0 \leq t \leq 4$, the nodes at level ℓ' are in state $\mathbf{c}_{\mathcal{X}}(\ell',t)$ at the end of round $\ell'+t$ during the execution of $\mathcal{X}$ with any message $\mu \in S$ on E_{12}. Thus, the precondition of Lemma 8 holds for $\ell_0 = \ell'$ and $t_0 = 4$. Hence, there is an algorithm with property $P(k-2)$. By definition of k, no such algorithm exists. Thus, $|S| \leq w_{k-3}$.

Recall that $|U| + |U'| = w_{k-1} + 1$. Since $|U|, |U'| \leq w_{k-3}$, it follows that $|U| + |U'| \leq w_{k-3} + w_{k-3} = w_{k-3} + w_{k-4} + w_{k-6}$. Note that, $w_0, w_1, \ldots$, is a non-decreasing sequence and, hence, $w_{k-6} \leq w_{k-5}$. Thus, $|U| + |U'| \leq w_{k-3} + w_{k-4} + w_{k-5} = w_{k-2} + w_{k-4} = w_{k-1}$. This is a contradiction. Thus, we conclude the proof of Theorem 2.

6 Future Work

In this paper, we presented an improved algorithm to broadcast a message from a finite set of values that works on the beeping model. We proved an exactly matching lower bound in the radio broadcast model with collision detection. This shows that the ability to send arbitrarily long messages instead of just a collision signal does not improve the round complexity.

Our algorithm relies on an encoding mechanism that requires the set of possible messages to be finite and known in advance. One way to extend our algorithm to handle an infinite set of possible messages is to have the source node first broadcast r using another algorithm (for example, using *Beep Waves* to broadcast a self-delimiting encoding of r [12]). A natural question is whether this approach or a recursive version of this approach is optimal.

A variant of the broadcast problem is the acknowledged broadcast problem in which the source node needs to eventually be informed that all nodes have learned the message. Chlebus, Gąsieniec, Gibbons, Pelc, and Rytter [6] provided a deterministic acknowledged broadcast algorithm in the radio broadcast model with collision detection, assuming that each node has a distinct identifier. A possible approach to solve the acknowledged broadcast problem in the anonymous beeping model is to extend our approach by modifying the algorithm as follows. The extension relies on the fact that, in our algorithm, all nodes at each level learn the last bit of the message at the same round, which is one round later than the nodes in the previous level. Each node v that is not connected to a node at a higher level terminates immediately after learning the last bit of the message. Now consider any node v at level $\ell \geq 1$ that is connected to a node at level $\ell+1$. Node v starts executing the acknowledgment process one round after relaying the last bit of the message to its neighbours at level $\ell+1$. It repeatedly executes phases of three rounds until it receives an acknowledgement from nodes at level $\ell+1$, which causes it to terminate. In the first round, v beeps to inform the nodes at level $\ell-1$ that it has not yet received an acknowledgement. Once v has received an acknowledgement, it terminates, so it does not beep in the first round of this phase. This serves as an acknowledgement for the nodes at level $\ell-1$. In the second round, v listens to detect whether the nodes at level $\ell+1$ have received an acknowledgement. Specifically, if v receives nothing, it means that the nodes at level $\ell+1$ have received the acknowledgement. Otherwise, v receives a beep, in which case it idles for one round and, then, starts executing the next phase. An interesting extension of our paper would be to see if this modified algorithm has optimal round complexity for the acknowledged broadcast problem in anonymous radio networks with collision detection.

In addition to round complexity, energy complexity has also received considerable attention in radio broadcast models. Energy complexity is generally defined as the maximum, over all nodes, of the number of rounds in which the node transmits or listens. Chang, Dani, Hayes, He, Li, Pettie [4] provided broadcast algorithms in the radio broadcast model with collision detection. In their model, they assumed that nodes have distinct identifiers and are given the number of nodes in the network, the maximum number of neighbours of any node in the network, and the maximum distance between any two nodes in the network. By modifying our encoding mechanism, we can improve the energy complexity of our algorithm. In particular, instead of using set W_i we can use the set W_i', which is defined as follows. Let $W_1' = \{0\}$, $W_2' = \{1\}$, $W_3' = \{0,1\}$, $W_4' = \{00, 01, 1\}$ and, for all $i \geq 5$, $W_i' = 0 \cdot W_{i-1}' \cup 1 \cdot W_{i-2}'$. For all $i \geq 1$, let $w_i' = |W_i|$. Then, $w_1' = w_2' = 1$ and, for all $i \geq 3$, $w_i' = w_{i-1}' + w_{i-2}'$. Note that, $w_1', w_2', w_3', \ldots$ is the Fibonacci sequence. The optimal energy complexity of the broadcast problem remains open.

Acknowledgments. I would like thank my supervisor, Faith Ellen, for her patience and support throughout this project. Her insightful feedback and guidance brought my work to a much higher level. I would also like to thank the anonymous reviewers for their time and helpful comments.

References

1. Alon, N., Bar-Noy, A., Linial, N., Peleg, D.: A lower bound for radio broadcast. J. Comput. Syst. Sci. **43**(2), 290–298 (1991)
2. Bar-Yehuda, R., Goldreich, O., Itai, A.: On the time-complexity of broadcast in multi-hop radio networks: An exponential gap between determinism and randomization. J. Comput. Syst. Sci. **45**(1), 104–126 (1992)
3. Bu, G., Potop-Butucaru, M., Rabie, M.: Wireless broadcast with short labels. In: Georgiou, C., Majumdar, R. (eds.) NETYS 2020. LNCS, vol. 12129, pp. 146–169. Springer, Cham (2021). https://doi.org/10.1007/978-3-030-67087-0_10
4. Chang, Y.J., Dani, V., Hayes, T.P., He, Q., Li, W., Pettie, S.: The energy complexity of broadcast. In: Proceedings of the 2018 ACM Symposium on Principles of Distributed Computing, pp. 95–104 (2018)
5. Chlamtac, I.: The wave expansion approach to broadcasting in multihop radio networks. IEEE Trans. Commun. **39**(3), 426–433 (1991)
6. Chlebus, B., Gąsieniec, L., Gibbons, A., Pelc, A., Rytter, W.: Deterministic broadcasting in ad hoc radio networks. Distrib. Comput. **15**(1), 27–38 (2002)
7. Chlebus, B.S., Gçasieniec, L., Östlin, A., Robson, J.M.: Deterministic radio broadcasting. In: Montanari, U., Rolim, J.D.P., Welzl, E. (eds.) ICALP 2000. LNCS, vol. 1853, pp. 717–729. Springer, Heidelberg (2000). https://doi.org/10.1007/3-540-45022-X_60
8. Chrobak, M., Gasieniec, L., Rytter, W.: Fast broadcasting and gossiping in radio networks. J. Algorithms **43**(2), 177–189 (2002)
9. Clementi, A., Monti, A., Silvestri, R.: Distributed broadcast in radio networks of unknown topology. Theoret. Comput. Sci. **302**(1–3), 337–364 (2003)
10. Cloitre, B.: The online encyclopedia of integer sequences (2002). http://oeis.org
11. Czumaj, A., Davies, P.: Deterministic communication in radio networks. SIAM J. Comput. **47**(1), 218–240 (2018)
12. Czumaj, A., Davies, P.: Communicating with beeps. J. Parallel Distrib. Comput. **130**, 98–109 (2019)
13. Czumaj, A., Rytter, W.: Broadcasting algorithms in radio networks with unknown topology. J. Algorithms **60**(2), 115–143 (2006)
14. Ellen, F., Gilbert, S.: Constant-length labelling schemes for faster deterministic radio broadcast. In: Proceedings of the 32nd ACM Symposium on Parallelism in Algorithms and Architectures, pp. 213–222 (2020)
15. Ellen, F., Gorain, B., Miller, A., Pelc, A.: Constant-length labeling schemes for deterministic radio broadcast. ACM Trans. Parallel Comput. **8**(3), 1–17 (2021)
16. Ghaffari, M., Haeupler, B.: Near optimal leader election in multi-hop radio networks. In: Proceedings of the Twenty-Fourth Annual ACM-SIAM Symposium on Discrete Algorithms, pp. 748–766. SIAM (2013)
17. Ghaffari, M., Haeupler, B., Khabbazian, M.: Randomized broadcast in radio networks with collision detection. Distrib. Comput. **28**(6), 407–422 (2014). https://doi.org/10.1007/s00446-014-0230-7
18. Hounkanli, K., Pelc, A.: Deterministic broadcasting and gossiping with beeps. arXiv preprint arXiv:1508.06460 (2015)

19. Kowalski, D., Pelc, A.: Broadcasting in undirected ad hoc radio networks. Distrib. Comput. **18**(1), 43–57 (2005)
20. Kowalski, D., Pelc, A.: Optimal deterministic broadcasting in known topology radio networks. Distrib. Comput. **19**(3), 185–195 (2007)
21. Kushilevitz, E., Mansour, Y.: An $\omega(d \cdot \log \frac{n}{d})$ lower bound for broadcast in radio networks. SIAM J. Comput. **27**(3), 702–712 (1998)
22. Nanah Ji, K.: Exactly optimal deterministic radio broadcasting with collision detection. arXiv preprint arXiv:2202.06375 (2022)
23. Pandita, N.: Ganita Kaumudi (1356)

Lower Bounds on Message Passing Implementations of Multiplicity-Relaxed Queues and Stacks

Edward Talmage(✉)

Bucknell University, Lewisburg, PA, USA
elt006@bucknell.edu

Abstract. A *multiplicity*-relaxed queue or stack data type allows multiple *Dequeue* or *Pop* operations to return the same value if they are concurrent. We consider the possible efficiency of message-passing implementations of such data types. We show that both the worst case and amortized time cost for *Dequeues* and *Pops* are nearly as high as upper bounds for their worst-case time in unrelaxed queues and stacks. Relaxed data types are of interest since they can in some cases trade off some of data types' ordering guarantees for increased performance or easier implementation. The multiplicity relaxation, in particular, is interesting as it has been shown to be less computationally complex than unrelaxed queues and stacks. Our results explore a different aspect of these data types, considering communication time complexity in a message passing system and showing limits on possible improved time performance.

Keywords: Distributed Data Structures · Message Passing · Relaxed Data Types · Lower Bounds

1 Introduction and Related Work

Organized access to data is a fundamental necessity for efficient computation. In a distributed setting, this access is non-trivial to implement, as physically-separated computing entities may have different data, and sharing that data requires explicit coordination between processes. Such coordination is typically costly, since the time delays inherent to communication across large distances are much greater than those in accessing local storage devices. Other issues, such as failures of participants in an algorithm and breakdowns of the communication network, can also interfere with efforts to coordinate data access.

This work continues exploration of a topic of recent interest: the notion of relaxing [1,5], or weakening, data type specifications in carefully defined ways to improve the performance of algorithms for distributed data structures. Past work has shown that traditional data types are inherently expensive [2,6,13], but certain relaxations can decrease the amortized cost of data structure implementations [11].

M. Parter (Ed.): SIROCCO 2022, LNCS 13298, pp. 253–264, 2022.
https://doi.org/10.1007/978-3-031-09993-9_14

Parallel to the exploration of the time costs of relaxed data structure implementations, various researchers have been exploring the computational power of relaxed data types. [10] and [12] computed the consensus numbers, a standard measurement of a data type's computational strength, of several relaxations of queues. [4] continued from there to consider a different relaxation, multiplicity, which allows some concurrent actions to be oblivious to each other. They showed that in a shared-memory system, queues and stacks with multiplicity are implementable, either in a wait-free or non-blocking way, from read/write registers. This implies that their computational strength is relatively low, and that they are achievable in systems without strong primitive data operations, despite providing nearly the full guarantees of first-in, first out queues and last-in first-out stacks.

Given the potential usefulness of multiplicity in shared memory systems, we ask how useful they would be in message passing systems, as it would be helpful to have relaxations which are useful in many settings. Further, intuition suggests that multiplicity might increase performance, since *Dequeue* and *Pop* instances no longer need to know about concurrent instances, so may be able to determine a return value more quickly. In this paper, we analyze the time cost of message-passing implementations of relaxed queues and stacks. We give lower bounds which show that the multiplicity relaxation cannot avoid the bulk of the cost of waiting for messages about concurrent instances, as compared to implementations of unrelaxed queues and stacks.

While we show a limited amount of possible performance gain, our bounds are not quite tight against the best known upper bounds [13], leaving open the question of whether multiplicity might still carry some performance advantages. There is also work to be done moving from the idealized model we consider in this paper to a more realistic model, since multiplicity may tolerate failures or decreased synchrony better than an unrelaxed type.

2 Model and Definitions

2.1 System Model

We consider a standard partially-synchronous, fault-free message passing model, based on that in [7]. We assume that each of the n processes participating in a shared memory implementation provides an interface by which users may invoke operations and the processes may generate responses to those invocations. Each process is sequential, so a user may not invoke an operation at a particular process until any previous instances at that process have returned. Beyond that constraint, invocations may arrive at any time. Each process p_i has a local clock, which runs at the same rate as real time, but may be offset by a constant c_i. We assume, for the sake of focusing on the communication costs inherent in distributed systems, that local computation is instantaneous. Each of the processes in the system is a state machine where steps may be sending or receiving a message, performing local computation, setting a timer, or a timer expiration.

Each step has an associated time, which is the current value of the local clock at the process where the step took place.

A *run* of an algorithm is a set of sequences, one for each process, of valid state machine steps at that process. Every process' sequence is infinite or ends with no messages to that process which have been sent and not received and no timers at that process which have not yet expired. A run is *admissible* with respect to parameters d, u, and n if there is a bijection from message sends to later receives, each message is received at least $d - u$ and no more than d time after it is sent, and the maximum difference (known as *skew*) between local clock offsets is at most $\epsilon \leq (1 - 1/n)u$ (shown as optimal and achievable in [8]). We also assume that algorithms are *eventually quiescent*, which means that if all users cease invoking operations, after a finite time the system will reach a state with no timers set and no messages in transit and remain in that state until another invocation.

We require that any implementation of a data type satisfy a *liveness* property: every operation invocation has a *matching* response, and every response follows a matching invocation. This invocation-response pair is an *operation instance.* We are interested in the real time which elapses between operation invocations and responses. Specifically, we focus on the worst-case time, $|OP|$, and amortized time, $|OP|_{am}$, for an operation OP. Worst-case time is the greatest time which may elapse, in any admissible run, between the invocation and response of any single instance of OP. Amortized time is the maximum, over all admissible runs, of the sum of the time from invocation to response of every instance of OP, divided by the number of instances of OP in the run.

A *set-sequential specification* of a data type T defines a set of operations, with their argument and return types, and a set of sequences of sets of instances of those operations which are legal. Each element in a legal sequence is a set of operation instances, each with an argument and return value. A run of an algorithm A implementing T is *set-linearizable* if there is a way to gather operation instances in the run into sets and order those sets in a legal sequence which respects the order of non-overlapping instances–if an operation instance *op* responds before another *op*′ is invoked, *op* precedes *op*′ in the sequence. We say that A is a set-linearizable implementation of T if every admissible run of A is set-linearizable.

2.2 Multiplicity

Intuitively, a *queue with multiplicity* is a First-In, First-Out queue structure, but with an explicit allowance for abnormal behavior in the presence of concurrency. In a traditional FIFO queue, each element may only be removed (by the *Dequeue* operation) one time. But when multiple users may invoke *Dequeue*, each of their *Dequeue* instances must communicate to collectively determine which instance returns which value to prevent duplicate returns. *Multiplicity* weakens the queue definition to (hopefully) reduce this communication requirement by loosening the unique-value requirement on *Dequeue* in the presence of concurrency. When a set of *Dequeue* instances are all concurrent, they are allowed (though not required) to return the same value. That value must still be the first value placed in the queue in the set-linearization of the run which has not yet been returned, maintaining FIFO ordering.

Stacks with multiplicity are defined similarly, with *Pop* taking the place of *Dequeue*. Concurrent instances of *Pop* may return the same value, as long as that is the most recently-added value in the stack in the set-linearization of the run.

Since these definitions explicitly refer to concurrency, they cannot be expressed as a sequential specification, which is the standard way to define data types, even in concurrent systems. Instead, we must use a concurrent specification to formally define a queue or stack with multiplicity. Following [4], we give set-sequential specifications for queues and stacks with multiplicity.

Definition 1. A queue with multiplicity *is a data type with two operations: Enqueue and Dequeue. Enqueue accepts one value from some set V of possible values as an argument and returns the constant ACK. Dequeue takes no argument and returns one value from $V \cup \{\bot\}$, where $\bot$ is a special symbol denoting empty. A sequence of sets of Enqueue and Dequeue instances is legal if*

- *Every Enqueue instance is in a singleton set.*
- *Each set of Dequeue instances ordered in the sequence contains only Dequeue instances with the same return value.*
- *Each Dequeue instance d returns the argument of the first Enqueue instance before d in the sequence which has not been returned by another Dequeue instance before d in the sequence. If there is no such Enqueue instance, Dequeue returns $\bot$.*

Definition 2. A stack with multiplicity *is a data type with two operations: Push and Pop. Push accepts one value from some set V of possible values as an argument and returns the constant ACK. Pop takes no argument and returns one value from $V \cup \{\bot\}, \bot \notin V$. A sequence of sets of Push and Pop instances is legal if*

- *Every Push instance is in a singleton set.*
- *Each set of Pop instances ordered in the sequence contains Pop instances with the same return value.*
- *Each Pop instance p returns the argument of the last Push instance before p in the sequence which has not been returned by another Pop instance before p in the sequence. If there is no such Push instance, Pop returns $\bot$.*

2.3 Shifting Arguments

For some of our results, we will use shifting arguments to show that runs are indistinguishable [3,8]. Shifting is a technique which takes advantage of the uncertainty in message delays and the skew between local clock values to adjust the real times when events occur. Given a run R, and a real-valued *shift vector* s of length n, we can create a new run R' by adjusting the real times at which all events at each process i occur by $s[i]$, the corresponding component of the shift vector. Thus, if an event e occurred at process 0 at real time t_r, it will now occur at real time $t_r + s[0]$. We also adjust p_i's clock offset c_i by $-s[i]$ so that events at each process occur at the same local clock value. In our example, e still occurs at the same local time, since local time is determined by real time and offset: $(t_r + s[0]) + (c_i - s[0])$.

If we can verify that the maximum skew between local clocks in R' is no more than ϵ and all message delays are still in the range $[d - u, d]$, then R' is also an admissible run. Since all events occur at the same local clock values and all messages arrive at the same times, the runs R and R' are indistinguishable to every process.

Such arguments are useful for proving lower bounds on linearizable and, as we do in this paper, set-linearizable implementations of data types. The primary way we use them is in showing that a run with overlapping operation instances at different processes is indistinguishable from another run where some operation instances are not concurrent. When they are not concurrent, we can draw conclusions about the order in which they must appear to occur and thus their return values. By carrying those conclusions back to the original run with concurrent instances, we can deduce time bounds.

3 Worst-Case Lower Bounds

3.1 *Dequeue* and *Pop*

We prove a lower bound on the worst-case cost of *Dequeue* in our relatively well-behaved partially synchronous, crash-free model. While this model may not be the most realistic, because of its strong assumptions, a lower bound in this model also applies in a less well-behaved model, such as one with asynchrony or faults. As a result, we know that it is impossible to implement a queue with multiplicity more efficiently in a realistic model than we can here. The proof of this theorem does not change if we consider *Pop* on a stack with multiplicity instead of a *Dequeue*, so we have the same lower bound for *Pop*.

To prove this bound, we carefully construct runs in which we can predict what value a *Dequeue* must return, by the data type semantics. We then alter these runs slightly to change the behavior of certain instances, then argue what values they return. We do this in a couple of different ways. First, we use the shifting technique to prove that a pair of overlapping *Dequeue* instances must return different values, as they cannot tell that they are concurrent, since the run is undetectably shifted from one in which they are not concurrent. We then create a third run with slightly different, but still admissible, message delays and consider how this affects when processes have and can have knowledge of events in the system which might change their behavior. This allows us to argue that if a certain *Dequeue* instance returned before a certain point in time, then it would necessarily return an illegal value, such as the same value as another, non-concurrent, *Dequeue* instance. This proves that these particular instances must have a certain minimum duration, proving the lower bound on the worst-cast time for *Dequeue*.

Theorem 1. *Any set-linearizable implementation of a queue with multiplicity on a system with at least three processes must have* $|Dequeue| \geq \min\left\{\frac{2d}{3}, \frac{d+u}{2}\right\}$.

Proof. We will construct a set of runs and show that a particular *Dequeue* instance in one of these admissible runs must return an incorrect value unless it takes at least $\min\left\{\frac{2d}{3}, \frac{d+u}{2}\right\}$ time.

For all of our runs, we first invoke the sequence of *Enqueue* instances

$$Enqueue(0) \cdot Enqueue(1) \cdot Enqueue(2) \cdot Enqueue(3)$$

sequentially at p_0. We invoke no more operations until the system becomes quiescent. Call the quiescent time t.

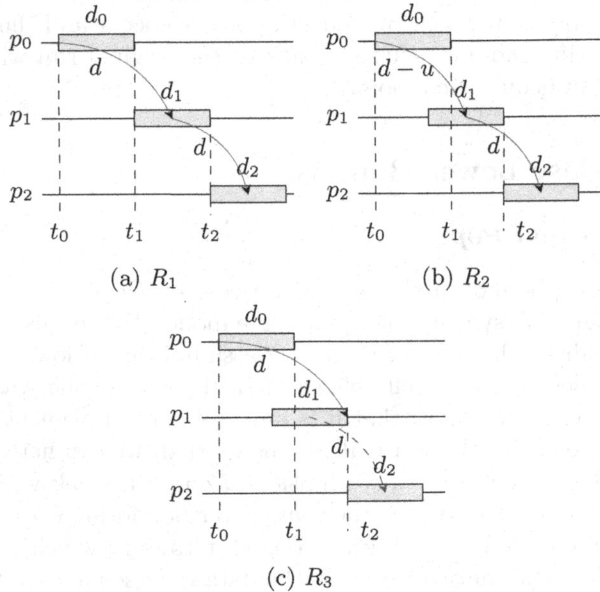

Fig. 1. Runs used in the proof of Theorem 1. Time increases to the right. Rectangles are operation instances. Curved red lines are messages, labeled with their delays; dashed blue indicates the missing information whose absence p_2 can detect to differentiate R_2 and R_3. (Color figure online)

Define run R_1 as follows: All messages from p_0 have delay d, all messages to p_0 have delay $d-u$, and all other messages have delay d. p_0's local clock value is $\epsilon \leq 2u/3$ lower than those of all other processes at the same real time. At time $t_0 > t$, process p_0 invokes a *Dequeue*, d_0. Let t_1 be the real time when d_0 returns. Immediately after t_1, p_1 invokes a *Dequeue*, d_1, which returns at real time t_2. Immediately after t_2, process p_2 invokes a *Dequeue*, d_2. (See Fig. 1a.) Since none of these *Dequeue* instances are concurrent, they must set-linearize in singleton sets as the sequence $\{d_0\} \cdot \{d_1\} \cdot \{d_2\}$, and must return the values $0, 1, 2$, respectively.

Define run R_2 (Fig. 1b) by shifting run R_1 by the shift vector $\langle 0, -u, -u, \ldots, -u\rangle$, shifting all processes but p_0 back in real time. Messages from p_0 now have delay $d - u$, since their receive events are sooner, and those to p_0 now have delay d, since their send events are sooner. Other messages' delays are unaffected, since their senders and receivers are shifted identically. p_0's local clock is now ahead of those of p_1 and p_2 (and all other processes) by $u/3 < \epsilon$. Thus, R_2 is admissible, by the bounds on message delay and clock skew. Since this run is a shift of R_1, no process can distinguish them, so the *Dequeue* instances have the same return values, despite d_0 and d_1 now being concurrent.

Finally, we define R_3 from R_2 (Fig. 1c) by increasing the delay for messages from p_0 to p_1 to d. We are not shifting the run, so processes may be able to distinguish the result from R_2, but we can determine when they will first be distinguishable. There are two possibilities for when the earliest possible message from p_0 to p_1 announcing the invocation of d_0 may arrive and from these we obtain the two terms of our lower bound. Call this hypothetical message m_0.

First, suppose that m_0 arrives before d_1 returns. Then we can directly calculate a bound on the response time of *Dequeue*, since two *Dequeue* instances which only overlap by u must take more than one message delay. That is, $2|Dequeue| - u > d$, giving $|Dequeue| > \frac{d+u}{2}$.

If $|Dequeue|$ is smaller than this bound, it follows that m_0 must arrive after d_1 returns. Then p_1 must have chosen a return value for d_1 without any knowledge of d_0. In this case, p_1 cannot distinguish, until after d_1 returns, R_3 from a run in which p_0 and p_2 take no action. In such a run, set-linearization implies that d_1 must return 0, so it must do the same in R_3.

It is legal by the specification of a queue with multiplicity for both d_0 and d_1 to return the same value in R_3, since the two instances are concurrent. However, d_1 now has a different return value than in R_2 and thus affects the behavior of d_2. The earliest time when p_2 can distinguish R_3 from R_2 is if p_1 sent a message m_1 to p_2 immediately on receiving m_0 in R_2. Since m_0 arrives later in R_3, p_1 may send a different or no message at the local time it received m_0 in R_2 but not in R_3. m_0 would have arrived at real time $t_0 + (d-u) + d$ in R_2, so $t_0 + 2d - u$ is the earliest real time when p_2 can distinguish R_2 and R_3. If d_2 returns before this time, then it must return the same value in both runs, but if it does so and returns 2 in R_3, there is no valid set-linearization, since no *Dequeue* instance returns 1, and thus none may return 2. Thus, d_2 cannot return until at least real time $t_0 + 2d - u$.

Now, we can set up an equation for the duration of the *Dequeue* instances. The three instances are end-to-end, with the exception of d_0 and d_1 overlapping by u. Thus, we have $3|Dequeue| - u \geq 2d - u$. This reduces to $|Dequeue| > \frac{2d}{3}$, giving us the second part of the lower bound.

Theorem 2. *Any set-linearizable implementation of a stack with multiplicity on a system with at least three processes must have* $|Pop| \geq \min\left\{\frac{2d}{3}, \frac{d+u}{2}\right\}$.

3.2 Amortized Cost Lower Bounds

While we've shown that the worst-case cost of *Dequeue* and *Pop* is relatively high, even with multiplicity, there is still the possibility of improved amortized complexity, as was the case for other relaxations in [11]. However, we next show that this cannot happen. For amortized cost to be lower than worst case, there must be instances which have lower cost than the worst case. The scenario which gave us our lower bound on worst-case cost does not depend on there being previous *Dequeue* or *Pop* instances in a run, so if the run terminates immediately after the worst-case instances, there are no cheaper instances over which we can amortize the cost. This means the same lower bound must hold on amortized cost as we have shown on worst-case cost, with only slight modifications to the argument.

Further, we can argue that we could recreate the same scenario from the proof of Theorem 1 over and over in a run. Thus, the lower bound on amortized cost is not limited to runs with only a small number of *Dequeue* instances, but applies for runs with any number (divisible by three) of instances.

Corollary 1. *Any implementation of a queue with multiplicity on at least 3 processes must have the amortized cost of Dequeue* $|Dequeue|_{am} \geq \min\left\{\frac{2d}{3}, \frac{d+u}{2}\right\}$. *Any implementation of a stack with multiplicity on at least 3 processes must have the amortized cost of Pop,*

$$|Pop|_{am} \geq \min\left\{\frac{2d}{3}, \frac{d+u}{2}\right\}$$

Proof. Consider the runs R_1, R_2, R_3 from the proof of Theorem 1, as shown in Fig. 1. Recall that one of two cases must hold in R_3: either m_0 arrives at p_1 before d_1 returns or d_2 cannot return until after message m_1 sent by p_1 when it receives m_0 in R_2 would have arrived.

In the first case, two *Dequeue* instances which overlapped by only u time took d time to complete. We here have no information on how long d_2 took to return, but since it is not invoked until after d_0 and d_1 return, there is another run in which p_0 and p_1 behave exactly the same and p_2 does not invoke d_2. In this run, the total cost of all *Dequeue* instances is $d + u$ and there are only two instances, so the amortized cost is $\frac{d+u}{2}$.

In the second case, we can bound the total time which three *Dequeue* instances, overlapping by only u, take to return. Here, two message delays must elapse, so we have the total cost of all three *Dequeue* instances in the run is $2d - u + u = 2d$. This gives an amortized cost of $\frac{2d}{3}$.

Since we do not know which case occurs, the overall bound is

$$|Dequeue|_{am} \geq \min\left\{\frac{d+u}{2}, \frac{2d}{3}\right\}$$

3.3 Bound on Sum of Operations

We note that a simple argument will also give a lower bound for the sum of an *Enqueue* and a *Dequeue* or the sum of a *Push* and a *Pop*. In an empty queue or stack, an *Enqueue* immediately before a *Dequeue* or a *Push* immediately before a *Pop* will determine the second operation instance's return value. This is similar to results from [9,13] and elsewhere on the sum of the costs of certain types of operations, but since we are in a set-linearizable model, the existing proofs for linearizability do not apply directly. Specifically, since overlapping operations may return the same value, it is much harder to force incorrect return values in shifting arguments. We present the simple results relevant to our current topic and leave generalizations for future work.

Theorem 3. *Any set-linearizable implementation of a queue with multiplicity must have* $|Enqueue| + |Dequeue| \geq d$.

Proof. Consider the following run, pictured in Fig. 2: At time 0, p_0 invokes *Enqueue*(x). Immediately after that instance returns, p_1 invokes *Dequeue*. Since these instances are not concurrent, the *Dequeue* instance must be set-linearized after *Enqueue*(x), and thus must return x, by the specification of a queue with multiplicity since x is the only, and thus first, previously-enqueued value. For p_1 to choose x as the return value for the *Dequeue* instance, it must be aware of the *Enqueue* instance. Otherwise, p_2 would not be able to distinguish this run from one where it is the only process which acts, in which case the *Dequeue* instance would return $\perp$. p_1 may not learn of *Enqueue*(x) until d real time after p_0's invocation, since any message containing information about the invocation may take that long to arrive. Thus, the *Dequeue* instance cannot return until at least d real time after *Enqueue*(x)'s invocation, so the two instances' durations must sum to at least d.

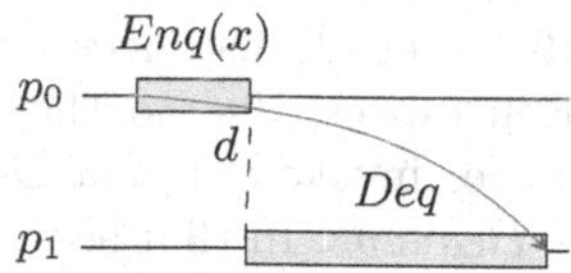

Fig. 2. Figure illustrating the run in the proof of Theorem 3 (Color figure online)

As with Theorems 1 and 2, the argument here works without further alteration if we substitute *Push* for *Enqueue* and *Pop* for *Dequeue*, so we have the below bound for stacks, as well. In fact, the argument for stacks is slightly simpler, since any *Push* will affect the return value of an immediately-following *Pop*, not only when the structure is empty.

Theorem 4. *Any set-linearizable implementation of a stack with multiplicity must have* $|Push| + |Pop| \geq d$.

4 Tightness

An algorithm in the idealized model we use in this paper is of limited practical value, since real systems are typically less well-behaved, but is still of theoretical interest, since it can tell us if our lower bound is tight. The best upper bound in the literature is that in [13], which provides a linearizable implementation of unrelaxed queues with $|Enqueue| = \epsilon$[1] and $|Dequeue| = d + \epsilon$, one maximum message delay plus clock skew. Since a linearizable implementation is a set-linearizable implementation, the bound carries over to the model we consider here. Similarly, an unrelaxed queue implementation also satisfies the specification of a queue with multiplicity, where no concurrent *Dequeue* instances happen to return the same value, so the data type implemented there is sufficient to give us an upper bound.

Corollary 2 (from Theorem 6 in [13]). *There is a set-linearizable implementation of a queue with multiplicity with* $|Dequeue| = d + \epsilon$ *and* $|Enqueue| + |Dequeue| = d + 2\epsilon$. *There is a set-linearizable implementation of a stack with multiplicity with* $|Pop| = d + \epsilon$ *and* $|Push| + |Pop| = d + 2\epsilon$.

Intuitively, it seems that it should be possible to achieve better performance with a set-linearizable implementation than a fully linearizable implementation, since there are fundamentally weaker constraints on the ordering, and thus on the knowledge of previous and concurrent operation instances. However, multiplicity only allows *Dequeue* (or *Pop*) instances to be combined into sets for linearization, not *Enqueue* (or *Push*) instances. Thus, participating processes must still totally order *Enqueue* (*Push*) instances, which can force *Dequeue* and *Pop* instances to take longer to return, even if *Enqueue* and *Push* return quickly.

We could easily reduce the delay from invocation to response for *Dequeue* instances to d and still correctly order *Dequeue* instances, since then each instance can be guaranteed to know the return value of any strictly preceding *Dequeue* instance and choose a different value. However, existing algorithms (such as those in [11] and [13] order all instances by timestamp. If we continue to use timestamps to order instances, due to the skew between local clocks, there may be *Enqueue* instances invoked after a *Dequeue* instance but with a smaller timestamp than that *Dequeue*. In that scenario, timestamp order forces the delay of $d + \epsilon$ for *Dequeue* to learn about such *Enqueue* instances. While that delay has not been proven to be necessary for all implementations, since we could use other (set-)linearizations than timestamp order, constructing such a set-linearization is far more complex. Thus, it remains an open question whether it may be possible to reduce the cost of *Dequeue* to d.

There is still work to do to determine tight bounds on implementing queues and stacks with multiplicity in the partially synchronous model we consider, but there is a fairly small window (roughly, between $d/3$ and $d/2$) for possible improvement in the complexity of *Dequeue* and *Pop*.

[1] Recall that ϵ is the clock skew, the maximum difference between any two processes' local clocks.

5 Conclusion

Multiplicity is a natural relaxation, since it captures the fact that concurrent actions may be unaware of each other. Since [4] showed that it is also computationally simple, it seems a natural candidate for widespread use in systems which can trade complete uniqueness of dequeued values for performance. We have begun exploring the feasibility of using queues and stacks with multiplicity in message passing systems, from the perspective of the time cost required to communicate updates to the structures.

We show lower bounds on queues and stacks with multiplicity in an idealized model, both for worst-case and amortized time. Our results, while not yet tight, imply that multiplicity gives at best a small space for possible increases in performance. While concurrent *Dequeue* and *Pop* operation instances do not need to know about each other, the requirements that they know about all preceding instances and totally order *Enqueue* and *Push* instances still impose significant minimum delays. It remains an open question whether a faster set-linearizable algorithm for these structures is possible, or if another proof could yield larger lower bounds.

Our results also show that, in terms of amortized cost, other relaxations may be more efficient. [11] showed that queues with relaxed ordering constraints can have an amortized cost inversely proportional to a degree-of-relaxation parameter. For large relaxations, that amortized cost is lower than the lower bound we showed on queues with multiplicity. Thus, in our idealized model, those order-based relaxations appear to be more efficient, in an amortized sense.

The lower bounds we show in this paper also apply to any less well-behaved model, so we can conclude that real-world implementations must also have costs above our bounds. Since these lower bounds are close to upper bounds for unrelaxed queues in the same model, we are inclined to guess that these relaxations provide little time benefit in a real-world message-passing data type implementation. More work is required in models closer to the real world, however, to determine whether unrelaxed types or other relaxations may have higher lower bounds in systems with less timing information or which must tolerate failures. It is still possible that multiplicity may result in performance gains in such a model if it suffers less from weakening the computation model.

References

1. Afek, Y., Korland, G., Yanovsky, E.: Quasi-Linearizability: Relaxed Consistency for Improved Concurrency. In: Lu, C., Masuzawa, T., Mosbah, M. (eds.) OPODIS 2010. LNCS, vol. 6490, pp. 395–410. Springer, Heidelberg (2010). https://doi.org/10.1007/978-3-642-17653-1_29
2. Attiya, H., Guerraoui, R., Hendler, D., Kuznetsov, P., Michael, M.M., Vechev, M.T.: Laws of order: expensive synchronization in concurrent algorithms cannot be eliminated. In: Ball, T., Sagiv, M. (eds.) Proceedings of the 38th ACM SIGPLAN-SIGACT Symposium on Principles of Programming Languages, POPL 2011, Austin, TX, USA, January 26–28, 2011. pp. 487–498. ACM (2011). https://doi.org/10.1145/1926385.1926442

3. Attiya, H., Welch, J.: Distributed Computing. Wiley (2004). https://doi.org/10.1002/0471478210.ch6
4. Castañeda, A., Rajsbaum, S., Raynal, M.: Relaxed queues and stacks from read/write operations. In: Bramas, Q., Oshman, R., Romano, P. (eds.) 24th International Conference on Principles of Distributed Systems, OPODIS 2020, December 14–16, 2020, Strasbourg, France (Virtual Conference). LIPIcs, vol. 184, pp. 13:1–13:19. Schloss Dagstuhl - Leibniz-Zentrum für Informatik (2020). https://doi.org/10.4230/LIPIcs.OPODIS.2020.13
5. Henzinger, T.A., Kirsch, C.M., Payer, H., Sezgin, A., Sokolova, A.: Quantitative relaxation of concurrent data structures. In: Giacobazzi, R., Cousot, R. (eds.) The 40th Annual ACM SIGPLAN-SIGACT Symposium on Principles of Programming Languages, POPL 2013, Rome, Italy - January 23–25, 2013. pp. 317–328. ACM (2013). https://doi.org/10.1145/2429069.2429109
6. Kosa, M.J.: Time bounds for strong and hybrid consistency for arbitrary abstract data types. Chic. J. Theor. Comput. Sci. 1999 (1999). http://cjtcs.cs.uchicago.edu/articles/1999/9/contents.html
7. Lundelius, J., Lynch, N.A.: A new fault-tolerant algorithm for clock synchronization. In: Kameda, T., Misra, J., Peters, J.G., Santoro, N. (eds.) Proceedings of the Third Annual ACM Symposium on Principles of Distributed Computing, Vancouver, B. C., Canada, August 27–29, 1984. pp. 75–88. ACM (1984). https://doi.org/10.1145/800222.806738
8. Lundelius, J., Lynch, N.A.: An upper and lower bound for clock synchronization. Inf. Control **62**(2/3), 190–204 (1984)
9. Mavronicolas, M., Roth, D.: Linearizable read/write objects. Theor. Comput. Sci. **220**(1), 267–319 (1999). https://doi.org/10.1016/S0304-3975(98)90244-4
10. Shavit, N., Taubenfeld, G.: The computability of relaxed data structures: queues and stacks as examples. Distributed Comput. **29**(5), 395–407 (2016). https://doi.org/10.1007/s00446-016-0272-0
11. Talmage, E., Welch, J.L.: Improving average performance by relaxing distributed data structures. In: Kuhn, F. (ed.) DISC 2014. LNCS, vol. 8784, pp. 421–438. Springer, Heidelberg (2014). https://doi.org/10.1007/978-3-662-45174-8_29
12. Talmage, E., Welch, J.L.: Anomalies and similarities among consensus numbers of variously-relaxed queues. Computing **101**(9), 1349–1368 (2019). https://doi.org/10.1007/s00607-018-0661-2
13. Wang, J., Talmage, E., Lee, H., Welch, J.L.: Improved time bounds for linearizable implementations of abstract data types. Inf. Comput. **263**, 1–30 (2018). https://doi.org/10.1016/j.ic.2018.08.004

Fixed Points and 2-Cycles of Synchronous Dynamic Coloring Processes on Trees

Volker Turau(✉)

Institute of Telematics, Hamburg University of Technology, Hamburg, Germany
turau@tuhh.de

Abstract. This paper considers synchronous discrete-time dynamical systems on graphs based on the threshold model. It is well known that after a finite number of rounds these systems either reach a fixed point or enter a 2-cycle. The problem of finding the fixed points for this type of dynamical system is in general both NP-hard and #P-complete. In this paper we give a surprisingly simple graph-theoretic characterization of fixed points and 2-cycles for the class of finite trees. Thus, the class of trees is the first nontrivial graph class for which a complete characterization of fixed points exists. This characterization enables us to provide bounds for the total number of fixed points and pure 2-cycles. It also leads to an output-sensitive algorithm to efficiently generate these states.

1 Introduction

Synchronous discrete-time dynamical systems for information spreading received a lot of attention in recent years. Often the following model is used: Let G be a graph with an initial configuration, where each node is either black or white. In discrete-time rounds, all nodes simultaneously update their color based on a predefined local rule. The rule is local in the sense that the color associated with a node in round t is determined by the colors of the neighboring nodes in round $t-1$. The main focus of the research so far has been on the stabilization time of this process [19] and the dominance problem, e.g., how many nodes must initially be black so that eventually all nodes are black [14]. These questions have been considered for various classes of graphs. These discrete-time dynamical systems are often based on the *threshold model*. In a simple version of this model a node becomes black if at least a fraction of α of its neighbors are black and white otherwise, $\alpha \in (0, 1)$ is a parameter of the model. In more elaborate versions edges have weights and the local rules are based on the weighted fraction of neighbors. The main property of these dynamical systems is that assuming symmetric weights, the system has period 1 or 2 [8,15]. This means that such a system eventually reaches a stable configuration or it toggles between two configurations. Fogelman et al. proved that the stabilization time is in $O(n^2)$ [5]. Frischknecht et al. showed that this bound is tight, up to some poly-logarithmic factor [6].

M. Parter (Ed.): SIROCCO 2022, LNCS 13298, pp. 265–282, 2022.
https://doi.org/10.1007/978-3-031-09993-9_15

In this paper we analyze a different aspect of discrete-time dynamical systems: The number and structure of fixed points and 2-cycles. This research is motivated by applications of so called Boolean networks (BN) [10], i.e., discrete-time dynamical systems, where each node (e.g., gene) takes either 0 (passive) or 1 (active) and the states of nodes change synchronously according to regulation rules given as Boolean functions. An example for a regulation rule is the majority rule, i.e., $\alpha = 0.5$. Since the problem of finding the fixed points of a BN is in general both NP-hard and #P-complete [3] (see Sect. 2), it is interesting to find graph classes, for which the number of fixed points can be determined efficiently. We regard our work as a step in this direction. Interest in the set of fixed points of BNs was also sparked by a result of Milano and Roli [12]. They use BNs to solve the satisfiability problem (SAT) by defining a mapping between a SAT instance and a BN and prove that BN fixed points correspond to SAT solutions.

This paper provides a characterization of the set of stable configurations (a.k.a. fixed points) and the set of states of period 2 (a.k.a. 2-cycles) for a given finite tree based on its edge set. We do this for two versions of the threshold model: minority and majority process. While the stabilization times for the majority and minority process can differ considerably for a given graph (see Fig. 1), the sets of stable configurations of a tree turn out to be closely related for both process types. Our main contributions are as follows:

1. We identify a subset $E_{fix}(T)$ of the power set of the edge set of a tree T and show that the elements of $E_{fix}(T)$ correspond one-to-one with the fixed points of T. $E_{fix}(T)$ is defined by a set of simple linear inequalities over the node degrees. The fixed point corresponding to an element of $E_{fix}(T)$ can be defined in simple terms. $E_{fix}(T)$ has the hereditary property, i.e., if $X \in E_{fix}(T)$ then all subset of X are also elements of $E_{fix}(T)$. This property allows to define a simple output-sensitive algorithm $\mathcal{A}_{\mathcal{M}}$ to explicitly generate all fixed points. This allows to prove upper bounds for the number of fixed points. We also show that elements of $E_{fix}(T)$ correspond to solutions of a system of linear diophantine inequalities.
2. We characterize the configurations of period 2, where each node changes its color in every round (a.k.a. pure configurations). As above we identify a subset $E_{pure}(T)$ of the power set of the edge set of T such that the elements of $E_{pure}(T)$ correspond one-to-one with the pure configurations of T. As above the definition of $E_{pure}(T)$ is based on simple linear inequalities and it has the hereditary property. The 2-cycle corresponding to an element of $E_{pure}(T)$ is also defined in simple terms. Again this allows to define a simple algorithm enumerating all 2-cycles and to prove upper bounds for their number. Interestingly, $E_{pure}(T)$ is a subset of $E_{fix}(T)$.
3. Finally we look at general configurations with period 2. We show that for each configuration c of this type each tree decomposes into subtrees, such that c induces either a fixed point or a pure configuration on each subtree. The subtrees allow to define a hyper structure of a tree, called the *block tree*. As in previous cases we identify a subset $E_{block}(T)$ of the power set of the

edge set of a tree T and show that the elements of $E_{block}(T)$ correspond one-to-one with the block trees of T. $E_{block}(T)$ is a subset of $E_{fix}(T)$. Since a tree can have several pure colorings, a block tree does not uniquely define a coloring. We define a subclass of 2-cycles called canonical colorings and prove that there is a direct correspondence between $E_{block}(T)$ and canonical colorings. The characterization of $E_{block}(T)$ is not as simple as in the above cases, since $E_{block}(T)$ does not have the hereditary property.

All results are obtained for the minority and the majority model. A long version of the paper including all proofs is available [17].

2 State of the Art

Most research on discrete-time dynamical systems on graphs consecrates oneself to bounds of the stabilization time. Good overviews for the majority resp. the minority process can be found in [19] resp. [13]. Rouquier et al. study the minority process in the asynchronous model, i.e., not all nodes update their color concurrently [16]. They show that the stabilization time strongly depends on the topology and observe that the case of trees is non-trivial.

The analysis of fixed points of the majority or minority process received only some attention. Královič determined the number of fixed points of a complete binary tree for the majority process [11]. Agur et al. did the same for ring topologies [2]. In both cases the number of fixed points is an exponentially small fraction of all configurations.

Boolean networks have been extensively used as models for the dynamics of gene regulatory networks. A gene is modeled by binary values, 0 or 1, indicating two transcriptional states, either active or inactive, respectively. Each network node operates by the same nonlinear majority rule, i.e., majority processes are a particular type of BN [18]. The set of fixed points is an important feature of the dynamical behavior of such networks [4]. The number of fixed points is a measure for the general memory storage capacity of a system. Many fixed points imply that a system can store a large amount of information, or, in biological terms, has a large phenotypic repertoire [1]. However, the problem of finding the fixed points of a Boolean network is in general both NP-hard and #P-complete [3]. There are only a few theoretical results to efficiently determine this set [9]. Aracena determined the maximum number of fixed points in a particular class of BN called regulatory Boolean networks [4].

3 Synchronous Discrete-Time Dynamical Systems

Let $G(V, E)$ be a finite, undirected graph. A coloring c assigns to each node of G a value of $\{0, 1\}$ with no further constraints on c. Denote by $\mathcal{C}(G)$ the set of all colorings of G, i.e., $|\mathcal{C}(G)| = 2^{|V|}$. A transition process $\mathcal{M}$ describes the transition of one coloring to another, i.e., it is a mapping $\mathcal{M} : \mathcal{C}(G) \longrightarrow \mathcal{C}(G)$. Given an initial coloring c, a transition process produces a sequence of colorings $c, \mathcal{M}(c), \mathcal{M}(\mathcal{M}(c)), \ldots$. We consider two transition processes: *Minority* and *Majority* process and denote the corresponding mappings by $\mathcal{MIN}$ and $\mathcal{MAJ}$. They are local mappings in the sense that the new color of a node is based on the current colors of its neighbors. To determine $\mathcal{M}(c)$ the local mapping is executed concurrently by all nodes. The transition from c to $\mathcal{M}(c)$ is called a *round*. In the minority (resp. majority) process each node adopts the minority (resp. majority) color among all neighbors. In case of a tie the color remains unchanged. Formally, the minority process is defined for a node v as follows:

$$\mathcal{MIN}(c)(v) = \begin{cases} c(v) & \text{if } \left|N^{c(v)}(v)\right| \leq \left|N^{1-c(v)}(v)\right| \\ 1 - c(v) & \text{if } \left|N^{c(v)}(v)\right| > \left|N^{1-c(v)}(v)\right| \end{cases}$$

$N^i(v)$ denotes the set of v's neighbors with color i $(i = 0, 1)$. The definition of $\mathcal{MAJ}$ is similar, only the binary operators $\leq$ and $>$ are reversed. Sometimes a result holds for both processes. To simplify matters in these cases we use the symbol $\mathcal{M}$ as a placeholder for $\mathcal{MIN}$ and $\mathcal{MAJ}$. Figure 1 depicts a sequence of colorings for $\mathcal{MIN}$.

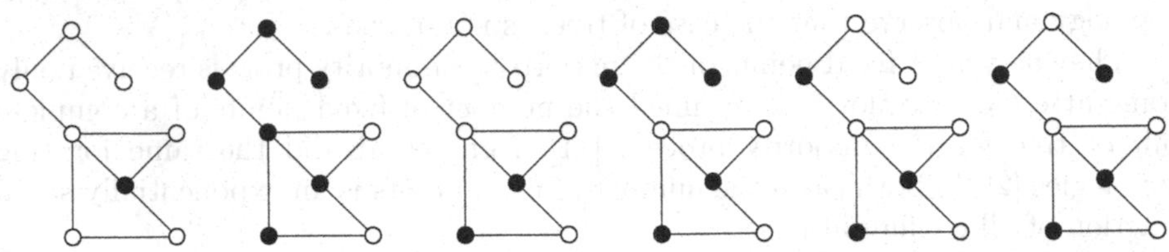

Fig. 1. For the initial coloring on the left $\mathcal{MIN}$ reaches after five rounds the coloring shown on the right. $\mathcal{MAJ}$ reaches for the same initial coloring after one round a monochromatic coloring.

In this paper we are interested in colorings with specific properties. Let $c \in \mathcal{C}(G)$. If $\mathcal{M}(c) = c$ then c is called a *fixed point*. It is called a *2-cycle* if $\mathcal{M}(c) \neq c$ and $\mathcal{M}(\mathcal{M}(c)) = c$. A 2-cycle is called *pure* if $\mathcal{M}(c)(v) \neq c(v)$ for each node v of G. c is called *monochromatic* if all nodes have the same color, i.e., $c(v) = c(w)$ for all $v, w \in V$. c is called *independent* if the color of each node is different from the colors of all its neighbors. Clearly, a monochromatic (resp. independent) coloring is a fixed point for the majority (resp. minority) process. An edge (v, w) is called *monochromatic* for c if $c(v) = c(w)$ otherwise it is called *multichromatic*.

For a mapping $\mathcal{M}$ denote by $\mathcal{F}_\mathcal{M}(G)$, $\mathcal{C}^2_\mathcal{M}(G)$, and $\mathcal{P}_\mathcal{M}(G)$, the set of all $c \in \mathcal{C}(G)$ that constitute a fixed point, a 2-cycle, or a pure coloring for $\mathcal{M}$. If c belongs to one of these sets the complementary coloring and $\mathcal{M}(c)$ also belong to this set. To cope with this fact we also define the sets $\mathcal{F}_\mathcal{M}(G)^+$, $\mathcal{C}^2_\mathcal{M}(G)^+$, and $\mathcal{P}_\mathcal{M}(G)^+$ as the subsets of those colorings of the corresponding sets which assign to a globally distinguished node v^* color 0. Hence, if $c \in \mathcal{F}_\mathcal{M}(G)$ then either c or the complement of c is in $\mathcal{F}_\mathcal{M}(G)^+$.

3.1 Notation

Let $T(V, E)$ be a finite, undirected tree with $n = |V|$. For $F \subseteq E$ let $\mathcal{C}_T(F)$ be the set of connected components of $T \backslash F$. We define a tree $\mathcal{T}_F$ with nodes $\mathcal{C}_T(F)$ and edges F. An edge of $(u, w) \in F$ connects components $T_1, T_2 \in \mathcal{C}_T(F)$ if and only if $u \in T_1$ and $w \in T_2$. For $F \subseteq E$ and $v \in V$ denote the number of edges in F incident to v by F_v.

The nodes of a nontrivial tree T can be uniquely partitioned into two subsets, such that the nodes of each subset form an independent set. In the following we denote these independent subsets by $\mathcal{I}_0(T)$ and $\mathcal{I}_1(T)$. To enforce unambiguousness when dealing with these subsets we demand that v^* is contained in $\mathcal{I}_0(T)$. A star graph is a tree with $n-1$ leaves. The maximal degree of a tree is denoted by Δ. We denote the nth Fibonacci number by F_n, i.e., $F_0 = 0, F_1 = 1$, and $F_n = F_{n-1} + F_{n-2}$. For a set S we denote by $\mathcal{P}(S)$ the power set of S, i.e., the set of all subsets of S.

4 Fixed Points

In this section we provide a characterization of $\mathcal{F}_{\mathcal{M}}(T)$ with respect to subsets of E. In particular; we identify a set $E_{fix}(T) \subset \mathcal{P}(E)$ and define a bijection $\mathcal{B}_{fix}$ between $E_{fix}(T)$ and $\mathcal{F}_{\mathcal{M}}(T)^+$. $E_{fix}(T) \neq \emptyset$ since $\emptyset \in E_{fix}(T)$. This shows that every tree has at least one fixed point. The definition of $\mathcal{B}_{fix}$ is different for $\mathcal{MIN}$ and $\mathcal{MAJ}$. These results allow to characterize the fixed points of paths. In the second subsection we prove an upper bound for $|\mathcal{F}_{\mathcal{M}}(T)|$ in terms of n and Δ. For the case of paths we give the exact numbers. In the last part we provide an output-sensitive algorithm to enumerate all fixed points.

4.1 The Bijection $\mathcal{B}_{fix}$

It is easy to see that for $c \in \mathcal{F}_{\mathcal{MIN}}(T)$ nodes adjacent to edges monochromatic for c have degree at least two, moreover at most one half of the adjacent edges of each node are monochromatic for c. Surprisingly the inverse of this statement is true in general and forms the basis for defining the bijection $\mathcal{B}_{fix}$: If F is a subset of the edges of T such that nodes adjacent to edges in F have degree at least two and at most one half of the adjacent edges of each node are in F then F uniquely defines a fixed point of T for $\mathcal{M}$.

Lemma 1. *Let T be a tree, $c \in \mathcal{F}_{\mathcal{M}}(T)$, and F the set of monochromatic (resp. multicolored) edges $(u, w) \in E$ if $\mathcal{M} = \mathcal{MIN}$ (resp. $\mathcal{M} = \mathcal{MAJ}$). If $(u, w) \in F$ then $deg_T(u) \geq 2$ and $deg_T(w) \geq 2$. Furthermore, $F_v \leq deg_T(v)/2$ for each node v of T.*

Proof. Assume $\mathcal{M} = \mathcal{MIN}$, the other case is proved similarly. Then $\left|N_T^{1-c(u)}(u)\right| \geq \left|N_T^{c(u)}(u)\right| \geq 1$ for $(u, w) \in F$. Thus, $deg_T(u) = \left|N_T^{c(u)}(u)\right| + \left|N_T^{1-c(u)}(u)\right| \geq 2$. Similarly $deg_T(w) \geq 2$. Let $v \in V$. Then $\left|N_T^{c(v)}(v)\right| \leq \left|N_T^{1-c(v)}(v)\right|$ since $c \in \mathcal{F}_{\mathcal{M}}(T)$, i.e., $deg(v) \geq 2\left|N_T^{c(v)}(v)\right| = 2F_v$. □

The last lemma motivates the following definition of $E_{fix}(T)$. Note that $E_{fix}(T)$ satisfies the hereditary property

Definition 1. *Let T be a tree. $E^2(T)$ denotes the set of edges of T where each end node has degree at least two. $F \subseteq E^2(T)$ is called* legal *if $F_v \le deg(v)/2$ for each node v. $E_{fix}(T)$ denotes the set of all legal subsets of a tree T.*

Theorem 1. *For a tree T there exists a bijection $\mathcal{B}_{fix}$ between $E_{fix}(T)$ and $\mathcal{F}_{\mathcal{M}}(T)^+$.*

Proof. Assume $\mathcal{M} = \mathcal{MIN}$. Let $F \in E_{fix}(T)$. We define a coloring $c_F \in \mathcal{F}_{\mathcal{MIN}}(T)$. Let $T^* \in \mathcal{C}_T(F)$ with $v^* \in T^*$. Let $c_F(v^*) = 0$ and extend c_F to an independent coloring of T^*, e.g., by using breadth-first search. This uniquely defines c_F on T^*. We extend c_F successively to a coloring with $c_F \in \mathcal{F}_{\mathcal{MIN}}(T)^+$. While there exists an already colored node u that has an uncolored neighbor do the following. Let $T_1 \in \mathcal{C}_T(F)$ with $u \in T_1$, $N_1 = N_{T_1}(u)$, and $N_2 = N_T(u) \setminus N_1$. All nodes in N_1 have color $1 - c_F(u)$ and $F_u = |N_2|$. No node of N_2 has yet been assigned a color. By assumption we have $|N_2| \le deg_T(u)/2$. Hence, $|N_2| \le |N_1|$. Set $c_F(w) = c_F(u)$ for all $w \in N_2$. For each $w \in N_2$ let $T_w \in \mathcal{C}_T(F)$ with $w \in T_w$. Extend c_F to an independent coloring on each T_w. Then $\left|N_T^{c_F(u)}\right| \le \left|N_T^{1-c_F(u)}\right|$. Clearly this uniquely defines c_F and $c_F \in \mathcal{F}_{\mathcal{MIN}}(T)^+$. Now we can define $\mathcal{B}_{fix}(F) = c_F$ for each $F \in E_{fix}(T)$.

Let $F_1 \ne F_2 \in E_{fix}(T)$ and $e = (u,w) \in F_1 \setminus F_2$. Then $c_{F_1}(w) = c_{F_1}(u)$ and $c_{F_2}(w) \ne c_{F_2}(u)$. Hence, $c_{F_1} \ne c_{F_2}$. Thus, $\mathcal{B}_{fix}(F)$ is injective. Next, we prove that $\mathcal{B}_{fix}$ is surjective, i.e., for every $c \in \mathcal{F}_{\mathcal{MIN}}(T)^+$ there exists $F_c \in E_{fix}(T)$ such that $\mathcal{B}_{fix}(F_c) = c$. For $c \in \mathcal{F}_{\mathcal{MIN}}(T)^+$ let $F_c = \{(u,w) \in E \mid c(u) = c(w)\}$. Then $F_c \in E_{fix}(T)$ by Lemma 1. By the first part of this proof we have $\mathcal{B}_T(F_c) \in \mathcal{F}_{\mathcal{MIN}}(T)^+$. Let $v \in T^*$ and $u \in N_{T^*}(v)$. Then $c(u) \ne c(v)$, otherwise $u \notin T^*$. Hence, $\mathcal{B}_T(F_c)$ is for T^* independent. Since $c_{F_c}(v^*) = c(v^*) = 0$ we have $\mathcal{B}_T(F_c)(v) = c(v)$ for all $v \in T^*$. Next we repeat this argument for all $\hat{T} \in \mathcal{C}_T(F_c)$. Thus, c and $\mathcal{B}_T(F_c)$ define the same coloring of T, i.e., $\mathcal{B}_T(F_c) = c$. □

Theorem 1 implies the following two results.

Corollary 1. *Let T be a tree. The minority process has an independent fixed point. It has a non-independent fixed point if and only if T has at least two inner nodes. The majority process has a monochromatic fixed point. It has a non-monochromatic fixed point if and only if T has at least two inner nodes.*

Corollary 2. *A coloring of a path is a fixed point of the minority (resp. majority) process if and only if each node has at least one neighbor with a different (resp. same) color.*

4.2 Counting Fixed Points

Theorem 1 allows to compute the number of fixed points in specific cases. If $\Delta = n - 1$ (resp. $\Delta = n - 2$) then $|\mathcal{F}_{\mathcal{MIN}}(T)| = 2$ (resp. $|\mathcal{F}_{\mathcal{MIN}}(T)| = 4$).

Furthermore, $|\mathcal{F}_{\mathcal{MIN}}(T)| \leq 8$ if $\Delta = n-3$. To get more general results we describe an algorithm $\mathcal{A}_{\mathcal{M}}$ to generate all fix points of a given tree T. We start with node v^* and color it with 0. Algorithm $\mathcal{A}_{\mathcal{M}}$ is recursive and extends a partial coloring by coloring all uncolored neighbors of an already colored node. In this context a partial coloring is a coloring of a subset of the nodes of T with the following property: Let v be an already colored node. Firstly, all nodes on the path from v^* to v in T are also colored. Secondly, if a neighbor of v other than the one closer to v^* is colored, then all neighbors of v are colored.

The details of a recursive call for the minority process, i.e., $\mathcal{A}_{\mathcal{MIN}}$ are as follows. Given a partial coloring c, a single invocation generates several extensions of c, all of them are again partial colorings covering more nodes. Let v be an already colored node that has an uncolored neighbor. First, each uncolored neighbor of v that is a leaf gets the complementary color of v. Then v has $r = deg(v) - |N^0(v)| - |N^1(v)|$ uncolored neighbors. Let U be the set of the uncolored neighbors of v, note none of them is a leaf. We color $\hat{N}_0$ (resp. $\hat{N}_1$) of these r neighbors with color 0 (resp. 1), i.e., $r = \hat{N}_0 + \hat{N}_1$. In order to produce a fixed point the following inequality must be satisfied:

$$\left|N^{c(v)}(v)\right| + \hat{N}_{c(v)} \leq \left|N^{1-c(v)}(v)\right| + \hat{N}_{1-c(v)} = \left|N^{1-c(v)}(v)\right| + r - \hat{N}_{c(v)}.$$

Hence,

$$\hat{N}_{c(v)} \leq \frac{r + |N_{1-c(v)}(v)| - |N_{c(v)}(v)|}{2}. \tag{1}$$

Let

$$r_0 = \min\left(\lfloor (r + |N_{1-c(v)}(v)| - |N_{c(v)}(v)|)/2 \rfloor, r\right). \tag{2}$$

For $i = 0, \ldots, r_0$ we extend c by coloring a subset S of U of size i with color $c(v)$ and the remaining nodes $U \setminus S$ with color $c(v) - 1$. This way we get $\sum_{i=0}^{r_0} \binom{r}{i}$ extended partial colorings. $\mathcal{A}_{\mathcal{MIN}}$ is applied to each of these extensions and terminates when all nodes are colored. Clearly, the resulting colorings are fixed points and all fixed points are generated this way. Algorithm $\mathcal{A}_{\mathcal{MAJ}}$ differs only in two places. Firstly, uncolored neighbors of v that are leaves gets the same color as v. Secondly, in Eq. (1) operator $\geq$ must be replaced by $\leq$ and the assignment of colors to nodes in U is reversed.

Next we prove an upper bound for $|F_{\mathcal{M}}(T)|$. According to Corollary 1 each tree has at least two fixed points. A star graph is an extreme case, because it only has two fixed points. The other extreme are paths as shown in this section.

Lemma 2. *Let T be a tree with a path v_0, v_1, v_2, v_3 such that $deg(v_0) = 1$ and $deg(v_1) = deg(v_2) = 2$. Let $T^0 = T \setminus v_0$ and $T^1 = T^0 \setminus v_1$. Then $|\mathcal{F}_{\mathcal{M}}(T)| = |\mathcal{F}_{\mathcal{M}}(T^0)| + |\mathcal{F}_{\mathcal{M}}(T^1)|$.*

Theorem 2. *Let T be a tree and P a path. Then $|\mathcal{F}_{\mathcal{M}}(T)| \leq 2F_{n-\lceil \Delta/2 \rceil}$ and $|\mathcal{F}_{\mathcal{M}}(P)| = 2F_{n-1}$.*

Figure 2 shows that the bound of Theorem 2 is not sharp. Let B_h be a binary tree of depth h. The equation $|\mathcal{F}_{\mathcal{M}}(B_h)| = |\mathcal{F}_{\mathcal{M}}(B_{h-1})| \left(|\mathcal{F}_{\mathcal{M}}(B_{h-1})| + 2\,|\mathcal{F}_{\mathcal{M}}(B_{h-2})|^2\right)$ already contained in [11] directly follows from Theorem 1.

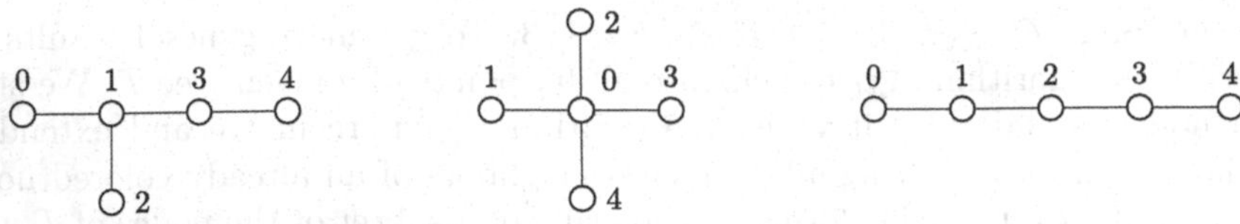

Fig. 2. Three trees with five nodes having 4, 2, and 6 fixed points for $\mathcal{MIN}$.

4.3 Generating Fixed Points

The fixed points of a tree T can be generated by iterating over all subsets of $E^2(T)$ and outputting the legal ones. The algorithm exploits the fact that $E_{fix}(T)$ has the hereditary property, i.e., if $X \in E^2(T)$ is legal, all subset of X are also legal. Algorithm 1 describes an output-sensitive algorithm running in time $O(n + |\mathcal{F}_{\mathcal{M}}(T)| \times |E^2(T)|)$. Since $|E^2(T)| \leq n$ the running time is in $O(n\,|\mathcal{F}_{\mathcal{M}}(T)|)$. If $E^2(T) = \{e_1, \ldots, e_l\}$ then the edges $\{e_1, \ldots, e_i\}$ for $i = 0, \ldots, l$. The inner **foreach**-loop always iterates over the list *fixedPoints* beginning at the first entry.

Algorithm 1: Algorithm to generate a list of all fixed points of a tree $T(V, E)$

```
E² := {(u, w) ∈ E | deg(u) ≥ 2 and deg(w) ≥ 2};
fixedPoints := ∅; fixedPoints.append(∅);
foreach e ∈ E² do
    count := fixedPoints.size();
    foreach X ∈ fixedPoints do
        if {e} ∪ X is legal then
            fixedPoints.append({e} ∪ X);
        count := count − 1;
        if count == 0 then
            break;
return fixedPoints;
```

Theorem 3. *Algorithm 1 computes all $|\mathcal{F}_{\mathcal{M}}(T)|$ fixed points of a tree T in time $O(n + |\mathcal{F}_{\mathcal{M}}(T)| \times |E^2(T)|)$ using $O(|E^2(T)| \times |\mathcal{F}_{\mathcal{M}}(T)|)$ memory.*

Proof. By Theorem 1 each legal subset of $E^2(T)$ uniquely corresponds to a fixed point of T. If a subset S of $E^2(T)$ is not legal, then no superset of S is legal and if S is legal then all subsets of S are legal. Therefore, the algorithm generates all legal subsets of $E^2(T)$. Let $l = |E^2(T)|$. Denote by S_i the set of elements of the list $fixedPoints$ at the beginning of the ith outer **foreach**-loop and S_{l+1} the elements of $fixedPoints$ after the last execution. Then $|S_1| = 1$ and $|\mathcal{F}(T)^+| = |S_{l+1}|$.

Next we prove that $(4/5)|S_{i+1}| \geq |S_i|$ for $i = 1,\ldots,l$. Let $e = (u,w) \in E^2(T)$. For $X \in S_i$ denote the number of edges in X that are incident with a node v by X_v. Let $\bar{S} = S_i$ and $\hat{S} = \emptyset$. Let $X \in \bar{S}$ with $X_u + 1 > deg(u)/2$ and $X_w + 1 > deg(w)/2$. Let e_u (resp. e_w) be an edge of X that is incident with u (resp. w). Then we remove $X, X \setminus \{e_u, e_w\}, X \setminus \{e_u\}$, and $X \setminus \{e_w\}$ from $\bar{S}$ and insert $X, X \setminus \{e_u, e_w\}, X \setminus \{e_u\}, X \setminus \{e_w\}$, and $X \setminus \{e_u, e_w\} \cup \{e\}$ into $\hat{S}$. We repeat this process until there is no X in $\bar{S}$ with the above property. Next, let $X \in \bar{S}$ with $X_u + 1 > deg(u)/2$ and $X_w + 1 \leq deg(w)/2$. Let e_u be an edge of X that is incident with u. Then we remove X, and $X \setminus \{e_u\}$ from $\bar{S}$ and insert $X, X \setminus \{e_u\}, X \setminus \{e_u\} \cup \{e\}$ into $\hat{S}$. We repeat this process until there is no X in $\bar{S}$ with the above property. Finally, for the remaining $X \in \bar{S}$ we insert $X, X \cup \{e\}$ into $\hat{S}$. Assume, that S_i contains n_1, n_2 resp. n_3 elements according to the above classification, then $|S_i| = 4n_1 + 2n_2 + n_3$ and $\left|\hat{S}\right| = 5n_1 + 3n_2 + 2n_3$. Since $S_{i+1} = \hat{S}$ we have $(4/5)|S_{i+1}| \geq |S_i|$. The overall number of executions of the inner **foreach**-loop is $\sum_{i=1}^{l} |S_i|$. Thus,

$$\sum_{i=1}^{l} |S_i| \leq (4/5) \sum_{i=2}^{l+1} |S_i| = (4/5) \sum_{i=1}^{l} |S_i| + (4/5)(|S_{l+1}| - 1).$$

Hence, $\sum_{i=1}^{l} |S_i| \leq 4(|S_{l+1}| - 1) < 4|\mathcal{F}(T)^+|$. In time $O(n)$ we provide the degrees of all nodes in an array. Also the test whether $X \cup e$ is legal and append the entry to the list can be performed in time $O(|X|)$. □

The bound $(4/5)|S_{i+1}| \geq |S_i|$ for all i can be used to prove the lower bound of $((5/4)^l$ with $l = \left|E^2(T)\right|$ for $|\mathcal{F}_{\mathcal{M}}(T)|$. We conjecture that a more detailed analysis of the relation between $|S_{i+1}|$ and $|S_i|$ leads to a better bound.

Finally, we sketch an alternative approach for computing all fixed points. The elements of $E_{fix}(T)$ correspond to the solutions of a system of linear diophantine inequalities $Ax \leq b$. Here, A is a binary $\left|E^2(T)\right| \times n$ matrix, where $a_{i,j} = 1$ if node i is incident with edge j of $E^2(T)$ and $b_i = \lfloor deg_T(i)/2 \rfloor$. Thus, by Theorem 1 the set of fixed points corresponds to the solutions of $Ax \leq b$. Unfortunately there isn't much work available for solving systems of linear diophantine inequalities [7].

5 General 2-Cycles

In this section we analyze the structure of $\mathcal{C}^2_{\mathcal{M}}(T)$. First we collect general results about colorings from $\mathcal{C}^2_{\mathcal{MIN}}(T)$. In the second subsection we consider the set $c \in \mathcal{P}_{\mathcal{M}}(T)$ of all pure colorings. We first prove properties of c and use these to define the set $E_{pure}(T)$ and define a bijection $\mathcal{B}_{pure}$ between $E_{pure}(T)$ and $\mathcal{P}_{\mathcal{M}}(T)^+$. Since $E_{pure}(T) \neq \emptyset$ this shows that every tree has pure coloring. These results immediately lead to a simple characterization pure coloring of paths. In the third subsection we derive from $\mathcal{B}_{pure}$ an upper bound for $|\mathcal{P}_{\mathcal{M}}(T)|$ in terms of n. Finally we consider the general case of 2-cycles. We prove that T decomposes into subtrees, such that c is either a fixed point or a pure coloring on each of these subtrees. These subtrees provide the basis to define a hyper structure of a tree, called the *block tree*. After analyzing properties of block trees we define a set $E_{block}(T)$ of subsets of the edge set of a tree T and show in Theorem 6 that the elements of $E_{block}(T)$ correspond one-to-one with the block trees of T. Since $E_{block}(T)$ does not have the hereditary property, we cannot use the approach of Algorithm 1 to enumerate all block trees.

5.1 General Results

Let $c \in \mathcal{C}^2_{\mathcal{M}}(T)$. We separate the nodes of T in two groups. A node u is called a *fixed node* for c if $\mathcal{M}(c)(u) = c(u)$; it is called a *toggle node* for c if $\mathcal{M}(c)(u) \neq c(u)$. Note that in any case $\mathcal{M}(\mathcal{M}(c))(u) = c(u)$. Denote by $N^i_f(u)$ (resp. $N^i_t(u)$) the number of neighbors of u with color i that are fixed (resp. toggle) nodes for c.

First, we provide a simple characterization of fixed and toggle nodes for $\mathcal{MIN}$, a corresponding result holds for $\mathcal{MAJ}$.

Lemma 3. *Let T be a tree and $c \in \mathcal{C}^2_{\mathcal{MIN}}(T)$. A node u of T is a fixed node of c if and only if $\left|N_t^{1-c(u)}(u) - N_t^{c(u)}(u)\right| \leq N_f^{1-c(u)}(u) - N_f^{c(u)}(u)$ and a toggle node of c if and only if $\left|N_f^{c(u)}(u) - N_f^{1-c(u)}(u)\right| < N_t^{c(u)}(u) - N_t^{1-c(u)}(u)$.*

5.2 Pure 2-Cycles

If $c \in \mathcal{P}_{\mathcal{M}}(T)$ then each node of T is a toggle node. In Theorem 4 we give a characterization $\mathcal{P}_{\mathcal{M}}(T)$, it allows to generate all pure 2-cycles and compute $|\mathcal{P}_{\mathcal{M}}(T)|$.

Lemma 4. *Let T be a tree, $c \in \mathcal{C}_{\mathcal{M}}(T)$. Then $c \in \mathcal{P}_{\mathcal{MIN}}(T)$ (resp. $c \in \mathcal{P}_{\mathcal{MAJ}}(T)$) if and only if $N^{c(u)}(u) > N^{1-c(u)}(u)$ (resp. $N^{c(u)}(u) < N^{1-c(u)}(u)$) for each node u.*

As in Sect. 4.1 we use properties of monochromatic edges to characterize pure 2-cycles. Corollary 3 is similar to Lemma 1 and is used to define the set $E_{pure}(T)$.

Lemma 5. *Let T be a tree, $c \in \mathcal{P}_{\mathcal{M}}(T)$, and $e = (u, w) \in E$ with $c(u) \neq c(w)$ if $\mathcal{M} = \mathcal{MIN}$ and $c(u) = c(w)$ if $\mathcal{M} = \mathcal{MAJ}$. Let T_u (resp. T_w) be the subtree of $T \setminus e$ that contains u (resp. w). Then u and w have degree at least 3, T_u and T_w contain at least 3 nodes, and c induces a pure 2-cycle on both subtrees.*

Proof. We state the proof for $\mathcal{M} = \mathcal{MIN}$. Since c is pure we have $N_T^{c(u)}(u) > N_T^{1-c(u)}(u)$ and since $c(u) \neq c(w)$ we also have $N_T^{1-c(u)}(u) \geq 1$. Hence, $deg(u) = N_T^{c(u)}(u) + N_T^{1-c(u)}(u) \geq 3$. Similarly $deg(w) \geq 3$. Let $v \in T_u$. If $v \neq u$ then all neighbors of v in T are in T_u and thus $\left|N_{T_u}^{c(u)}(u)\right| > \left|N_{T_u}^{1-c(u)}(u)\right|$. Next consider the case $v = u$. Since c is pure, there exists in $N(u)$ at least one more node with color $c(u)$ than with color $c(w)$. Thus, u has at least two neighbors in T_u, hence T_u contains at least three nodes. Since $\left|N_{T_u}^{c(u)}(u)\right| = \left|N_T^{c(u)}(u)\right| > \left|N_T^{1-c(u)}(u)\right| = \left|N_{T_u}^{1-c(u)}(u)\right| + 1$ we have $\left|N_{T_u}^{c(u)}\right| > \left|N_{T_u}^{1-c(u)}(u)\right|$. Hence, Lemma 4 implies that c induces a pure 2-cycle for $\mathcal{MIN}$ on T_u. The same is true for T_w. □

Corollary 3. *Let T be a tree. If $c \in \mathcal{P}_{\mathcal{MIN}}(T)$, $F_c = \{(u,w) \in E \mid c(u) \neq c(w)\}$, and $\hat{T} \in \mathcal{C}_T(F_c)$ then $\left|\hat{T}\right| \geq 3$ and c induces a monochromatic coloring on $\hat{T}$. If $c \in \mathcal{P}_{\mathcal{MAJ}}(T)$, $F_c = \{(u,w) \in E \mid c(u) = c(w)\}$, and $\hat{T} \in \mathcal{C}_T(F_c)$ then $\left|\hat{T}\right| \geq 3$ and c induces an independent coloring on $\hat{T}$. Furthermore, $(F_c)_v < deg_T(v)/2$ for $v \in V$.*

Corollary 3 motivates the following definition of $E_{pure}(T)$. Note that $E_{pure}(T)$ satisfies the hereditary property and $E_{pure}(T) = E_{fix}(T)$ if all degrees of T are odd.

Definition 2. *Let T be a tree. $E^3(T)$ denotes the set of all edges of T where each end node has degree at least three. $F \subseteq E^3(T)$ is called* legal *if $F_v < deg(v)/2$ for each node v. $E_{pure}(T)$ denotes the set of all legal subsets of $E^3(T)$.*

Theorem 4. *For a tree T there exists a bijection $\mathcal{B}_{pure}$ between $E_{pure}(T)$ and $\mathcal{P}_{\mathcal{M}}(T)^+$.*

Proof. Let $F \in E_{pure}(T)$. We uniquely partition the nodes of $\mathcal{T}_F$ into two independent subsets $\mathcal{I}_0$ and $\mathcal{I}_1$ with $v^* \in \mathcal{I}_0$. Assume $\mathcal{M} = \mathcal{MIN}$. Define a mapping $C_F : \mathcal{C}_T(F) \rightarrow \{0,1\}$ by setting $C_F(\hat{T}) = i$ if $\hat{T} \in \mathcal{I}_i$. Based on C_F we define a coloring c_F of T as follows $c_F(v) = C_F(\hat{T})$ if $v \in \hat{T}$. Note that $c_F(v^*) = 0$. F uniquely defines c_F, since for each node v there is a unique $\hat{T} \in \mathcal{C}_T(F)$ that contains v. First, we prove that $c_F \in \mathcal{P}_{\mathcal{MIN}}(T)^+$. For $v \in V$ let $\hat{T} \in \mathcal{C}_T(F)$ with $v \in \hat{T}$. Then $N(v) \cap \hat{T} = N_T^{c_F(v)}(v)$. Since $F \in E_{pure}(T)$ we have $\left|N_T^{c_F(v)}(v)\right| > deg(v)/2$. Thus, $2\left|N_T^{c_F(v)}(v)\right| > \left|N_T^{c_F(v)}(v)\right| + \left|N_T^{1-c_F(v)}(v)\right|$ and hence, $\left|N_T^{c_F(v)}(v)\right| > \left|N_T^{1-c_F(v)}(v)\right|$ for all v. Hence, $c_F \in \mathcal{P}_{\mathcal{MIN}}(T)^+$ by Lemma 4. Now we can define $\mathcal{B}_{pure}(F) = c_F$ for each $F \in E_{pure}(T)$. Let $F_1 \neq F_2 \in E_{pure}(T)$ and $e = (u,w) \in F_1 \setminus F_2$. Then $c_{F_1}(w) \neq c_{F_1}(u)$ and $c_{F_2}(w) = c_{F_2}(u)$. Hence, $c_{F_1} \neq c_{F_2}$, i.e., $\mathcal{B}_{pure}(F)$ is injective. Next, we prove that $\mathcal{B}_{pure}$ is surjective, i.e., for every $c \in \mathcal{P}_{\mathcal{MIN}}(T)^+$ there exists $F_c \in E_{pure}(T)$ with $\mathcal{B}_{pure}(F_c) = c$. For $c \in \mathcal{P}_{\mathcal{MIN}}(T)^+$ define $F_c = \{(u,w) \in E \mid c(u) \neq c(w)\}$. By Lemma 5 we have $F_c \in E^3(T)$. Let $v \in V$. Since c is a pure 2-cycle we have $\left|N_T^{c(v)}(v)\right| > \left|N_T^{1-c(v)}(v)\right|$, i.e., $deg(v) > 2\left|N_T^{1-c(v)}(v)\right|$. Since, $(F_c)_v = \left|N_T^{1-c(v)}(v)\right|$ we have $deg(v)/2 > (F_c)_v$. This yields $F_c \in E_{pure}(T)$. By the first part of this proof we have $\mathcal{B}_{pure}(F_c) \in \mathcal{P}_{\mathcal{MIN}}(T)^+$. By Corollary 3 $\mathcal{B}_{pure}(F_c)$ is for each tree $\hat{T} \in \mathcal{C}_T(F_c)$ a monochromatic coloring with

$\mathcal{B}_{pure}(F_c)(v) = c(v)$ for all $v \in \hat{T}$. Hence, c and $\mathcal{B}_{pure}(F_c)$ define the same coloring of T, i.e., $\mathcal{B}_{pure}(F_c) = c$.

The proof for the case $\mathcal{M} = \mathcal{MAJ}$ is similar. The main differences are that we define c_F such that it induces an independent coloring on each $\hat{T} \in \mathcal{C}_T(F)$ and in the second part we define $F_c = \{(u, w) \in E \mid c(u) = c(w)\}$. □

Corollary 4. *Every tree T has a pure coloring for the minority and the majority process. T has a non-monochromatic (resp. non-independent) pure coloring for the minority (resp. majority) process if and only if there exist an edge $(u, w) \in T$ such that $deg(u) \geq 3$ and $deg(w) \geq 3$.*

Proof. We provide the proof for $\mathcal{M} = \mathcal{MIN}$. The result follows from Theorem 4. Since $\emptyset \in E_{pure}(T)$ we have $c_\emptyset \in \mathcal{F}_\mathcal{M}(T)^+$. $c_\emptyset$ is a monochromatic coloring. T has a non-monochromatic pure coloring if and only if $E_{pure}(T) \neq \emptyset$. This is equivalent to having an edge with the stated properties. □

Corollary 5. *Let P be a path and $c \in \mathcal{C}(P)$. Then $c \in \mathcal{P}_{\mathcal{MIN}}(P)$ (resp. $c \in \mathcal{P}_{\mathcal{MAJ}}(P)$) if and only if $c(v) = c(w)$ (resp. $c(v) \neq c(w)$) for each edge (v, w) of P.*

Since $E_{pure}(T) \subseteq E_{fix}(T)$ we have $\mathcal{P}_{\mathcal{MAJ}}(T) \subseteq \mathcal{F}_{\mathcal{MIN}}(T)$ and $\mathcal{P}_{\mathcal{MIN}}(T) \subseteq \mathcal{F}_{\mathcal{MAJ}}(T)$. Figure 3 shows that there are trees T where $\mathcal{P}_{\mathcal{MAJ}}(T) \subset \mathcal{F}_{\mathcal{MIN}}(T)$ and $\mathcal{P}_{\mathcal{MIN}}(T) \subset \mathcal{F}_{\mathcal{MAJ}}(T)$.

Fig. 3. The left coloring is in $\mathcal{F}_{\mathcal{MAJ}}(T) \setminus \mathcal{P}_{\mathcal{MIN}}(T)$, the right one is in $\mathcal{F}_{\mathcal{MIN}}(T) \setminus \mathcal{P}_{\mathcal{MAJ}}(T)$.

5.3 Counting Pure 2-Cycles

Theorem 4 allows to determine the pure 2-cycles of a tree T, and thus, $|\mathcal{P}_\mathcal{M}(T)|$. Since $E_{pure}(T) \subseteq E_{fix}(T)$ we have $|\mathcal{P}_\mathcal{M}(T)| \leq |\mathcal{F}_\mathcal{M}(T)|$ and $|\mathcal{P}_\mathcal{M}(T)| \leq 2F_{n-\lceil\Delta/2\rceil}$ by Theorem 2. To generate all pure 2-cycles Algorithm 1 can be adopted, note that $E_{pure}(T)$ has the hereditary property. The difference is that it uses $E^3(T)$ and the corresponding notion of legal. The algorithm works in time $O(n + |\mathcal{P}_\mathcal{M}(T)|\,|E^3(T)|)$. Next we provide a better upper bound for $|\mathcal{P}_\mathcal{M}(T)|$. Let $e_T = |E^3(T)|$.

Lemma 6. *Let T be a tree, then $e_T \leq (n-4)/2$.*

The last lemma implies $|\mathcal{P}_\mathcal{M}(T)| \leq 2^{1+(n-4)/2}$. This bound is purely based on the bound for $|E^3(T)|$. By utilizing the constraints imposed by $E_{pure}(T)$ better bounds may be derived. The tree H_n with $n \equiv 0(2)$ that consists of a path of length $(n+2)/2$ and a single node attached to each inner node of the path (see Fig. 4) shows that the bound of Lemma 6 is sharp, but there is large gap between $|E^3(H_n)|$ and $|E_{pure}(H_n)|$.

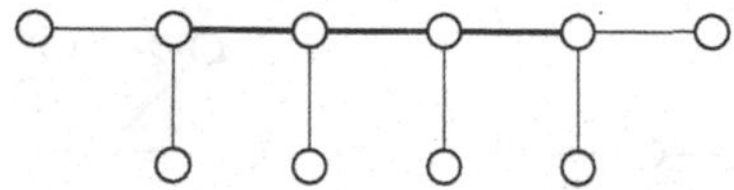

Fig. 4. The graph H_{10}, the three edges belonging to $E^3(H_{10})$ are depicted by solid lines. In general we have $\left|E^3(H_n)\right| = 2^{(n-4)/2}$ and $|E_{pure}(H_n)| = F_{n/2}$.

5.4 Block Trees of 2-Cycles

In this section we consider general 2-cycles, i.e., those that have both fixed and toggle nodes. We characterize the coarse grain structure of $\mathcal{C}^2_{\mathcal{M}}(T)$, called the block tree of T.

Definition 3. *Let T be a tree and $c \in \mathcal{C}^2_{\mathcal{M}}(T)$. Let V_f (resp. V_t) be the set of fixed (resp. toggle) nodes of c and T^f (resp. T^t) the subgraph of T induced by V_f (resp. V_t).*

The next result shows that a 2-cycle c induces a structure on T that allows to define a hypertree $\mathcal{B}_c(T)$.

Lemma 7. *Let T be a tree, $c \in \mathcal{C}^2_{\mathcal{M}}(T)$, and T' a connected component of T^f (resp. T^t). Then c induces a fixed point (resp. a pure 2-cycle) on T'.*

Proof. We assume $\mathcal{M} = \mathcal{MIN}$. Let T' be a connected component of T^f and u a node of T'. With respect to T we have $\left|N_t^{1-c(u)}(u) - N_t^{(u)}(u)\right| \leq N_f^{1-c(u)}(u) - N_f^{c(u)}(u)$ by Lemma 3. Restricting c to T' gives $N_{T'}^{c(u)}(u) = N_f^{c(u)}(u)$ and $N_{T'}^{1-c(u)}(u) = N_f^{1-c(u)}(u)$. Thus, $N_{T'}^{1-c(u)}(u) \geq N_{T'}^{c(u)}(u)$ and u is a fixed node of T' for c. Hence, c is a fixed point for T'. The result about components of T^t is proved similarly. □

Lemma 7 provides the base to define the *block tree* of a coloring $c \in \mathcal{C}^2_{\mathcal{M}}(T)$.

Definition 4. *Let T be a tree, $c \in \mathcal{C}^2_{\mathcal{M}}(T)$, and $T_1, \ldots, T_s$ the connected components of T^f and T^t. The block tree $\mathcal{B}_c(T)$ of T for c is a tree with nodes $\{T_1, \ldots, T_s\}$, nodes T_i and T_j are connected if there exists $(u, w) \in E$ with $u \in T_i$ and $w \in T_j$. A node T_i is called a* fixed block *(resp.* toggle block*) of $\mathcal{B}_c(T)$ if T_i is a connected component of T^f (resp. T^t).*

Obviously $\mathcal{B}_c(T)$ is a tree. $\mathcal{B}_c(T)$ is uniquely defined, but different 2-cycles can induce the same block tree (see Fig. 5). Each edge e of $\mathcal{B}_c(T)$ connects a fixed block with a toggle block, e uniquely corresponds to an edge of T. For convenience we denote this edge also by e. If T_i is a toggle block then obviously $|T_i| \geq 2$, since all neighboring blocks are fixed blocks. Fixed blocks can consist of a single node only (see Fig. 6).

Fig. 5. Two colorings leading to the same block tree. For the minority process both colorings define the same block tree. The left block node is a toggle node while the right is a fixed point.

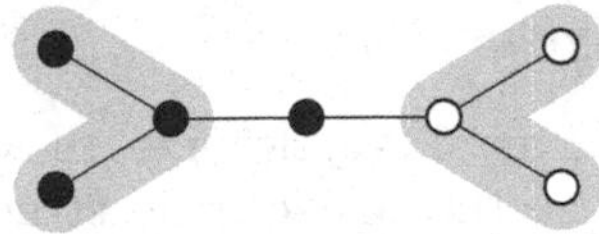

Fig. 6. A block tree consisting of two toggle blocks and one fixed block with a single node.

The goal of this section is to present a characterization of the set of all block trees for a given tree T similar to Theorem 4, i.e., the trees T_B for which there exists $c \in \mathcal{C}^2_{\mathcal{M}}(T)$ such that $T_B = \mathcal{B}_c(T)$. The following theorem summarizes properties of 2-cycles.

Theorem 5. *Let T be a tree, $c \in \mathcal{C}^2_{\mathcal{M}}(T)$, and $e = (u, w)$ an edge of $\mathcal{B}_c(T)$. Then*

1. *If $deg_T(u) = 2$ then u is a fixed node.*
2. $\min(deg_T(u), deg_T(w)) \geq 2$ *and* $\max(deg_T(u), deg_T(w)) \geq 3$.
3. *If T_0 is a node of $\mathcal{B}_c(T)$, $v \in T_0$, $deg_{T_0}(v) = 1$ and $deg_T(v) \equiv 0(2)$ then v is a fixed node and T_0 is a fixed block.*
4. *If $T_0 = \{v\}$ is a node of $\mathcal{B}_c(T)$ then v is a fixed node, T_0 is a fixed block, and $deg_T(v)$ is even.*

Proof. Assume $\mathcal{M} = \mathcal{MIN}$, the proof for $\mathcal{MAJ}$ is similar. Assume that u is toggle node. Then $|N^{c(u)}(u)| > |N^{1-c(u)}|$. Thus, if $|N^{1-c(u)}| > 0$ then $deg_T(u) \geq 3$. Therefore, $|N^{1-c(u)}| = 0$ and $|N^{c(u)}(u)| = 2$. Since u is toggle node, both neighbors must change their color, i.e., both are toggle nodes. This yields that w is a toggle node. Contradiction, since $e(u, w)$ is an edge of $\mathcal{B}_c(T)$.

WLOG we assume that u is a fixed node while w toggles its color. Assume that $\min(deg(u), deg(w)) = 1$. If $deg(u) = 1$ then u cannot be a fixed node because w toggles its color. Similarly, w cannot have degree 1. Hence, $\min(deg(u), deg(w)) \geq 2$. Assume that $deg(u) = deg(w) = 2$. Then by the first part, both nodes are fixed nodes. Contradiction. Assume that v is a toggle node. Then $N_t^{c(v)} = 1$ and $N_t^{1-c(v)} = 0$. Hence, by Lemma 3 we have $N_f^{1-c(v)} = N_f^{c(v)}$ thus, $deg_T(v) = 1 + 2N_f^{c(v)} \equiv 1(2)$. Contradiction. Let $T_0 = \{v\}$. If v is a toggle node then all neighbors are fixed nodes. Hence, v is also a fixed node. Contradiction. Lemma 3 yields that $deg_T(v)$ is even. □

The last theorem list properties of $\mathcal{B}_c(T)$ for $c \in \mathcal{C}^2_{\mathcal{M}}(T)$. As before we take these properties to identify a set of edges F_c such that $\mathcal{T}_{F_c} = \mathcal{B}_c(T)$. The following two definitions provide a formal framework for this purpose.

Definition 5. *Let T be a tree. $E^{2.5}(T)$ denotes the set of edges of T, where one end node has degree at least two and the other has degree at least 3. For $F \subseteq E^{2.5}(T)$ a component $\hat{T} \in \mathcal{C}_T(F)$ is called* fixed *if $\left|\hat{T}\right| = 1$ or if there exists $v \in \hat{T}$ such that $deg_T(v) \equiv 0(2)$ and $deg_{\hat{T}}(v) = 1$. $Fix(T, F)$ denotes the set of all fixed components of $\mathcal{C}_T(F)$.*

Definition 6. *Let T be a tree. $F \subseteq E^{2.5}(T)$ is called* legal *if all components of $Fix(T, F)$ are fully contained in $\mathcal{I}_0(\mathcal{T}_F)$ and if $T_0 \in \mathcal{C}_T(F)$ with $T_0 = \{v\}$ then $deg_T(v) \equiv 0(2)$. $E_{block}(T)$ denotes the set of all legal subsets of $E^{2.5}(T)$.*

The next result reveals the significance of $E_{block}(T)$ for block trees.

Lemma 8. *Let T be a tree, $c \in \mathcal{C}^2_{\mathcal{M}}(T)$, and F_c the edges of $\mathcal{B}_c(T)$. Then $F_c \in E_{block}(T)$.*

Proof. Note that $\mathcal{T}_{F_c} = \mathcal{B}_c(T)$. By Theorem 5.2 we have $F_c \subseteq E^{2.5}(T)$. By construction of $\mathcal{B}_c(T)$ and Theorem 5.4 and 5.3 we have $Fix(T, F_c) \subseteq \mathcal{I}_0(\mathcal{T}_{F_c})$. Theorem 5.4 completes the proof. □

Definition 7. *Let T be a tree. A coloring $c \in \mathcal{C}^2_{\mathcal{MIN}}(T)$ is called* canonical *if c induces a monochromatic (resp. independent) coloring on each connected component of T^t (resp. T^f). A coloring $c \in \mathcal{C}^2_{\mathcal{MAJ}}(T)$ is called* canonical *if c induces an independent (resp. monochromatic) coloring on each connected component of T^t (resp. T^f).*

The next result lays the groundwork for our characterization of block trees.

Lemma 9. *Let T be a tree and $F \in E_{block}(T)$. There exits $c \in \mathcal{C}^2_{\mathcal{M}}(T)$ with $\mathcal{B}_c(T) = \mathcal{T}_F$ such that c is canonical and $\mathcal{I}_0$ (resp. $\mathcal{I}_1$) is the set of fixed (resp. toggle) nodes of c.*

Proof. Assume $\mathcal{M} = \mathcal{MIN}$, $\mathcal{M} = \mathcal{MAJ}$ is similar. The proof is by induction on $|F|$. The case $|F| = 0$ is obvious, c is the monochromatic coloring. Let $|F| > 0$. Let $L \in \mathcal{T}_F$ be a leaf and $e = (u, w) \in F$ such that $w \in L$. Then $|L| \geq 2$ if $L \in \mathcal{I}_0(\mathcal{T}_F)$ and $|L| \geq 3$ if $L \in \mathcal{I}_1(\mathcal{T}_F)$. Remember that $\mathcal{I}_0(\mathcal{T}_F)$ contains the fixed components of $\mathcal{T}_F$. By the definition of $E_{block}(T)$ we have to consider four cases.

Case 1: $L \in \mathcal{I}_0(\mathcal{T}_F)$ and $|L| > 2$. We construct a tree $\tilde{T}$ as follows: Remove from T all nodes of L except w and add a new neighbor v to w. Then $\left|\tilde{T}\right| < |T|$. Then $deg_T(u) > 2$ otherwise L would not be in $\mathcal{I}_0(\mathcal{T}_F)$. Hence, $F \subseteq E^{2.5}(\tilde{T})$. Denote the leaf of $\mathcal{C}_{\tilde{T}}(F)$ consisting of v and w by $\tilde{L}$. Thus, $\tilde{L} \in Fix(\tilde{T}, F)$ and $Fix(\tilde{T}, F) = Fix(T, F) \cup \tilde{L} \setminus L \subseteq \mathcal{I}_0(\mathcal{T}_F)$. Let $T_0 = \{v\} \in \mathcal{C}_{\tilde{T}}(F)$. Then, $T_0 \in \mathcal{C}_T(F)$. Hence, $deg_T(v) \equiv 0(2)$ by assumption. Since $T_0 \in \mathcal{I}_0(\mathcal{T}_F)$ we also have $deg_{\tilde{T}}(v) \equiv 0(2)$. This shows that $\tilde{T}$ and F satisfy the theorem's assumption.

Hence, by induction there exists a canonical coloring $\tilde{c} \in \mathcal{C}^2(\tilde{T})$ with $\mathcal{B}_{\tilde{c}}(\tilde{T}) = \mathcal{T}_F$ satisfying all properties. We can extend $\tilde{c}$ to a coloring $c \in \mathcal{C}^2(T)$ by setting $c(x) = \tilde{c}(x)$ for all nodes $x \in T \setminus L$, $c(w) = \tilde{c}(w)$, and color the remaining nodes of L in the canonical way for a fixed point.

Case 2: $L \in \mathcal{I}_0(\mathcal{T}_F)$ and $|L| = 2$. Let $\tilde{F} = F \setminus e$. Let $v \in L$ be a neighbor of w and set $\tilde{T} = T \setminus v$. Let $T_u \in \mathcal{C}_T(F)$ with $u \in T_u$. Then $T_u \in \mathcal{I}_1(\mathcal{T}_F)$ and thus, $|T_u| > 1$, $deg_{T_u}(u) \geq 1$ and $deg_T(u) \geq 3$. Let $\tilde{T}_u \in \mathcal{C}_{\tilde{T}}(\tilde{F})$ with $u \in \tilde{T}_u$. Then $w \in \tilde{T}_u$, $\tilde{T}_u \in \mathcal{I}_1(\mathcal{T}_F)$ and $T_u \subset \tilde{T}_u$. Clearly, $\tilde{F} \subseteq E^{2.5}(\tilde{T})$. Let $T_0 = \{v_0\} \in \mathcal{C}_{\tilde{T}}(\tilde{F})$ with $|T_0| = 1$. Then $T_0 \in \mathcal{C}_T(F)$, thus $deg_T(v_0) \equiv 0(2)$. Hence, $deg_{\tilde{T}}(v_0) \equiv 0(2)$. Let $\hat{T} \in \mathcal{C}_{\tilde{T}}(\tilde{F})$ and $v_0 \in \hat{T}$ with $deg_{\hat{T}}(v_0) = 1$, $deg_{\tilde{T}}(v_0) \equiv 0(2)$. Assume $\hat{T} = \tilde{T}_u$. Then $v_0 \neq w$ since $deg_{\tilde{T}}(w) = 1 \not\equiv 0(2)$. Thus, $\hat{T} = \tilde{T}_u$ if $v_0 \in \hat{T}$ with $deg_{\hat{T}}(v_0) = 1$ for some $v_0 \neq w$. Hence, $\hat{T} \in Fix(\tilde{T}, \tilde{F}) = Fix(T, F) \subseteq \mathcal{I}_0(\mathcal{T}_F) = \mathcal{I}_0(\mathcal{T}_{\tilde{F}})$.

Therefore, $\tilde{T}$ and $\tilde{F}$ satisfy the theorem's assumption. By induction there exists a canonical coloring $\tilde{c} \in \mathcal{C}^2(\tilde{T})$ with $\mathcal{B}_{\tilde{c}}(\tilde{T}) = \mathcal{T}_{\tilde{F}}$ satisfying all properties. Since $\tilde{T}_u \in \mathcal{I}_1(T)$ all nodes of $\tilde{T}_u$ have the same color, thus $N_t^{1-\tilde{c}(u)}(u) = 0$ and $\tilde{c}(u) = \tilde{c}(w)$. By Lemma 3 we have $\left|N_f^{\tilde{c}(u)}(u) - N_f^{1-\tilde{c}(u)}(u)\right| < N_t^{\tilde{c}(u)}(u)$.

We change $\tilde{c}$ to a coloring c of T as follows. First, we set $c(x) = \tilde{c}(x)$ for all $x \notin \{w, v\}$. We apply Lemma 3 to prove that u is still a toggle node for c.

If $N_f^{\tilde{c}(u)}(u) > N_f^{1-\tilde{c}(u)}(u)$ we set $c(w) = 1 - \tilde{c}(w)$ and $c(v) = \tilde{c}(w)$. If $N_f^{\tilde{c}(u)}(u) < N_f^{1-\tilde{c}(u)}(u)$ we set $c(w) = \tilde{c}(w)$ and $c(v) = 1 - \tilde{c}(w)$. At last consider the case $N_f^{\tilde{c}(u)}(u) = N_f^{1-\tilde{c}(u)}(u)$. If $N_t^{\tilde{c}(u)}(u) = 2$ then $N_t^{c(u)}(u) = 1$, i.e., $deg_{T_u}(u) = 1$. Also $deg_{\tilde{T}}(u) = 2N_f^{\tilde{c}(u)}(u) + 2$, i.e., $deg_T(u) \equiv 0(2)$. Hence, $T_u \in \mathcal{I}_0(\mathcal{T}_F)$. Contradiction and thus $N_t^{\tilde{c}(u)}(u) > 2$. Set $c(w) = 1 - \tilde{c}(w)$ and $c(v) = \tilde{c}(w)$. Then $N_t^{c(u)}(u) > 1$ and thus, $\left|N_f^{c(u)} - N_f^{1-c(u)}\right| = 1 < N_t^{c(u)}(u)$. Therefore, c has the desired properties. □

Theorem 6. *For a tree T there exists a bijection $\mathcal{B}_{block}$ between $E_{block}(T)$ and the set of block trees of T of the minority and the majority process.*

Corollary 6. *Let T be a tree where all nodes have odd degree. Then $E_{block}(T) = \{F \subseteq E^3(T) \mid \mathcal{C}_T(F)$ does not contain a component of size $1\}$. Let P be a path. Then $\mathcal{C}^2_{\mathcal{M}}(P) = \mathcal{P}_{\mathcal{M}}(P)$.*

5.5 Counting Block Trees

The concept of Algorithm 1 can not be used to generate all elements of $E_{block}(T)$ because $E_{block}(T)$ does not have the hereditary property (see example in [17]). Since $E_{block}(T) \subseteq E_{fix}(T)$ each upper bound for $|\mathcal{F}_{\mathcal{M}}(T)|$ is also an upper bound for $\left|\mathcal{C}^2_{\mathcal{M}}(T)\right|$. A naive way to generate all block trees of a tree is to iterate over the set $E_{fix}(T)$ and test, whether an element is legal according to Definition 6.

6 Conclusion and Open Problems

In this paper we provided characterizations of several categories of colorings of trees for the minority and majority process in terms of subsets of the tree edges. This means that the class of trees is the first nontrivial graph class for which a complete characterization of fixed points for the minority/majority process exists. This includes an algorithm to enumerate all fixed points and upper bounds for the number of fixed points.

There are several open questions that are worth pursuing. Firstly, is it possible to characterize fixed points and pure colorings for other graph classes? Clearly, the results for trees do not hold for general graphs, e.g. for cycles. But, it might be possible to use the same approach, i.e., find suitable subsets of the edge set similar to E_{fix}.

Furthermore, the current work for trees can be improved. It would be interesting to find better general upper bounds for $|\mathcal{F}_{\mathcal{M}}(T)|$ and $|\mathcal{P}_{\mathcal{M}}(T)|$ for trees. Also, we believe that the run-time of Algorithm 1 can be improved. Moreover, an algorithm to enumerate all block trees is an open problem. Finally, a full characterization of all 2-cycles with the help of a subset of the power set of the tree edges is still missing.

Another line of research is to consider random trees and compute the expected number of fixed points and pure colorings. Using our results, it suffices to compute the expected sizes of $|E_{fix}(T)|$ and $|E_{pure}(T)|$ for these trees.

References

1. Agur, Z.: Fixed points of majority rule cellular automata with application to plasticity and precision of the immune system. Complex Syst. **5**(3), 351–357 (1991)
2. Agur, Z., Fraenkel, A., Klein, S.: The number of fixed points of the majority rule. Discret. Math. **70**(3), 295–302 (1988)
3. Akutsu, T., Kuhara, S., Maruyama, O., Miyano, S.: A system for identifying genetic networks from gene expression patterns produced by gene disruptions and overexpressions. Genome Inform. **9**, 151–160 (1998)
4. Aracena, J.: Maximum number of fixed points in regulatory boolean networks. Bull. Math. Biol. **70**(5), 1398 (2008)
5. Fogelman, F., Goles, E., Weisbuch, G.: Transient length in sequential iteration of threshold functions. Discret. Appl. Math. **6**(1), 95–98 (1983)
6. Frischknecht, S., Keller, B., Wattenhofer, R.: Convergence in (Social) influence networks. In: Afek, Y. (ed.) DISC 2013. LNCS, vol. 8205, pp. 433–446. Springer, Heidelberg (2013). https://doi.org/10.1007/978-3-642-41527-2_30
7. Gao, C., Dong, Y.: ABS algorithm for solving a class of linear Diophantine inequalities and integer LP problems. J. Appl. Math. Inf. **26**(12), 349–353 (2008)
8. Goles, E., Olivos, J.: Periodic behaviour of generalized threshold functions. Discret. Math. **30**(2), 187–189 (1980)
9. Irons, D.: Improving the efficiency of attractor cycle identification in boolean networks. Physica D **217**(1), 7–21 (2006)
10. Kauffman, S., et al.: The Origins of Order: Self-organization and Selection in Evolution. Oxford University Press, USA (1993)

11. Královič, R.: On majority voting games in trees. In: Pacholski, L., Ružička, P. (eds.) SOFSEM 2001. LNCS, vol. 2234, pp. 282–291. Springer, Heidelberg (2001). https://doi.org/10.1007/3-540-45627-9_25
12. Milano, M., Roli, A.: Solving the satisfiability problem through boolean networks. In: Lamma, E., Mello, P. (eds.) AI*IA 1999. LNCS (LNAI), vol. 1792, pp. 72–83. Springer, Heidelberg (2000). https://doi.org/10.1007/3-540-46238-4_7
13. Papp, P., Wattenhofer, R.: Stabilization time in minority processes. In: 30th Int. Symp. on Algorithms & Computation, volume 149 of LIPIcs, pp. 43:1–43:19, Dagstuhl (2019)
14. Peleg, D.: Local majorities, coalitions and monopolies in graphs: a review. Theoret. Comput. Sci. **282**(2), 231–257 (2002)
15. Poljak, S., Sura, M.: On periodical behaviour in societies with symmetric influences. Combinatorica **3**(1), 119–121 (1983)
16. Rouquier, J., Regnault, D., Thierry, É.: Stochastic minority on graphs. Theoret. Comput. Sci. **412**(30), 3947–3963 (2011)
17. Turau, V.: Fixed points and 2-cycles of synchronous dynamic coloring processes on trees (2022). arXiv:2202.01580
18. Veliz-Cuba, A., Laubenbacher, R.: On the computation of fixed points in boolean networks. J. Appl. Math. Comput. **39**, 145–153 (2012)
19. Zehmakan, A.: On the Spread of Information Through Graphs. Ph.D. thesis, ETH Zürich (2019)

Foremost Non-stop Journey Arrival in Linear Time

Juan Villacis-Llobet[1,2], Binh-Minh Bui-Xuan[1(✉)], and Maria Potop-Butucaru[1]

[1] LIP6 (CNRS – Sorbonne Université), Paris, France
{buixuan,maria.potop-butucaru}@lip6.fr

[2] Institut Polytechnique de Paris, Palaiseau, France
juan.villacisllobet@ip-paris.fr

Abstract. A journey in a temporal graph is a sequence of adjacent and dated edges preserving the increasing order of arrival dates to the consecutive edges. When a journey never visits a vertex twice it is also called a temporal path. Given a pair of source and target vertices, a journey connecting them is foremost if the arrival date at the target vertex is the earliest. Like in the static case, there always exists a foremost journey which is also a temporal path because it is useless to circle around an intermediary vertex. It is therefore equivalent to compute the arrival date of a foremost journey or a foremost temporal path.

A non-stop journey is a journey where every pair of consecutive edges must also fulfill a maximum waiting time constraint. Foremost non-stop journeys can achieve strictly earlier arrival date than foremost non-stop temporal paths. We present a linear time algorithm computing the earliest arrival date of such a non-stop journey connecting any two given vertices in a given temporal graph.

Keywords: temporal graph · foremost journey · non-stop journey

1 Introduction

In a static graph, both ShortestWalk and ShortestPath ask for the same computation, that is, a path joining two given vertices with the least number of edges. There is no need to make a distinction between walks and paths because a shortest walk never visits a vertex twice, hence, is also a path. Moreover, shortest paths fulfill a very convenient local optimisation property called prefix preservation: any prefix of a shortest path is itself a shortest path. Exploiting prefix preservation, popular greedy algorithms such as Dijkstra or Bellman-Ford algorithms can be used to compute shortest path in polynomial time [4].

Generalising to the temporal case, given a temporal graph whose edges are weighted with cost function c and two vertices s and t, a *journey* from s to t is a sequence of dated edges $(d_1, s = v_1, v_2), (d_2, v_2, v_3), \ldots, (d_p, v_p, v_{p+1} = t)$ satisfying some condition of *realizability* over the dates d_i's. Furthermore, when a journey never visits a vertex twice, it is called a *temporal path*. A fundamental

M. Parter (Ed.): SIROCCO 2022, LNCS 13298, pp. 283–301, 2022.
https://doi.org/10.1007/978-3-031-09993-9_16

realizability constraint we will impose on all journeys appearing in this paper is being *timely increasing*, that is, $d_i + c(d_i, v_i, v_{i+1}) \leq d_{i+1}$ for every $1 < i \leq p$. Finding the earliest arrival date at destination $d_p + c(d_p, v_p, v_{p+1} = t)$ of a journey, resp. temporal path, satisfying the timely increasing property, or outputting a negative answer when such a date does not exist, is called the ForemostJourneyArrival, resp. ForemostTemporalPathArrival, problem. This helps modelling both ground traffic [5] and TCP/IP transmission [10], where a vehicle or a TCP/IP package need to be at successive checkpoints in increasing arrival dates. Like in the static case, there is no need here to make a distinction between journeys and temporal paths because removing from any foremost journey a cycle around an intermediary vertex does not modify the arrival date at its final destination. Furthermore, prefix preservation can also be retrieved for such journeys after a topological sort over the vertices [1]. From that point, the Dijkstra or Bellman-Ford approach can be extended to compute a foremost journey in polynomial time. Prefix preservation also plays a crucial role in obtaining algorithmic solutions for other path problems in temporal graphs [2,9].

In addition to the timely increasing property, a stronger realizability condition is to also have *non-stop transit*, that is, we also have the inequality the other way around $d_{i+1} \leq d_i + c(d_i, v_i, v_{i+1})$ for every $1 < i \leq p$. Here, $c(d_i, v_i, v_{i+1})$ represents exactly the time it takes for sailing from one vertex to another, where the journey *must continue* without delays. This helps dealing with physical constraints when the traversal is performed by an aircraft or a boat [6]: while a TCP/IP package can be retained at a vertex for an unlimited delay, an aircraft can not perform a stationary flight at a vertex waiting for a better wind condition. In the present paper, we address the slightly more general condition of (α, β)*-transit* which allows for d_{i+1} to depart within a time window, that is, $d_i + c(d_i, v_i, v_{i+1}) + \alpha(v_{i+1}) \leq d_{i+1} \leq d_i + c(d_i, v_i, v_{i+1}) + \beta(v_{i+1})$ for every $1 < i \leq p$. This helps modeling disease spreading where an infection is supposed not to stay on any infected individual v_{i+1} for more than $\beta(v_{i+1}) = 7$ days. With α being constantly equal to 0 and β constantly equal to 7 days, solving the corresponding ForemostJourneyArrival under (α, β)-transit would let us know if the destination vertex would be at risk of contamination whenever the source vertex is infected.

On the theoretical side, not only ForemostJourneyArrival and ForemostTemporalPathArrival strictly differ under (α, β)-transit, but it is also unclear how to retrieve the prefix preservation property, as the topological sort approach does not seem to give satisfying results. It is even more unfortunate that ForemostTemporalPathArrival under (α, β)-transit is *NP*-hard [3]. The situation is better for ForemostJourneyArrival, where to the best of our knowledge, it can be solved in $O(n + m \log m)$ time under (α, β)-transit, with n being the number of vertices and m the number of dated edges in the input temporal graph following a recent result in Ref. [8]. Therein, the main idea is to slice the input temporal graph into graphs G_d containing arcs of the input temporal graph dated with d, plus some well selected arcs with arrival date equal to d. Then, by sliding the value of d over the time dimension, one can decompose the original problem into a computation of journeys ending before d and what will be remaining. Using a Dijkstra approach to solve the former part, one can achieve a global $O(n + m \log m)$ time solution for FOREMOSTJOURNEYARRIVAL under (α, β)-transit, along with a larger class of path-like problems [8].

We present a linear time solution for FOREMOSTJOURNEYARRIVAL under (α, β)-transit. Unlike the previous decomposition approach, we focus in reducing a set of relevant arcs into one arc which achieves the desired foremost journey arrival date. It is divided into four stages which will be called sequentially. In a nutshell, we first compute a representation for the arc set of a large gadget digraph (first two stages), then traverse it (third stage) before a last scan to filter the output (fourth stage). Considering implementation matters, we devise the first two stages and the fourth stage using the filter-map-reduce programming paradigm. Additionally, these stages can easily be batch-performed in a distributed setting. Our third stage computation is a graph traversal and it is very unclear whether this stage can be parallelized. Nevertheless, we leave open the question whether the third stage can also be implemented using functional programming, which would be interesting among other things for reuse matters.

In order to achieve linear time complexity, we cope with the gadget digraph using an implicit representation. Originally, each vertex of the gadget graph is a pair (d, u) denoting that it is possible in the input temporal graph to leave u at d (to any destination). There is an arc from (d, u) to (d', v) if it is possible to leave u at d to arrive to v, *and* leave again v at d' while fulfilling the (α, β)-transit condition at v. We show that the gadget graph allows for encoding every information we need to solve FOREMOSTJOURNEYARRIVAL. However, its size is very large. For every fixed v, we then exploit the total order of the time dimension, and regroup only relevant values of d' into disjoint-sets using a restricted version of disjoint-set data structure [7]. Finally, we show a constant upper bound for the out-going degree of the leftover implicit representation of the gadget graph, and use it to prove the global linear time complexity.

Our paper is organised as follows. We formalise in Sect. 2 problem FOREMOSTJOURNEYARRIVAL under (α, β)-transit. In Sect. 3 we present the main structures to be computed before solving this problem. We show in Sect. 4 how to compute all these structures in linear time. In Sect. 5 we close the paper with concluding remarks and open perspectives for further research.

2 Journey in a Temporal (di)graph

In this paper, digraphs are simple loopless directed graphs. This encompasses the case of simple loopless undirected graphs, whose formalism is equivalent to that of symmetric digraphs. We denote $V \otimes V = V \times V \setminus \{(v, v) : v \in V\}$ for any finite set V. A *temporal digraph* is a tuple $G = (\tau, V, A, c)$ where:

- $\tau \in \mathbb{N}$ is an integer called the *timespan* of G. We define interval $T = [\![0, \tau - 1]\!]$ as the set of time instants used in G.
- V is a finite set called the *vertex set* of G.
- $A \subseteq T \times V \otimes V$ is called the *arc set* of G.
- $c : A \to \mathbb{N}$ represents the traversal time of every arc, it is called the *cost function* for G.

For every arc $a = (d, s, t) \in A$, we denote $s(a) = s$ the *source vertex* of the arc, $t(a) = t$ its *target vertex*, and $d(a) = d$ its *departure time*. The traversal of arc a departs from s towards t at departure time d and arrives to t at arrival time $d + c(a)$.

Remark 1. With this formalism, if $a = (d, s, t)$ belongs to A and $c(a) > 1$, it is still not necessarily the case that $a' = (d + 1, s, t)$ belongs to A. If both a and a' belong to A, the formalism allows for $c(a)$ and $c(a')$ to differ arbitrarily. This helps modeling the routing condition from s to t according to the moment the arc is traversed.

We define journeys with waiting time constraints following the formalism given in [8]. Let $s, t \in V$ be two distinct vertices of G. Let $\alpha, \beta : V \to \mathbb{N}$ be two functions representing the minimum and maximum waiting time at every vertex. An (α, β)-*journey* from s to t is a sequence of arcs $J = (a_1, a_2, \ldots, a_p) \in A^p$, where $s(a_1) = s$, $t(a_p) = t$, and for every $1 \leq i < p$ we have both $t(a_i) = s(a_{i+1})$ and $d(a_i) + c(a_i) + \alpha(t(a_i)) \leq d(a_{i+1}) \leq d(a_i) + c(a_i) + \beta(t(a_i))$. For $1 \leq i < p$, the traversal of arc a_i begins from source vertex $s(a_i)$ at departure time $d(a_i)$, it takes $c(a_i)$ time steps to arrive at target vertex $t(a_i)$, where the journey has to be delayed for at least $\alpha(t(a_i))$ and at most $\beta(t(a_i))$ time steps before pursuing with the traversal of arc a_{i+1}. The *arrival date* of J is defined as $d(a_p) + c(a_p)$. A journey is called *foremost* when its arrival date is minimum.

On input a temporal graph $G = (\tau, V, A, c)$ with transit functions (α, β) and two vertices s, t in G, the problem of computing the minimum value of arrival date $d(a_p) + c(a_p)$ taken over every (α, β)-journey $J = (a_1, a_2, \ldots, a_p)$ from s to t is called the ForemostJourneyArrival under (α, β)-transit problem. When both α and β are constantly equal to 0, such a $(0, 0)$-journey J is called a *non-stop* journey. Figure 1 exemplifies such journeys.

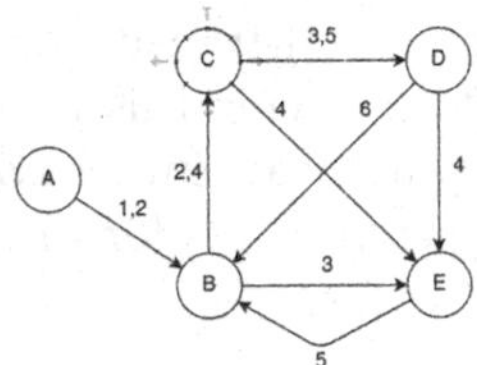

Fig. 1. A temporal digraph, the labels on the arcs denote the time instants where the arcs are active. If the cost function is uniformly unitary, then the following journey from A to E is foremost: $A \xrightarrow{1} B \xrightarrow{3} E$, while the journey $A \xrightarrow{1} B \xrightarrow{2} C \xrightarrow{3} D \xrightarrow{4} E$ is the foremost $(0, 0)$-journey, also known as non-stop.

3 From Journey to Time Set

In this section we prepare the way for solving ForemostJourneyArrival under (α, β)-transit. Lemma 1 below is crucial because it helps us reduce a path-like problem down to a set problem. Then, Lemmas 2 and 4 introduce the structures that will be used in next Sect. 4 to solve FOREMOSTJOURNEYARRIVAL under (α, β)-transit. For the sake of clarity, the time complexity which will be given in both Lemmas 2 and 4 is inefficient. In the next section, we depict optimised replacements for these steps, reducing the time complexity down to linear.

Let $G = (\tau, V, A, c)$ be a temporal digraph. Let $\alpha, \beta : V \to \mathbb{N}$ be two functions representing the minimum and maximum waiting time constraints. For any source vertex $s \in V$, we define the set of *(α, β)-reachable arcs* from s as $R(s) = \{a_p \in A : \exists$ *(α, β)-journey* $J = (a_1, a_2, \ldots, a_p) \in A^p \wedge s(a_1) = s\}$. We show below how to compute, from the supposed knowledge of set $R(s)$, the minimum arrival date of an (α, β)-journey from s to any target vertex t.

Lemma 1 (Reachable arcs). *On input a temporal digraph G, a pair of source and target vertices s, t in G, two constraint functions $\alpha, \beta : V \to \mathbb{N}$, and the above defined (α, β)-reachable arc set $R(s)$, it is possible to output in linear time the minimum arrival date of an (α, β)-journey from s to t. More precisely, the minimum arrival date is equal to $\min\{d(a) + c(a) : a \in R(s) \wedge t(a) = t\}$, which can be reduced from $R(s)$ in linear time. Moreover, the reduction can be done using a functional programming approach as showed in Remark 2 below.*

Proof. We first address the case when there exists an (α, β)-journey from s to t. Let *mad* be the minimum arrival date of such a journey. Let $m = \min\{d(a) + c(a) : a \in R(s) \wedge t(a) = t\}$. We claim that $m = mad$.

Indeed, let $J = (a_1, a_2, \ldots, a_p) \in A^p$ be an (α, β)-journey from s to t minimising the arrival date. By definition, we firstly have that *mad* is the arrival date of J, that is, $mad = d(a_p) + c(a_p)$. Besides, it also follows from definition of J that $t(a_p) = t$. Moreover, J is also such that $s(a_1) = s$, therefore, we have from definition of $R(s)$ that $a_p \in R(s)$. Now, we have both $a_p \in R(s) \wedge t(a_p) = t$, therefore, by definition of m we have that $m \leq d(a_p) + c(a_p)$. Combining with earlier proven $mad = d(a_p) + c(a_p)$ we obtain $m \leq mad$.

Conversely, let $a \in R(s)$ such that $t(a) = t$. We claim that $mad \leq d(a) + c(a)$. By definition of $a \in R(s)$, there exists an (α, β)-journey $J = (a_1, a_2, \ldots, a_p) \in A^p$ such that $s(a_1) = s$ and $a_p = a$. Combining with $t(a) = t$ we have that J is exactly an (α, β)-journey from s to t. Since the arrival date of J is $d(a) + c(a)$, we have that $mad \leq d(a) + c(a)$ because *mad* is the minimal arrival date taken over all (α, β)-journeys from s to t. We have proven that $mad \leq d(a) + c(a)$ for every $a \in R(s) \wedge t(a) = t$. Hence, $mad \leq m$.

All in all we have just proved that $m = mad$. Therefore it is sufficient to compute m in order to output the value of *mad*. Finally, computing m from the input of $R(s)$, vertex t, and cost function c can be done in linear time by any standard streaming process.

When there is no (α, β)-journey from s to t then $mad = \infty$. In this case $R(s)$ contains no arc a such that $t(a) = t$ and therefore m would have the value ∞ after performing the map-reduce (as no arc satisfies both properties needed to be considered a suitable value for m). Therefore the result still holds for this particular case. □

Remark 2. Lemma 1 is proper to temporal digraphs in the sense that we can from input $G = (\tau, V, A, c)$ filter the set A to a smaller subset $R(s) \subseteq A$, then filter further to the set of arcs whose target vertex is t, and finally reduce the stream to find the minimum value $d(a)+c(a)$. As a comparison no shortest path algorithm on static (di)graphs allows for using filter-map-reduce programming in such a straightforward manner.

We now introduce an intermediary step to compute the arc set $R(s)$. We define the set of *valid transit departures* in an (α, β)-journey from s as $D(s) = \{(d, v) \in T \times V : \exists\ (\alpha, \beta)\text{-journey } J = (a_1, a_2, \ldots, a_p) \in A^p,\ \text{such that } s(a_1) = s \wedge s(a_p) = v \wedge d(a_p) = d\}$.

Lemma 2 (Valid transit departures). *On input a temporal digraph G, a pair of source and target vertices s, t in G, two constraint functions $\alpha, \beta : V \to \mathbb{N}$, and the above defined set $D(s)$ of valid transit departures in an (α, β)-journey from s, it is possible to output in polynomial time the set $R(s)$ of (α, β)-reachable arcs from s.*

Proof. A naive way to output $R(s)$ from $D(s)$ is as follows. We initialize a boolean table `R` indexed by the elements of A. For any $a \in A$ with $a = (d, u, v)$, we scan $D(s)$ and check if $(d, u) \in D(s)$. If this is the case we set `R[a]` to `true`. At the end of the process, we scan `R` and output every index a where `R[a]` has value `true`. □

In the sequel we show how to compute in polynomial time the set $D(s)$ from the input of G. We first define a static digraph associated to temporal digraph G, then we perform a graph search on the thus defined static graph.

The (α, β)-*transit departure digraph* of G, that we call $G_D = (V_D, A_D)$, is defined as follows. First, $V_D = \{(d, v) : \exists a \in A, s(a) = v \wedge d(a) = d\}$ is the set of all possible transit departures, including those not necessarily valid w.r.t. any (α, β)-journey from s. In other words, $D(s) \subseteq V_D$, however, V_D could be much larger than $D(s)$. Then, for any pair of vertices $x = (d, u)$ and $y = (d', v)$ of V_D, we define $(x, y) \in A_D$ if and only if we have both that $a = (d, u, v)$ belongs to A and that $d + c(a) + \alpha(v) \leq d' \leq d + c(a) + \beta(v)$. Figure 2 exemplifies the construction of a transit departure digraph.

We capture in the following property our main computational purposes of defining G_D. It gives a reasonable upper bound for both $|V_D|$ and $|A_D|$.

Property 1. Let $G = (\tau, V, A, c)$ be a temporal digraph, $\alpha, \beta : V \to \mathbb{N}$ two functions representing the minimum and maximum waiting time constraints, and $G_D = (V_D, A_D)$ the (α, β)-transit departure digraph of G. Let $\gamma = \max\{\beta(v) - \alpha(v) + 1 : v \in V\}$. Then, $|V_D| \leq |A|$ and $|A_D| \leq \gamma \times |A|$.

Proof. Note that two naive upper bounds for G_D exist: $|V_D| \leq \tau \times |V|$ and $|A_D| = O(|V_D|^2)$. Furthermore, we can also note by definition $V_D = \{(d, v) : \exists a \in A, s(a) = v \wedge d(a) = d\}$ that $|V_D| \leq |A|$ because there will be at most one vertex in V_D for every arc in A. As a side note, it could be the case that $|V_D| < |A|$: if we have $(d, v, w) \in A$ and $(d, v, w') \in A$ for distinct vertices $w \neq w'$.

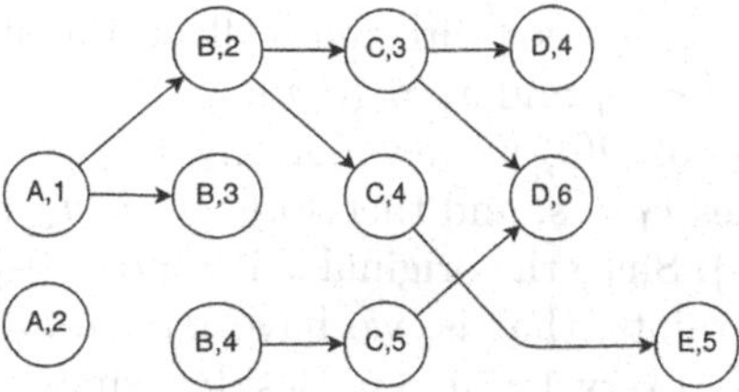

Fig. 2. The corresponding static representation G_D of the temporal graph in Fig. 1 taking 0 as minimum waiting time, 2 as the maximum waiting time and a traversal time of 1 for all arcs.

Now, let us examine $(x, y) \in A_D$ with $x = (d, u)$ and $y = (d', v)$. By definition, $a = (d, u, v)$ must belong to A, and d' must satisfy the waiting time constraints $d+c(a)+\alpha(v) \leq d' \leq d+c(a)+\beta(v)$. Let us define function $f : A_D \to A \times [\![0, \gamma-1]\!]$ as $f((d, u), (d', v)) = (d, u, v, d' - d - c(a))$. Then, we can check that f is a well-defined injective function, and therefore, deduce that $|A_D| \leq \gamma \times |A|$. □

Essentially, the size of thus defined graph G_D is not far from linear in $|A|$. We now would like to extract from V_D all vertices belonging to $D(s)$. By definition, we have the following closure property: $(x, y) \in A_D \wedge x \in D(s)$ implies $y \in D(s)$. Hence, $D(s)$ encompasses the set of vertices in G_D reachable from any vertex of the set $V_D \cap \{(d, s) : 0 \leq d < \tau\}$. Moreover, we show in the following lemma that $D(s)$ is exactly the latter set.

Lemma 3. *Let $G = (\tau, V, A, c)$ be a temporal digraph, $\alpha, \beta : V \to \mathbb{N}$ two functions representing the minimum and maximum waiting time constraints, $s \in V$, and $D(s)$ the set of valid transit departures in an (α, β)-journey from s. Let $G_D = (V_D, A_D)$ be the (α, β)-transit departure digraph of G. Then, $D(s)$ is exactly the set of vertices in G_D which are reachable from a (directed) path beginning at any vertex of the set $V_D \cap \{(d, s) : 0 \leq d < \tau\}$.*

Proof. We denote by $R_D(s)$ the set of vertices in G_D which are reachable from a (directed) path beginning at any vertex of the set $V_D \cap \{(d, s) : 0 \leq d < \tau\}$. By definition of G_D, we have the following closure property: if $(x, y) \in A_D$ and $x \in D(s)$ then $y \in D(s)$. Besides, whenever $(d, s) \in V_D$ for any $0 \leq d < \tau$, that is, whenever there exists $a \in A$ such that $s(a) = s$ and $d(a) = d$, we also have that $(d, s) \in D(s)$ by using the single-arc (α, β)-journey $J = (a)$ in the definition of $D(s)$. Now, we use the above mentioned closure property in order to deduce that $R_D(s) \subseteq D(s)$. Hence, the only thing left for us to show is that $D(s) \subseteq R_D(s)$.

Let $(d, v) \in D(s)$. We would like to prove that $(d, v) \in R_D(s)$. By definition of $D(s)$, there exists an (α, β)-journey $J = (a_1, a_2, \ldots, a_p) \in A^p$ such that $s(a_1) = s$, $s(a_p) = v$, and $d(a_p) = d$. Let us consider $J_q = (a_1, a_2, \ldots, a_q)$, for any $1 \leq q \leq p$. For convenience, we denote $d_q = d(a_q)$, $v_q = s(a_q)$, and $x_q = (d_q, v_q)$. Since $a_q \in A$ we have from definition of V_D that $x_q \in V_D$, for any $1 \leq q \leq p$.

We claim that $(x_1, x_2, \ldots, x_p)$ is a directed walk in the static digraph G_D, with $x_1 \in V_D \cap \{(d, s) : 0 \leq d < \tau\}$ and $x_p = (d, v)$.

Indeed, by definition of $D(s)$ we have for any $1 \leq q \leq p$ that $(d_q, v_q) \in D(s)$. When $q = 1$, this implies $v_1 = s$, and therefore $x_1 = (d_1, v_1) = (d_1, s)$ belongs to $V_D \cap \{(d, s) : 0 \leq d < \tau\}$. Since the original J is an (α, β)-journey, it must satisfy the waiting time constraints, that is, we have $d_q + c(a_q) + \alpha(t(a_q)) \leq d_{q+1} \leq d_q + c(a_q) + \beta(t(a_q))$, for every $1 \leq q < p$. Besides, since $t(a_q) = s(a_{q+1}) = v_{q+1}$, we have both $(d_q, v_q, v_{q+1}) = a_q \in A$ and $d_q + c(a_q) + \alpha(v_{q+1}) \leq d_{q+1} \leq d_q + c(a_q) + \beta(v_{q+1})$. This implies (x_q, x_{q+1}) belongs to A_D for every $1 \leq q < p$. In other words, $(x_1, x_2, \ldots, x_p)$ is a directed walk in G_D. Since $d_p = d(a_p) = d$ and $s_p = s(a_p) = v$, we also have $x_p = (d, v)$. We have shown a directed walk in G_D beginning from vertex $x_1 \in V_D \cap \{(d, s) : 0 \leq d < \tau\}$, and ending at vertex $x_p = (d, v)$. This also implies there exists a directed path in G_D from x_1 to $x_p = (d, v)$. Hence, $(d, v) \in R_D(s)$. We have proved for every $(d, v) \in D(s)$ that $(d, v) \in R_D(s)$. In other words, $D(s) \subseteq R_D(s)$. □

Lemma 4. *On input a temporal digraph G, a source vertex s in G, two constraint functions $\alpha, \beta : V \to \mathbb{N}$, it is possible to output in polynomial time the set $D(s)$ of valid transit departures in an (α, β)-journey from s.*

Proof. Let $G = (\tau, V, A, c)$, and $G_D = (V_D, A_D)$ its (α, β)-transit departure digraph. By Lemma 3, $D(s)$ can be computed by a graph search on G_D, that is, in $O(|V_D| + |A_D|)$ time from the knowledge of G_D. From Property 1, the size of V_D and A_D is polynomial in $|A|$, α, and β. Hence, it is straightforward to construct V_D in $O(\tau \times |V| \times |A|)$, then A_D in $O(|V_D|^2 \times |A|)$. □

All in all, we presented in Lemmas 1, 2 and 4 polynomial procedures for computing the minimum arrival date of an (α, β)-journey from s to t. Whereas the procedure presented in Lemma 1 requires linear time, the other two might take more time to terminate. The total time complexity is significantly worse than the recently known $O(|V| + |A| \log |A|)$ algorithm presented in [8]. In the next section we present an improvement in the way we construct the graph G_D and traverse it in linear time in $|A|$.

4 Foremost Non-stop Journey Arrival in Linear Time

In this section we show how to solve in linear time ForemostJourneyArrival under (α, β)-transit from a source vertex s to a target vertex t in a temporal digraph $G = (\tau, V, A, c)$. This encompass the case of foremost non-stop journeys when both α and β are constantly equal to 0. We do this by an implicit traversal of the (α, β)-transit departure digraph $G_D = (V_D, A_D)$ as defined in the previous section. We suppose the three functions c, α, β are given as tables, so that the cost for accessing $c(a)$ for every $a \in A$, and the cost for accessing $\alpha(v)$ and $\beta(v)$ for every $v \in V$ are constant.

Our algorithm is composed of four main stages, each one terminates in $O(\tau + |V| + |A|)$ time: first we construct V_D; then we construct a subset of A_D of representative arcs whose number is bounded by $2 \times |A|$; in a third stage we use the previously constructed structures to compute an implicit representation of the set $D(s)$ defined in the previous section; finally, we use this information and construct the set $R(s)$ defined in Lemma 1, and result as a byproduct in the earliest arrival date of an (α, β)-journey from s to t.

For use in Algorithm 1, we perform two linear bucket sorting processes (a.k.a. radix sorting) as follows. We first initialize two arrays containing τ buckets each. Each bucket is to contain a list of arcs initially empty. Then, we stream through A, where for every arc $a \in A$ we: first append a to the list present in the bucket numbered $d(a)$ in the first array of buckets; second append a to the list present in the bucket numbered $d(a) + c(a)$ in the second array of buckets. For later use, we also keep a variable counting the number of elements in every bucket. After the streaming process, we have filled two arrays of τ buckets each, where every bucket contains a list of arcs, as well as the number of arcs in the bucket. We now iterate over the buckets by increasing order and concatenate all the two times τ lists consecutively, resulting in a list of arcs sorted in increasing departure time from source vertex, and a list of arcs sorted in increasing arrival time to target vertex. Since the number of elements in each list is known, each concatenation can be done in constant time. The whole procedure is hence in $O(\tau + |A|)$.

From now on we suppose A is given twice: sorted by increasing departure time from source vertex, and sorted by increasing arrival time to target vertex.

Algorithm 1. Construction of V_D, the vertex set of G_D.

```
procedure GenerateVertices(G = (τ, V, A, c), α, β)
    Departures ← ∅                                  ▷ Set containing all the new vertices in G_D
    Initialize table VDep with |V| entries           ▷ Same set, fast track for later use
    Arrivals ← ∅
    Initialize table VArr with |V| entries
    for each vertex v ∈ V do
        VDep[v] ← ∅
        VArr[v] ← ∅
    for each arc a ∈ A in increasing departure time do
        Append (d(a), s(a)) to Departures if not already present
        Append d(a) to VDep[s(a)] if not already present
    for each arc a ∈ A in increasing arrival time do
        Append (d(a) + c(a), t(a)) to Arrivals if not already present
        Append d(a) + c(a) to VArr[t(a)] if not already present
    Either output VDep or Departures as the vertex set V_D
    For later use in Algorithm 2, also output VArr.
```

Stage 1: Construction of V_D. We stream through every element $a \in A$ in increasing departure time and append $(d(a), s(a))$ to a `Departures` list. Similarly, we stream through every element $a \in A$ in increasing arrival time and append $(d(a) + c(a), t(a))$ to an `Arrivals` list. By definition, the vertex set

V_D contains exactly the elements present in the `Departures` list. Furthermore, we will organise V_D in the following manner, for later use in Stage 2. Let $V = \{v_1, v_2, \ldots, V_{|V|}\}$. We create $|V|$ buckets numbered by these v_i's, each bucket is to contain a list of departure times initially empty. When streaming through every element $a \in A$, we also append $d(a)$ to the list present in the bucket numbered $s(a)$. After the streaming process, we keep the $|V|$ lists in a table named `VDep`, indexed by the v_i's. Thus, for every vertex $v \in V$, reading `VDep`$[v]$ gives a quick access to all the departure times d associated to that vertex, *i.e.* where $(d, v) \in V_D$. Since A is sorted by increasing departure time, it is also the case with list `Departures`, as well as with list `VDep`$[v]$, for every $v \in V$. For later use in Stage 2, we also organise `Arrivals` into a table named `VArr`, in a similar way. In reality, we do not use `Departures` and `Arrivals` in the rest of the manuscript. However, we keep them in the discussion for more clarity. We capture the pseudo-code in Algorithm 1, and result in the following lemma.

Lemma 5. *On input a temporal digraph $G = (\tau, V, A, c)$ and two constraint functions $\alpha, \beta : V \to \mathbb{N}$, Algorithm 1 correctly generates in time $O(|V| + |A|)$ all the vertices of G_D, the (α, β)-transit departure graph of G.*

Proof. The algorithm's correctness follows from definition. Lines 3,5,6–8 take $O(|V|)$ times while lines 9–14 take $O(|A|)$ time. □

Stage 2: Implicit representation of A_D in $O(|A|)$ space. If $\beta(v) - \alpha(v) = 1$ for every $v \in V$, then a similar argument as in the proof of Property 1 implies $|A_D| \leq 2 \times |A|$. Indeed, every arc $(d, u, v) = a \in A$ gives rise to at most two arcs in G_D: one from (d, u) to (d', v) with $d' = d + c(a) + \alpha(v)$ if the latter vertex (d', v) belongs to V_D; one from (d, u) to (d', v) with $d' = d + c(a) + \beta(v)$ if the latter vertex (d', v) belongs to V_D.

Now if $\beta(v) - \alpha(v)$ is an arbitrary integer, we remark the following organisation of A_D. Consider set $D' = \{d' : ((d, u), (d', v)) \in A_D\}$, then if both $d'_1 \leq d'_3$ belong to D' and d'_2 is such that we have both $d'_1 \leq d'_2 \leq d'_3$ and $(d'_2, v) \in V_D$, then d'_2 belong to D'. Moreover, in the (ordered) list `VDep`$[v]$, the elements of D' appear consecutively. Therefore, in order to represent all arcs of A_D of the form $((d, u), (d', v))$, for every given $(d, u, v) = a \in A$, we only need to store the arc from (d, u) to (d'_{min}, v) and the arc from (d, u) to (d'_{max}, v), where $d'_{min} = \min D'$ and $d'_{max} = \max D'$. All the other arcs of the form $((d, u), (d', v))$ can be obtained by enumerating from `VDep`$[v]$ all d' such that $d'_{min} \leq d' \leq d'_{max}$.

In order to implement this idea, we define `DPmin` and `DPmax` to be two tables, indexed by the elements of A. For every $(d, u, v) = a \in A$, we define `DPmin`$[(d, u, v)] = \min\{d' : ((d, u), (d', v)) \in A_D\}$ and `DPmax`$[(d, u, v)] = \max\{d' : ((d, u), (d', v)) \in A_D\}$. Then, we will use in Stage 3 and Stage 4 the input of `VDep`, `DPmin`, and `DPmax` as an implicit representation of $G_D = (V_D, A_D)$.

In order to avoid computing `DPmin` and `DPmax` in quadratic time in $|A|$, our main trick is to break down the total transit time cost into two parts: the traversal time represented by function $c : A \to \mathbb{N}$, and the delay time represented by functions $\alpha, \beta : V \to \mathbb{N}$. For this, we build auxiliary tables `FirstTransit` and `LastTransit` with the help of Algorithm 1 table `VArr`. We capture the pseudo-code for first computing the auxiliary tables, then DPmin and DPmax in Algorithm 2, and result in the following lemma.

Algorithm 2. Implicit representation of $G_D = (V_D, A_D)$ by VDep, DPmin, and DPmax.

---

procedure GENERATETABLES($G = (\tau, V, A, c), \alpha, \beta$)
  Call Algorithm 1 GENERATEVERTICES(G, α, β) and obtain VDep and VArr
  Initialize table FirstTransit with $|V|$ entries
  Initialize table LastTransit with $|V|$ entries
  for each vertex $v \in V$ **do** ▷ Find for each arrival to v the first/last departure
    $d_{arr} \leftarrow$ first element of VArr[v]
    Initialize table FirstTransit[v] with |VArr[v]| entries
    $d'_{first} \leftarrow$ first element of VDep[v]
    while d_{arr} is still an element of VArr[v] **do**
      while $\neg(d_{arr} + \alpha(v) \leq d'_{first})$ **do**
        $d'_{first} \leftarrow$ next element after d'_{first} in VDep[v]
      FirstTransit[v][d_{arr}]$\leftarrow d'_{first}$
      $d_{arr} \leftarrow$ next element after d_{arr} in VArr[v]
    $d_{arr} \leftarrow$ last element of VArr[v]
    Initialize table LastTransit[v] with |VArr[v]| entries
    $d'_{last} \leftarrow$ last element of VDep[v]
    while d_{arr} is still an element of VArr[v] **do**
      while $\neg(d'_{last} \leq d_{arr} + \beta(v))$ **do**
        $d'_{last} \leftarrow$ previous element before d'_{last} in VDep[v]
      LastTransit[v][d_{arr}]$\leftarrow d'_{last}$
      $d_{arr} \leftarrow$ previous element before d_{arr} in VArr[v]
  Initialize table DPmin with $|A|$ entries
  Initialize table DPmax with $|A|$ entries
  for each arc $(d, u, v) = a \in A$ **do** ▷ Take c into account: $d_{arr} = d + c(a)$
    DPmin[(d, u, v)]$\leftarrow$ FirstTransit[v][$d + c(a)$]
    DPmax[(d, u, v)]$\leftarrow$ LastTransit[v][$d + c(a)$]
  Output VDep, DPmin and DPmax as implicit representation of G_D.

Lemma 6. *On input a temporal digraph* $G = (\tau, V, A, c)$ *and two constraint functions* $\alpha, \beta : V \to \mathbb{N}$*, Algorithm 2 outputs in time* $O(|V| + |A|)$ *three tables called* VDep, DPmin, *and* DPmax. *From these tables one can generate* G_D*, the* (α, β)*-transit departure graph of* G*, in linear time (in the size of the input* G *and the output* G_D*).*

Proof. It is a standard exercise to prove that Algorithm 2 correctly computes DPmin[(d, u, v)] $= \min\{d' : ((d, u), (d', v)) \in A_D\}$ and DPmax[(d, u, v)] $= \max\{d' : ((d, u), (d', v)) \in A_D\}$, *e.g.* by induction on the total size of VArr.

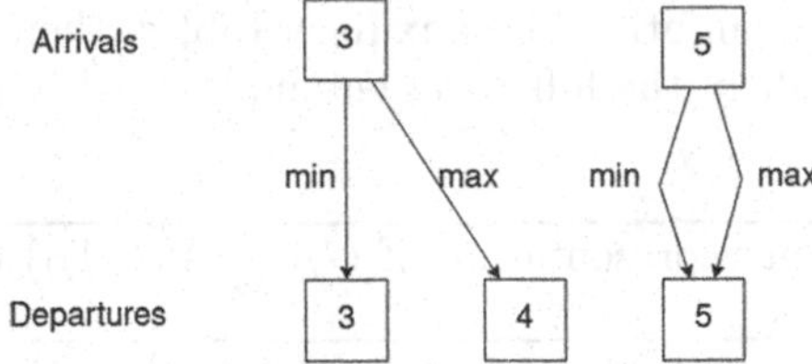

Fig. 3. Data structure used in Algorithm 2 for each vertex $v \in V$, where we suppose $\alpha(v) = 0$ and $\beta(v) = 2$. In this case it is done for the node c in the temporal graph of Fig. 1. List $\texttt{VArr}[v]$ is represented by the "Arrivals" row on top, it contains all times at which a path arrives to node c. List $\texttt{VDep}[v]$ is represented by the "Departures" row on the bottom, it contains all times at which a path departs from node c. For every element d_{arr} in $\texttt{VArr}[v]$, its associated value $\texttt{FirstTransit}[v][d_{arr}]$ in $\texttt{VDep}[v]$ is pointed to by the *min* arrow departing from d_{arr}, while its associated value $\texttt{LastTransit}[v][d_{arr}]$ in $\texttt{VDep}[v]$ is pointed to by the *max* arrow.

For complexity issues, the main point lies in the chasing *while* loops which happens twice, once at lines 9–13, and the other time at lines 17–21. Here, vertex $v \in V$ is already fixed. In the first case, lines 9–13, variables d_{arr} and d'_{first} can only move forward in lists $\texttt{VArr}[v]$ and $\texttt{VDep}[v]$, respectively. Accordingly, at the end of the process lines 9–13, every entry in $\texttt{VArr}[v]$ and $\texttt{VDep}[v]$ are visited exactly once. The case with lines 17–21 is similar, where variables d_{arr} and d'_{last} can only move backward. Figure 3 exemplifies for every entry of table $\texttt{VArr}[v][d_{arr}]$ the position of $\texttt{FirstTransit}[v][d_{arr}]$ and $\texttt{LastTransit}[v][d_{arr}]$ in $\texttt{VDep}[v]$.

Summing up over every vertex $v \in V$, at the end of the process lines 5–21, every elements in VArr and VDep are visited exactly twice. Since the total size of VArr is at most $|A|$, and so is the total size of VDep, *cf.* Property 1, Lemma 5 and Algorithm 1, we deduce that the total contribution of lines 9–13 and 17–21 to the *for* loop lines 5–21 is $O(|A|)$. The contribution of lines 7 and 15 is $O(|A|)$, and that of lines 6,8,14,16 is $O(|V|)$. To complete the complexity analysis for Algorithm 2, we note that line 2 takes $O(|V| + |A|)$ time (Lemma 5), lines 3–4 take $O(|V|)$ time, and lines 23–26 take $O(|A|)$ times.

In order to generate V_D, we scan over all elements of VDep. The enumeration of A_D can proceed as follows. For every arc $(d, u, v) = a \in A$, we scan from the element $\texttt{DPmin}[(d, u, v)]$ of $\texttt{VDep}[v]$ to its element equal to $\texttt{DPmax}[(d, u, v)]$ and enumerate all d' in between, generating an arc $((d, u), (d', v))$ for each of them. By definition of DPmin and DPmax, we have $A_D = \{((d, u), (d', v)) : (d, u, v) \in A \wedge (d', v) \in V_D \wedge \texttt{DPmin}[(\text{d,u,v})] \leq d' \leq \texttt{DPmax}[(\text{d,u,v})]\}$. □

Stage 3: Marking all vertices in V_D whether it belongs to $D(s)$. After the first two stages, every vertex of $G_D = (V_D, A_D)$ is constructed, as well as an implicit representation of the arcs of G_D. We now need to visit the vertices of G_D and mark them according to Lemma 3. Because the representation does not include all arcs of G_D, solving this step using standard graph searches such as Breadth-First

Search seems difficult. More precisely, our implicit representation only stores in entries DPmin[(a)] and DPmax[(a)] the earliest and the latest departure dates for a valid transit at vertex $v = t(a)$ w.r.t. the (α, β)-transit condition. Let us assume $(d, u) \in D(s)$ for $d = d(a)$ and $u = s(a)$, and let us consider $v = t(a)$: if we do test whether (d', v) belongs to $D(s)$ by naively scanning all potential values of d' s.t. DPmin[(a)] $\leq d' \leq$ DPmax[(a)], then the overall complexity will have a factor depending on $\max_{v \in V}$ |VArr[v]|, which is a sharp upper bound for DPmax[(a)] − DPmin[(a)]. This could lead to a non-linear complexity in the worst case.

Algorithm 3. Traverse the graph G_D

```
procedure TRAVERSE(G, α, β, s)
    Call Algorithm 2 GENERATETABLES(G, α, β) and obtain VDep, DPmin and DPmax
                    ▷ We refer to V_D as the concatenation of all entries in VDep
    Q ← a queue initially with all vertices in VDep[s]
    Traversed ← a list that stores all vertices that are visited during the traversal
    for each vertex u ∈ V do
        Make disjoint-set D_u    ▷ With .join(d, d') for union and .comp(d) for find
        for each vertex (d, u) ∈ V_D do
            B_u[d] = −1           ▷ Values in B_u will represent the date when vertex
                                  ▷ (d, u) ∈ V_D was visited before, −1 represents false
    while Q is not empty do
        (d, u) ← Q.pop()
        for each a = (d, u, v) ∈ A and d' = DPmin[(a)] do     ▷ Implicit iteration
                                                ▷ over ((d, u), (d', v)) ∈ A_D
            d'_max ← DPmax[(d, u, v)]
            d'_now ← d'
            d'_prev ← d'
            while d'_now ≤ d'_max do
                visited = B_v[D_v.comp(d'_now)]
                temp ← max(B_v[D_v.comp(d'_prev)], B_v[D_v.comp(d'_now)])
                D_v.join(d'_now, d'_prev)
                B_v[D_v.comp(d'_now)] ← temp
                if visited ≠ −1 then  ▷ Value in B_v states if the vertex was visited
                    d'_now ← B_v[D_v.comp(d'_now)]    ▷ Jump to the vertex with the
                                              ▷ largest time in that component
                else
                    Q.push((d'_now, v))
                    Traversed.append((d'_now, v))
                B_v[D_v.comp(d'_now)] = max(d'_now, B_v[D_v.comp(d'_now)])
                d'_prev ← d'_now
                (d'_now, v) ← next vertex after (d'_now, v) in VArr[v]
```

To achieve linear runtime, we use two main ideas. First, we use disjoint-set data structure to dynamically join the possible values of d' whenever we do the test whether (d', v) belongs to $D(s)$, for every $v \in V$. Generally, the use of disjoint-set data structure leads to a quasi-linear time complexity. In our case, note that the possible values of d' for every $v \in V$ are in reality values in `VArr[`v`]`, Furthermore, already after calling Algorithm 1 the original values present in `VArr[`v`]` are both known and totally ordered. Therefore, a faster case of disjoint-set data structure [7] can be used, with $O(1)$ amortised cost per operation. Hence, there is only a total contribution in time $O(\sum_{v \in V} |\texttt{VArr[}v\texttt{]}|) = O(|A|)$ for disjoint-set operations. The second idea is captured in the proof of Lemma 9. Roughly, we show with Lemma 9 that after joining dynamically the values of d' while testing whether (d', v) belongs to $D(s)$, we can control the out-degree of what is leftover in our implicit representation of G_D.

We formalize Stage 3 of our computation in Algorithm 3. Roughly, we initialize a visiting queue with all vertices in V_D associated to s, that is, of the form (d, s) for any d. At each iteration we remove the top element (d, u) of the queue, which is considered as a vertex of G_D. In this iteration, we would like to follow all arcs that the vertex (d, u) might have in G_D. For this purpose we make an implicit iteration: by iterating over every arc $a = (d, u, v) \in A$ and every possible value of d' s.t. $\texttt{DPmin[}(a)\texttt{]} \leq d' \leq \texttt{DPmax[}(a)\texttt{]}$, as explained in Algorithm 2 and Lemma 6. However, to control the global runtime, we only start with $d'_{now} \leftarrow d' = \texttt{DPmin[}(a)\texttt{]}$, and our plan is to check whether (d'_{now}, v) belongs to $D(s)$ while dynamically joining all previously visited $D_v.comp(d'_{now})$, for every possible value of d'_{now} between $d' = \texttt{DPmin[}(a)\texttt{]}$ and $d'_{max} = \texttt{DPmax[}(a)\texttt{]}$. These values are read from `VArr[`v`]`.

More precisely, for each unvisited vertex visited in this process the algorithm marks it as visited, adds it to the queue, joins the components for it and the previous vertex (itself in case it is the first one) in the disjoint-set D_v, sets as d'_{max} the reference to the largest visited vertex for its component and continue to the next vertex in increasing order of time in `VArr[`v`]`. When the algorithm reaches a vertex previously visited it goes to D_v and gets the identifier of the component it belongs to, jumps to the vertex with the maximum time that belongs to the component and continues to the next vertex. In total, the algorithm will visit the number of unvisited arcs in between d' and d'_{max} plus at most 3 previously visited vertices. The previous value 3 comes from the fact that when the algorithm gets for the first time to a visited vertex it will follow the link to the visited vertex with the largest time value in that component, then it continues iterating vertices until it arrives once more to a visited vertex and follows the link again. The link of the second visited vertex necessarily has to lead to a vertex with a time component at least as large as d'_{max}, and then the visiting process for this particular arc has finished. This is proven in Lemma 9. From the previous analysis it follows that over the whole traversal the algorithm will make at most $3|A| + |A|$ steps, where the first term comes from the bound obtained earlier and the second term is the maximum number of vertices in the graph.

Figure 4 exemplifies Algorithm 3. Therein, the upper and lower arrows indicate the links towards d' and d'_{max} at each time. Eventually, all vertices in between become part of the same component, which is represented by the same color, with the vertex in darker shade indicates the component with the largest time. As we move from left to right we can see the progression of the components as they are visited. On the leftmost column of the picture we can see the component derived from a path arriving to b whose minimum departure time is 3 and whose maximum departure time is 6 (denoted by the top and bottom arrows pointing to these nodes). As there were no previous components on the disjoint-set data structure at this iteration we create a component that contains all nodes within those times. On the second column we see the resulting components after a path with minimum departure time 8 and maximum departure time 9. As this times do not intersect with those of the previous components we get two disjoint components. The same process can be seen in the third column, where the red component is expanded to include the node $(7, b)$. On the last column we get the result of processing a path that has minimum departure time from b equal to 5 and maximum departure time from b equal to 8. The nodes comprised between these times contain elements from both sets in the disjoint-set data structure. Therefore they are merged, resulting in a single set of all vertices of the form (d', b) in V_D.

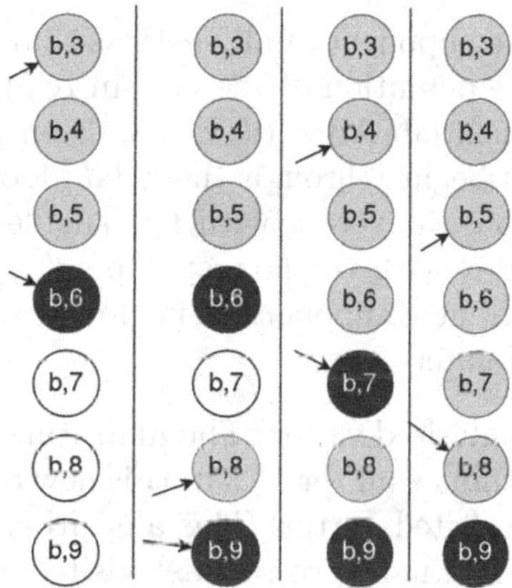

Fig. 4. The process of visiting all vertices in V_D associated with a single vertex b in V.

We recall that for each arc $a = (d, u, v) \in A$ a vertex $(d, u) \in V_D$ will have arcs towards all vertices in G_D associated to vertex v between time d' and d'_{max}. However, our representation of G_D does not include all these arcs: only (d', v) and (d'_{max}, v) are explicit. We will call all of these vertices associated to the same vertex $v \in V$ to which (d, u) has an arc in G_D cousins with respect to (d, u). This is defined formally in the following definition.

Definition 1. *Two vertices (d'_1, v) and $(d'_2, v) \in V_D$ associated to the same vertex $v \in V$ are cousins with respect to a vertex $(d, u) \in V_D$ if arc $(d, u, v) \in A$ exists, $d + c((d, u, v)) + \alpha(v) \leq d'_1 \leq d + c((d, u, v)) + \beta(v)$ and $d + c((d, u, v)) + \alpha(v) \leq d'_2 \leq d + c((d, u, v)) + \beta(v)$.*

Lemma 7. *There does not exist a vertex* $(d'_2, v) \in V_D$ *such that there exist two vertices* (d'_1, v) *and* $(d'_3, v) \in V_D$ *both associated to the same vertex* $v \in V$ *and* $D_v.comp(d'_1) = D_v.comp(d'_3) \neq D_v.comp(d'_2)$ *and* $d'_1 < d'_2 < d'_3$.

Proof. The only place where components are joined in the disjoint-set in Algorithm 3 is in line 21. As can be seen in this line of the algorithm, only components that are next to each other in the order given by `VArr[`v`]` can be merged. Lets suppose that d'_1 and d'_3 are part of the same component. As at the beginning all vertices belong to distinct components there should have been a moment when the components of d'_1 and d'_3 were merged. This implies that the components were next to each other and therefore there cannot be a value d'_2 in between these two components because otherwise the merge operation would not have been possible. This leads to the fact that all members of the same component are sequential in the order given by `VArr[`v`]` and proves the lemma. □

Lemma 8. *All vertices that are cousins with respect to a vertex* (d, u) *will be in the same component after* (d, u) *has been popped from the queue and its arcs traversed in Algorithm 3.*

Proof. Let C denote the component of all vertices that are cousins with respect to vertex (d, u). From the definition of the cousin relationship it is known that all vertices in C are sequential. After (d, u) has been popped from Q and the processing of all vertices begins through the *while* loop located in line 18 the algorithm starts processing the vertex with the smallest time d' and continues all the way until the vertex with the largest time d'_{max} each time going to the next vertex after the current component. The following will analyze each case the algorithm might encounter

- Reaches a previously unvisited vertex: The algorithm merges it with the previous component, by Lemma 7 all vertices in this new component are sequential.
- Reaches a previously visited vertex: The algorithm merges this component with the previous one and moves to the last visited vertex of this component. By Lemma 7 both merged components are sequential and therefore their merge produces a component with no gap.

Therefore all merged vertices will be sequential and because the process visits all components between d' and d'_{max} all previously existing components in that range will become one after the end of the process. □

Lemma 9. *For each arc* $(d, u, v) \in A$ *Algorithm 3 will visit* $e_{(d,u,v)} + 4$ *vertices where* $\sum_{a \in A} e_a = O(|A|)$.

Proof. From Algorithm 2, each arc in $(d, u, v) \in A$ leads to two explicit arcs $((d, u), (d', v))$ and $((d, u), (d'_{max}, v))$ of G_D, along with several implicit arcs of G_D, where d' = `DPmin[`(d, u, v)`]` and d'_{max} = `DPmax[`(d, u, v)`]`. After following one explicit arc it is possible to visit several other vertices by moving through the ordered list of vertices `VArr` created in Algorithm 1. From the inner *while* loop in line 18 of the algorithm it can be seen that the number of steps that

Algorithm 3 takes is bounded by the number of vertices visited by each explicit arc in G_D (by the *for* in line 13). What follows will analyze how many vertices will be visited in the traversal process, as this will be the number of steps that Algorithm 3 will perform. This can be seen in lines 18–31 of Algorithm 3 where for each iteration of the *while* loop there is a vertex that is processed. Therefore it is of interest to count the number of these aforementioned loops. The algorithm will loop until the value of d'_{now} is greater than that of d'_{max}. In lines 15–17 d'_{now} receives as initial value the minimum time d' such that the path that contains (d, u, v) can continue through an outgoing arc from v at time d', while d'_{max} receives as value the maximum at which the journey could continue through an outgoing arc from v. Both values where calculated in Algorithm 2. When the algorithm visits a vertex (d'_{now}, v) there are 2 possibilities, depending on whether (d'_{now}, v) has been visited before or not. What follows will analyze both cases. If the vertex has not been previously visited then it is marked as visited (line 22), its component is joined with that of the previous visited vertex (line 21) and the algorithm moves to the successor vertex. In this case the value of $e_{(d,u,v)}$ is increased by one. In this case vertex (d'_{now}, v) becomes the largest visited vertex of the component containing all vertices associated with v with times between d' and d'_{now} as can be deduced from the analysis of Lemma 8.

If the vertex has been visited previously then in line 24 we get the reference to the largest visited vertex of the component and move to the successor vertex. In what follows the argument that this can happen at most two times is presented. This is the same as showing that at most two previously visited components will be visited. By the definition of disjoint-set it is deduced that vertices in the first and second components of visited vertices cannot intersect because otherwise they would be part of the same set. By Lemma 7 it is also known that the vertex of the second component with the minimum time (let it be denoted by d'') should have a time bigger than that of d' because even if d' is the last vertex from the first visited set, vertex (d'', v) is at least the successor of the last vertex from the first visited set. Now following the link to the latest vertex of this component would lead to at least time `DPmax`$[(d'', u, v)]$ given that from Lemma 8 it follows that if this is the first vertex in the component it will have at least enough vertices to lead to the said time, which because $d'' > d'$ should be at least as big as `DPmax`$[(d', u, v)]$ and as such we have at least reached time d'_{max} and we finish the loop. □

Stage 4: Computing $R(s)$ and a foremost (α, β) -journey from s to t. Now that we have described the 3 main stages of the algorithm, we present them together in the form of the full procedure in Algorithm 4 and show that it is linear in Theorem 1.

Theorem 1. *On input a temporal digraph G, a pair of source and target vertices s, t in G, and two constraint functions $\alpha, \beta : V \to \mathbb{N}$, Algorithm 4 computes the arrival date of a foremost (α, β)-journey from s to t in linear time.*

Algorithm 4. Foremost (α, β)-journey arrival date in linear time

```
procedure FOREMOST(G, α, β, s, t)
    GENERATEVERTICES(G, α, β)                                  ▷ Algorithm 1
    GENERATETABLES(G, α, β)                                    ▷ Algorithm 2
    TRAVERSE(G, α, β, s)                                       ▷ Algorithm 3
    ArrivalDate = ∞
    for each arc (d, u, v) in A do
        if v = t and (d, u) in traversed then
            ArrivalDate = min(ArrivalDate, d + c((d, u, v)))
    return ArrivalDate              ▷ the time of the foremost journey from s to t
```

Proof. From Lemmas 5 and 6 it is known that Algorithms 1 and 2 have linear complexity. All that is left to do is show that Algorithm 3 is linear. From Lemma 9 it is known that for each unvisited vertex the value of $e_{(d,u,v)}$ is increased by one, and because there are at most $2|A|$ vertices in V then $\sum_{a \in A} e_a \leq 2|A|$. Because each arc in A produces at most one iteration of the *for* loop in line 13 of Algorithm 3 then from Lemma 9 it is deduced that the number of visited vertices over all Algorithm 3 is smaller than $|A| + 4|A|$ and therefore the number of visited vertices is linear in the size of $|A|$ for Algorithm 3. Now that the complexity of Algorithms 1, 2 and 3 are linear it follows that Algorithm 4's complexity is linear as it is a combination of the three aforementioned algorithms plus an iteration over all arcs in A. □

5 Conclusion and Perspectives

We addressed the problem of computing in a temporal graph the arrival date of a foremost journey under non-stop transit constraints. It is a polynomial time problem, contrary to the computation of a temporal path under the same constraints which is NP-hard. We then depict a linear time solution for finding the minimum arrival date of a foremost non-stop journey.

As for perspectives, it turns out that most our algorithmic steps follow the filter-map-reduce programming paradigm. In particular, we make intensive use of bucket sorting (a.k.a. radix sort) which can very naturally be implemented by lambdas. We believe that bucket sorting is important for processing historic data in general, and temporal graphs in particular. In this sense, we raise the open question whether our algorithm can be rewritten using a fully functional programming approach.

Acknowledgements. We are grateful to the anonymous reviewers for their helpful comments which greatly improved the paper.

References

1. Bui-Xuan, B.M., Ferreira, A., Jarry, A.: Computing shortest, fastest, and foremost journeys in dynamic networks. Int. J. Found. Comput. Sci. **14**(2), 267–285 (2003)

2. Casteigts, A., Flocchini, P., Mans, B., Santoro, N.: Shortest, fastest, and foremost broadcast in dynamic networks. Int. J. Found. Comput. Sci. **26**(4), 499–522 (2015)
3. Casteigts, A., Himmel, A.S., Molter, H., Zschoche, P.: Finding temporal paths under waiting time constraints. Algorithmica **83**, 2754–2802 (2021)
4. Cormen, T.H., Leiserson, C.E., Rivest, R.L., Stein, C.: Introduction to Algorithms. The MIT Press, Cambridge(1989)
5. Dibbelt, J., Pajor, T., Strasser, B., Wagner, D.: Connection scan algorithm. ACM J. Exp. Algorithm. **23**, 1–56 (2018)
6. Dupuy, M., d'Ambrosio, C., Liberti, L.: Optimal paths on the ocean (2021). https://hal.archives-ouvertes.fr/hal-03404586
7. Gabow, H., Tarjan, R.: A linear-time algorithm for a special case of disjoint set union. J. Comput. Syst. Sci. **30**, 209–221 (1985)
8. Himmel, A.S., Bentert, M., Nichterlein, A., Niedermeier, R.: Efficient computation of optimal temporal walks under waiting-time constraints. In: 8th International Conference on Complex Networks and Their Applications. SCI, vol. 882, pp. 494–506 (2019)
9. Rymar, M., Molter, H., Nichterlein, A., Niedermeier, R.: Towards Classifying the Polynomial-Time Solvability of Temporal Betweenness Centrality. In: Kowalik, Ł, Pilipczuk, M., Rzążewski, P. (eds.) WG 2021. LNCS, vol. 12911, pp. 219–231. Springer, Cham (2021). https://doi.org/10.1007/978-3-030-86838-3_17
10. Saramäki, J., Kivelä, M., Karsai, M.: Weighted temporal event graphs. Temporal Network Theory, pp. 107–128 (2019)

Author Index

Printed in the United States
by Baker & Taylor Publisher Services

Printed in the United States
by Baker & Taylor Publisher Services